NORTH CAROLINA
STATE BOARD OF COM
LIBRAR
ASHEVILLE-BUNCOMBE TECHNICAL COLLEGE

DISCARDED

DEC 1 1 2024

APPLIED AIR CONDITIONING AND REFRIGERATION

Second Edition

APPLIED AIR CONDITIONING AND REFRIGERATION

Second Edition

C. T. GOSLING

A.M.Inst.R., A.M.A.S.H.R.A.E., F.I.S.E.

*Ductwork Engineering Systems Limited,
Croydon, Surrey, England*

APPLIED SCIENCE PUBLISHERS LTD
LONDON

APPLIED SCIENCE PUBLISHERS LTD
RIPPLE ROAD, BARKING, ESSEX, ENGLAND

British Library Cataloguing in Publication Data

Gosling, Clifford Trevor
 Applied air conditioning and refrigeration.—
 2nd ed.
 1. Air conditioning
 I. Title
 697.9'3   TH7687

ISBN 0-85334-877-4

WITH 217 ILLUSTRATIONS AND 95 TABLES

© APPLIED SCIENCE PUBLISHERS LTD 1980

All rights reserved. No part of this publication may be reproduced, stored in a retrieval system, or transmitted in any form or by any means, electronic, mechanical, photocopying, recording, or otherwise, without the prior written permission of the publishers, Applied Science Publishers Ltd, Ripple Road, Barking, Essex, England

Printed in Great Britain by Galliard (Printers) Ltd, Great Yarmouth

FOREWORD

The first edition of this book stemmed from a series of articles written by Trevor Gosling a decade ago, to provide an up-to-date source of knowledge and information for all involved—consultants, architects, plant engineers and contractors—with the then very fast growing air conditioning business.

A major problem attendant upon this fast growth was the amount of expertise available for and from all interests; many people were entering an area largely foreign to their previous experience, and others, while involved in disciplines close to, or part of air conditioning, were having to tackle problems very different from those dealt with previously.

Today, air conditioning is on a much more gentle upward curve in terms of market growth, but has taken new directions, thanks to the economic, social and political changes of the past ten years, and there is as big a need as ever for a thoroughly up-to-date, authoritative source of information.

This edition recognises the changes, includes new chapters on heat pumps and energy cost comparisons for different types of air conditioning system, and has extended coverage of such topics as heat rejection, air cooled systems and variable air volume systems.

This book, like its first edition, will surely prove of benefit to experienced heating, ventilating, air conditioning and refrigeration engineers, specifiers and others with a close interest in air conditioning and its development over the past decade.

TERRY O'GORMAN,
Editor,
Refrigeration and Air Conditioning

CONTENTS

FOREWORD v

AUTHOR'S NOTES ON IMPERIAL AND SI UNITS . . . ix

Chapter
1 PSYCHROMETRICS 1
2 THE REFRIGERATION CYCLE 22
3 SITE SURVEY 32
4 LOAD ESTIMATING 40
5 ROOM AIR DISTRIBUTION 94
6 DUCT DESIGN 107
7 PARTIAL LOAD AND ZONING 120
8 EVAPORATORS 129
9 RECIPROCATING REFRIGERATION COMPRESSORS 141
10 CENTRIFUGAL REFRIGERATION COMPRESSORS 154
11 ABSORPTION CYCLE REFRIGERATION EQUIPMENT 170
12 CONDENSERS AND COOLING TOWERS . . 186
13 AIR MOVING DEVICES 225
14 TERMINAL DEVICES AND AIR CONDITIONING SYSTEMS 242
15 LIQUID CHILLING PACKAGES 309
16 PACKAGED EQUIPMENT 319
17 REFRIGERATION PIPEWORK 345
18 WATER PIPEWORK 358

19	HEAT PUMPS AND HEAT RECLAIM DEVICES	368
20	ENERGY COMPARISON OF AIR CONDITIONING SYSTEMS FOR MULTI-ROOM APPLICATIONS	392
INDEX		407

AUTHOR'S NOTES ON IMPERIAL AND SI UNITS

The basic units on which the International System (SI) is based are as follows:

Quantity	Unit	Symbol
Length	metre	m
Mass	kilogramme	kg
Time	second	s
Temperature	Kelvin	K

From the above the following units have been derived:

Force	newton	N	$= kgm/s^2$
Work, Energy, Heat	joule	$J = Nm$	$= kgm^2/s^2$
Power	watt	$W = J/s$	$= kgm^2/s^2$
Temperature	degree Celsius	°C	$= K + 273.15\,°C$

However, the use of basic SI units in air conditioning calculations often reflects either uncommonly large units or the necessity to use many places of decimals. Therefore, further derivations have been made using units which are to the power 10^3, i.e. kilo-, or 10^{-3}, i.e. milli-, to avoid cumbersome units. These derivations are reflected as follows:

$$\begin{aligned}
\text{mm (millimetre)} &= \text{m} \times 10^{-3} \\
\text{litre (litre)} &= \text{m}^3 \times 10^{-3} \\
\text{g (gramme)} &= \text{kg} \times 10^{-3} \\
\text{kN (kilonewton)} &= \text{N} \times 10^3 \\
\text{kJ (kilojoule)} &= \text{J} \times 10^3 \\
\text{kW (kilowatt)} &= \text{W} \times 10^3
\end{aligned}$$

CONVERSION FACTORS—IMPERIAL AND SI UNITS

(units derived from basic SI units in common use are shown in bold type)

Quantity	To convert Imperial units	Multiply by	Divide by	Equals SI units	SI symbol
Length	**inch**	**0·0254**	**39·37**	**metre**	**m**
	inch	25·4	0·039 37	millimetre	mm
	foot	**0·3048**	**3·2808**	**metre**	**m**
Area	**square inch**	**0·000 645**	**1550·0**	**square metre**	**m²**
	square foot	**0·0929**	**10·763**	**square metre**	**m²**
Volume	**cubic inch**	**0·000 016**	**61 010·0**	**cubic metre**	**m³**
	cubic inch	0·016 39	61·01	litre	litre
	cubic foot	**0·028 31**	**35·32**	**cubic metre**	**m³**
	cubic foot	28·31	0·035 32	litre	litre
Air flow	**cubic feet/minute**	**0·000 4719**	**2119·0**	**cubic metre/second**	**m³/s**
	cubic feet/minute	0·4719	2·119	litre/second	litre/s
Water flow	gallons/minute	0·000 075 77	13 198·00	cubic metre/second	m³/s
	gallons/minute	0·075 77	13·198	litre/second	litre/s
Mass	**pound**	**0·453 592**	**2·204**	**kilogramme**	**kg**
	pound	453·592	0·002 2	gramme	g
Pressure	**pound/square inch**	**6895·0**	**0·000 145**	**newton/square metre**	**N/m²**
	pounds/square inch	6·895	0·145	kilonewton/square metre	kN/m²
	inch w.g.	249·1	0·004 015	newton/square metre	N/m²
	inch Hg	**3386·4**	**0·000 295**	**newton/square metre**	**N/m²**
	inch Hg	3·3864	0·295 3	kilonewton/square metre	kN/m²

	Equals Imperial Units	Divide by	Multiply by	To convert SI Units	
Pressure drop per unit length	inch w.g./100 feet foot hd water/100 feet	8·176 98·112	0·1223 0·0102	newton/cubic metre newton/cubic metre	N/m³ N/m³
Weight	pound/square foot	4·883	0·2048	kilogramme/square metre	kg/m²
Density	pound/cubic foot	16·02	0·06243	kilogramme/cubic metre	kg/m³
Specific volume	cubic foot/pound	0·06243	16·02	cubic metre/kilogramme	m³/kg
Velocity	foot/minute	0·00508	196·85	metre/second	m/s
Heat output	BTU BTU BTU/hour	1·055 1055·0 0·2931	0·9479 0·000948 3·412	joule kilojoule watt	J kJ W
Refrigeration	ton R (12 000 BTU/h) ton R (12 000 BTU/h)	3516·0 3·516	0·000284 0·2844	watt kilowatt	W kW
Power	horsepower horsepower	745·7 0·7457	0·00134 1·341	watt kilowatt	W kW
Heat content	BTU/pound BTU/pound	0·002326 2·326	429·9 0·4299	joule/kilogramme kilojoule/kilogramme	J/kg kJ/kg
Specific heat	BTU/pound °F	0·004187	238·83	joule/kilogramme °C	J/kg °C
Heat transfer	BTU/h sq ft °F	5·678	0·1761	watt/square metre °C	W/m² °C
Thermal resistance	h sq ft °F/BTU	0·1761	5·678	square metre °C/watt	m² °C/W

The conversion table on pp. x–xi reflects both the use of pure or basic SI units and the derivatives used within the text of this book.

Wherever possible all examples have been provided in both Imperial and SI units, and these are shown with the SI units as a direct conversion and italicised after each Imperial unit. However, where examples have been made to show typical situations, often a direct conversion tends to mislead and to detract from the simplicity of the example. In such cases a separate worked example is provided, printed in italics, and it should be stressed that such instances do not reflect a true conversion.

Chapter 1

PSYCHROMETRICS

The fundamental tool of an air-conditioning engineer is the psychrometric chart and this chapter will examine this and the various processes which can be analysed from it.

Psychrometrics is the science involving thermodynamic properties of moist air, the definition must be broadened to include the effect of atmospheric moisture on human comfort and materials, and the method of controlling the thermal properties of moist air.

The psychrometric chart is basically a graph of Dry Bulb Temperature (that measured by a normal thermometer) plotted against the weight of water vapour in grains or pounds of moisture per pound of dry air (g/kg), the Specific Humidity or Moisture Content. At any given dry bulb temperature a pound (kg) of dry air (the basis for all psychrometric calculations) can only absorb a certain amount of moisture and this is found at the Saturation Line. The higher the dry bulb temperature, the greater the moisture content the air can absorb. At the saturation line the temperature is known as the Dew Point Temperature, i.e. the temperature at which condensation will take place if the air is cooled. The saturation line can be considered at 100 per cent Relative Humidity, which is the ratio of actual water vapour pressure of the air to the saturated water vapour pressure of the air at the same dry bulb temperature. Almost equal to the relative humidity is the Humidity Ratio: a ratio of the actual moisture content to the moisture content that could be absorbed at saturation at the same dry bulb temperature. Relative humidity lines are shown on the psychrometric chart as diverging lines almost parallel to the saturation line.

Wet Bulb Temperature is the temperature measured by a

thermometer with its bulb covered by a wick, wetted with distilled water exposed to a current of rapidly moving air. Constant wet bulb temperature lines are shown on the chart as diagonal lines, where at saturation, the dry bulb, wet bulb and dew point temperatures coincide. The drier the air, i.e. the lower the relative humidity, the more rapidly the moisture on the wetted wick will evaporate and the greater the depression from dry bulb temperature.

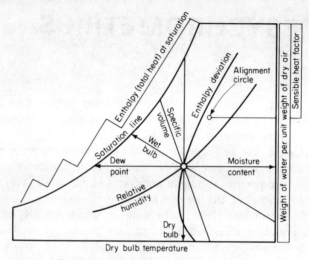

FIG. 1.1. Skeleton psychrometric chart.

As the dry bulb temperature increases so the air expands and the Specific Volume, the cubic feet of the mixture per pound of dry air, increases. Specific Volume lines are shown as almost vertical lines on the psychrometric chart, the greater the moisture content at a given dry bulb temperature the greater the specific volume.

Examination of the skeleton psychrometric chart (Fig. 1.1) shows that if any two of the air properties mentioned are known, the air condition can be plotted from which all other properties can be found.

Up to this point all psychrometric charts, of which there are many variations, are the same. The Enthalpy, a thermal property indicating the quantity of heat in the air above an arbitrary datum, is measured in British Thermal Units (BTU) per pound of dry air (kJ/kg). It is this datum and the presentation of enthalpy scale which differ amongst the various psychrometric charts available. The chart used in this chapter (Figs. 1.2 and 1.2a), that originally devised by Dr Willis

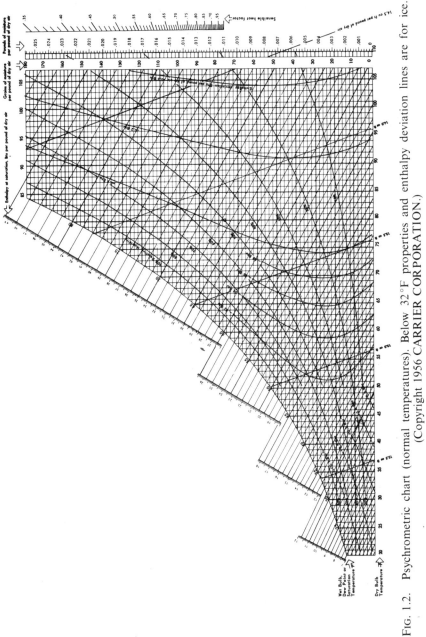

FIG. 1.2. Psychrometric chart (normal temperatures). Below 32°F properties and enthalpy deviation lines are for ice. (Copyright 1956 CARRIER CORPORATION.)

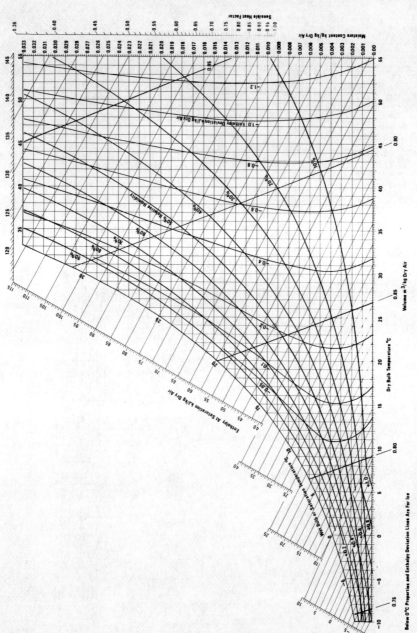

Fig. 1.2a. Psychrometric chart (normal temperatures). Barometric pressure 101·325 kN/m². (Copyright 1973 CARRIER CORPORATION).

H. Carrier in 1911, although subsequently revised to its present form, shows enthalpy at saturation and is found at the extension of the wet bulb lines beyond the saturation line. The datum used is 0 °F for dry air, and for the moisture content 32 °F water for the chart in Imperial units and *0 °C for dry air and moisture content on the SI chart*. Because the enthalpy shown is that at saturation it should be corrected by Enthalpy Deviation due to the air not being at the saturated state. The enthalpy deviation lines, shown almost vertically on the psychrometric chart, are measured in BTU per pound of dry air (kJ/kg), are of negative quantity and should be applied to the enthalpy at saturation to achieve the true enthalpy of the air considered. The deviation is normally applied when extreme accuracy is required and cannot often be applied to the enthalpy scale since the scale itself cannot be read so accurately. However, accuracy can be obtained when using the table of Enthalpy at Saturation (Tables 1.1 and 1.1a) in conjunction with the deviation lines from the psychrometric chart.

It can be seen that the wet bulb temperature occurs almost at constant enthalpy and if shown on the psychrometric chart occurs as almost parallel lines. The chart used by the American Society of Heating, Refrigeration and Air Conditioning Engineers (ASHRAE) shows such lines and care should be taken to ensure proper conditions are plotted.

The psychrometric chart produced by The Chartered Institute of

TABLE 1.1

ENTHALPY AT SATURATION, BTU PER POUND OF DRY AIR
(Extracted from Chapter 3, ASHRAE 1961 Guide and Data Book. Used by permission.)

Wet bulb (°F)	0	0·1	0·2	0·3	0·4	0·5	0·6	0·7	0·8	0·9	Wet bulb (°F)
40	15·23	15·28	15·32	15·37	15·42	15·47	15·51	15·56	15·61	15·65	40
41	15·70	15·75	15·80	15·84	15·89	15·94	15·99	16·04	16·08	16·13	41
42	16·17	16·22	16·27	16·32	16·37	16·42	16·46	16·51	16·56	16·61	42
43	16·66	16·71	16·76	16·80	16·85	16·90	16·95	17·00	17·05	17·10	43
44	17·15	17·20	17·25	17·30	17·35	17·40	17·45	17·50	17·55	17·60	44
45	17·65	17·70	17·75	17·80	17·85	17·91	17·96	18·01	18·06	18·11	45
46	18·16	18·21	18·26	18·32	18·37	18·42	18·47	18·52	18·58	18·63	46
47	18·68	18·73	18·79	18·84	18·89	18·95	19·00	19·05	19·10	19·16	47
48	19·21	19·26	19·32	19·37	19·43	19·48	19·53	19·59	19·64	19·70	48
49	19·75	19·81	19·86	19·92	19·97	20·03	20·08	20·14	20·19	20·25	49

TABLE 1.1—continued

Wet bulb (°F)	0	0.1	0.2	0.3	0.4	0.5	0.6	0.7	0.8	0.9	Wet bulb (°F)
50	20.30	20.36	20.41	20.47	20.53	20.58	20.64	20.69	20.75	20.81	50
51	20.86	20.92	20.98	21.03	21.09	21.15	21.21	21.26	21.32	21.38	51
52	21.44	21.49	21.55	21.61	21.67	21.73	21.78	21.84	21.90	21.96	52
53	22.02	22.08	22.14	22.20	22.26	22.32	22.38	22.44	22.50	22.56	53
54	22.62	22.68	22.74	22.80	22.86	22.92	22.98	23.04	23.10	23.16	54
55	23.22	23.28	23.34	23.41	23.47	23.53	23.59	23.65	23.72	23.78	55
56	23.84	23.90	23.97	24.03	24.10	24.16	24.22	24.29	24.35	24.42	56
57	24.48	24.54	24.61	24.67	24.74	24.80	24.86	24.93	24.99	25.06	57
58	25.12	25.19	25.25	25.32	25.38	25.45	25.52	25.58	25.65	25.71	58
59	25.78	25.85	25.92	25.98	26.05	26.12	26.19	26.26	26.32	26.39	59
60	26.46	26.53	26.60	26.67	26.74	26.81	26.87	26.94	27.01	27.08	60
61	27.15	27.22	27.29	27.36	27.43	27.50	27.57	27.64	27.70	27.78	61
62	27.85	27.92	27.99	28.07	28.14	28.21	28.28	28.35	28.43	28.50	62
63	28.57	28.64	28.72	28.79	28.87	28.94	29.01	29.09	29.16	29.24	63
64	29.31	29.39	29.46	29.54	29.61	29.62	29.76	29.83	29.91	29.98	64
65	30.06	30.14	30.21	30.29	30.37	30.45	30.52	30.60	30.68	30.75	65
66	30.83	30.91	30.99	31.07	31.15	31.23	31.30	31.38	31.46	31.54	66
67	31.62	31.70	31.78	31.86	31.94	32.02	32.10	32.18	32.26	32.34	67
68	32.42	32.50	32.59	32.67	32.75	32.84	32.92	33.00	33.08	33.17	68
69	33.25	33.33	33.42	33.50	33.59	33.67	33.75	33.84	33.92	34.01	69
70	34.09	34.18	34.26	34.35	34.43	34.52	34.61	34.69	34.78	34.86	70
71	34.95	35.04	35.13	35.21	35.30	35.39	35.48	35.57	35.65	35.74	71
72	35.83	35.92	36.01	36.10	36.19	36.28	36.38	36.47	36.56	36.65	72
73	36.74	36.83	36.92	37.02	37.11	37.20	37.29	37.38	37.48	37.57	73
74	37.66	37.76	37.85	37.95	38.04	38.14	38.23	38.33	38.42	38.51	74
75	38.61	38.71	38.80	38.90	39.00	39.09	39.19	39.28	39.38	39.47	75
76	39.57	39.67	39.77	39.87	39.97	40.07	40.17	40.27	40.37	40.47	76
77	40.57	40.67	40.77	40.87	40.97	41.08	41.18	41.28	41.38	41.48	77
78	41.58	41.68	41.79	41.89	42.00	42.10	42.20	42.31	42.41	42.52	78
79	42.62	42.73	42.83	42.94	43.05	43.16	43.26	43.37	43.48	43.58	79
80	43.69	43.80	43.91	44.02	44.13	44.24	44.34	44.45	44.56	44.67	80
81	44.78	44.89	45.00	45.11	45.23	45.34	45.45	45.56	45.67	45.79	81
82	45.90	46.01	46.13	46.24	46.36	46.48	46.59	46.70	46.81	46.93	82
83	47.04	47.16	47.28	47.39	47.51	47.63	47.75	47.87	47.98	48.10	83
84	48.22	48.34	48.46	48.58	48.70	48.82	48.95	49.07	49.19	49.31	84
85	49.43	49.55	49.68	49.80	49.92	50.04	50.17	50.29	50.41	50.54	85

TABLE 1.1a
ENTHALPY AT SATURATION, kJ/kg

Wet bulb (°C)	0	0·1	0·2	0·3	0·4	0·5	0·6	0·7	0·8	0·9	Wet bulb (°C)
4	16·69	16·88	17·08	17·27	17·47	17·66	17·85	18·05	18·24	18·44	4
5	18·63	18·83	19·03	19·23	19·43	19·63	19·83	20·03	20·23	20·43	5
6	20·63	20·84	21·05	21·25	21·46	21·67	21·87	22·08	22·29	22·50	6
7	22·70	22·92	23·13	23·34	23·56	23·77	23·99	24·20	24·41	24·63	7
8	24·84	25·06	25·28	25·51	25·73	25·95	26·17	26·39	26·61	26·83	8
9	27·05	27·28	27·51	27·74	27·96	28·19	28·42	28·65	28·88	29·11	9
10	29·33	29·57	29·81	30·05	30·28	30·52	30·76	30·99	31·23	31·47	10
11	31·70	31·95	32·20	32·44	32·69	32·93	33·18	33·42	33·67	33·92	11
12	34·16	34·41	34·67	34·92	35·18	35·43	35·68	35·94	36·19	36·44	12
13	36·70	36·96	37·23	37·49	37·76	38·03	38·29	38·56	38·82	39·09	13
14	39·35	39·63	39·90	40·17	40·44	40·72	40·99	41·26	41·53	41·81	14
15	42·08	42·37	42·66	42·95	43·23	43·52	43·81	44·10	44·39	44·68	15
16	44·97	45·26	45·55	45·85	46·14	46·44	46·73	47·02	47·32	47·61	16
17	47·90	48·21	48·52	48·83	49·13	49·44	49·75	50·06	50·36	50·67	17
18	50·98	51·30	51·62	51·94	52·27	52·59	52·91	53·23	53·55	53·87	18
19	54·19	54·53	54·86	55·19	55·53	55·86	56·19	56·53	56·86	57·20	19
20	57·52	57·87	58·22	58·57	58·91	59·26	59·61	59·95	60·30	60·65	20
21	61·00	61·36	61·72	62·08	62·44	62·80	63·16	63·52	63·88	64·24	21
22	64·61	64·99	65·37	65·75	66·12	66·50	66·88	67·26	67·64	68·02	22
23	68·40	68·80	69·19	69·58	69·98	70·37	70·76	71·16	71·55	71·94	23
24	72·34	72·75	73·16	73·57	73·99	74·40	74·81	75·22	75·64	76·05	24
25	76·46	76·89	77·32	77·75	78·17	78·60	79·03	79·46	79·89	80·32	25
26	80·74	81·19	81·64	82·09	82·54	82·99	83·44	83·89	84·34	84·79	26
27	85·24	85·71	86·18	86·65	87·12	87·59	88·06	88·53	89·00	89·47	27
28	89·94	90·43	90·91	91·40	91·88	92·37	92·86	93·34	93·83	94·32	28
29	94·80	95·32	95·83	96·34	96·85	97·37	97·88	98·39	98·90	99·42	29
30	99·93										

Building Services (CIBS) has no enthalpy lines, but increments around the perimeter of the chart which can be aligned by a rule to find the total heat at any given air condition. This avoids errors in plotting, but proves difficult to read accurately. The CIBS chart uses dry air at 32 °F (0 °C) as a datum.

In consideration of the different methods of measuring enthalpy from various charts, only differences in total heat should be referred to. All other conditions, of course, could be plotted irrespective of which chart is used.

The thermal properties of air can be split into two categories, Sensible Heat and Latent Heat. Sensible Heat is that added to moist air without change in moisture content, and Latent Heat is that added to moist air without change in dry bulb temperature. Total Heat is the summation of Sensible and Latent Heat. Whenever cooled and dehumidified air is introduced into an area to be conditioned the condition of that air must be able to satisfy the sensible and latent heat gain to the area (Room Sensible Heat and Room Latent Heat) in the correct proportions. This relationship is known as the Room Sensible Heat Factor (RSHF) and is defined as:

$$\text{RSHF} = \frac{\text{RSH}}{\text{RSH} + \text{RLH}} = \frac{\text{RSH}}{\text{RTH}}$$

where RSH = room sensible heat
RLH = room latent heat
RTH = room total heat

On the psychrometric chart there is a construction method from which the room sensible heat factor can be plotted. An alignment circle, located at 80 °F dry bulb (DB) and 50 per cent relative humidity on Fig. 1.2, and *24 °C dry bulb and 50 per cent relative humidity on Fig. 1.2a*, is used as a reference point and a line from this, to the scale extended beyond the specific humidity scale, produces the slope of the line to produce the correct ratio of sensible to latent heats. From this constructed line a parallel line from the desired room condition to be maintained can be made in the direction of the saturation line, and any air condition located on this line will satisfy the ratio of room sensible and latent heat (see Fig. 1.3).

The mixing process can be shown very easily on a psychrometric chart with sufficient accuracy for air conditioning problems. A common example of this is the mixture of return air at desired room conditions and ambient air used for ventilation purposes. Figure 1.4 shows a typical air-conditioning process and the mixture condition C will fall on a line connected from the room condition A, to the outside air condition B. The location of point C will depend on the quantities of air to be mixed. If 75 per cent room air is mixed with 25 per cent ventilation air then the point will fall 25 per cent along the line from point A. The best method of calculating this point is to use the dry bulb scale as a reference such that with a room condition of 68 °F (*20 °C*) and an ambient of 86 °F (*30 °C*), point C will occur where 68 °F + 0·25

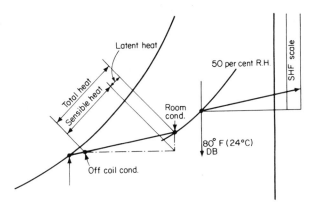

Fig. 1.3. Room sensible heat factor.

(86–68), or 72·5 °F *(20 °C + 0·25 (30–20), or 22·5 °C)*, and the constructed line AB meet.

Figure 1.4 also illustrates a further mixing process which relates to the performance of the cooling and dehumidifying coil. If the coil was 100 per cent efficient all the air would be cooled to the effective coil surface temperature point D, commonly referred to as the Apparatus Dew Point (ADP). However, the effectiveness of a coil will depend on its geometry, i.e. the number and arrangement of rows and the fin

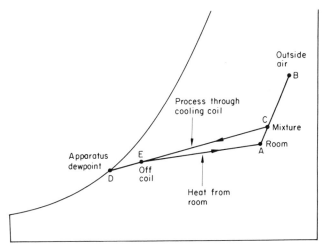

Fig. 1.4. Typical air conditioning process.

spacing, together with the air velocity through the coil. The effectiveness of the coil is referred to as the Contact Factor, and the inefficiency as the By-pass Factor, i.e. that proportion of air which can be considered not to have made contact with the coil and which remains at the entering condition.

$$\text{by-pass factor} = \frac{t_{la} - t_{adp}}{t_{ea} - t_{adp}}$$

where t_{adp} = apparatus dewpoint temperature (Point D)
t_{ea} = entering air temperature (Point C)
t_{la} = temperature leaving apparatus (Point E)

Therefore, point E will occur as a mixture between entering air at condition C, and Apparatus Dew Point, condition D. In commercial practice by-pass factors can be as low as 0·02 for eight-row coils with fins spaced at 14 per in and as high as 0·30 for three-row coils. Typical by-pass factors are discussed in Chapter 9.

Simplified factors, based on standard air, have been established; these can be used to simplify calculations and are as follows:

Sensible heat (BTU/h) = 1·08 × cfm × temperature difference (°F)
Latent heat (BTU/h) = 0·68 × cfm × moisture difference (grs/lb)
Total heat (BTU/h) = 4·45 × cfm × enthalpy difference
(BTU/lb air)

where $1·08 = 0·244 \times \dfrac{60}{13·34}$

$0·68 = \dfrac{1076 \times 60}{7000 \times 13·5}$

$4·45 = \dfrac{60}{13·5}$

and cfm = quantity of air (cubic feet per minute)
 0·244 = specific heat of moist air
   60 = minutes per hour
 13·34 = specific volume of dry air
 1076 = latent heat of 1 lb moisture
 7000 = grs/lb
 13·5 = specific volume of moist air

1 PSYCHROMETRICS

Sensible heat $(kW) = 1\cdot21 \times m^3/s \times$ *temperature difference* $(°C)$
$(W) = 1\cdot21 \times litres/s \times$ *temperature difference* $(°C)$
$(W) = 1210 \times m^3/s \times$ *temperature difference* $(°C)$
Latent heat $(kW) = 2\cdot98 \times m^3/s \times$ *moisture difference* (grs/kg)
$(W) = 2\cdot98 \times litres/s \times$ *moisture difference* (grs/kg)
$(W) = 2980 \times m^3/s \times$ *moisture difference* (grs/kg)
Total heat $(kW) = 1\cdot19 \times m^3/s \times$ *enthalpy difference* (kJ/kg)
$(W) = 1\cdot19 \times litres/s \times$ *enthalpy difference* (kJ/kg)
$(W) = 1190 \times m^3/s \times$ *enthalpy difference* (kJ/kg)

where $1\cdot21 = \dfrac{1\cdot021}{0\cdot832}$

$2\cdot98 = \dfrac{2503}{1000 \times 0\cdot841}$

$1\cdot19 = \dfrac{1}{0\cdot841}$

and $1\cdot021 =$ *specific heat of moist air* $(kJ/kg/°C)$
$0\cdot832 =$ *specific volume of dry air* (m^3/kg)
$0\cdot841 =$ *specific volume of moist air* (m^3/kg)
$2503 =$ *latent heat of 1 kg moisture* (kJ/kg)
$1000 = grs/kg$

The air quantity required to offset room sensible and latent heats can be calculated once point E the supply air condition (Fig. 1.4) has been established. Since, by construction, any point falling along the line AE will satisfy both sensible and latent gains in the correct proportion, then only the sensible heat content need be considered to establish the supply air quantity, which may be calculated from the following equation:

$$\text{cfm}_{sa} = \frac{\text{RSH(BTU/h)}}{1\cdot08(t_{rm} - t_{la})°F} \qquad m^3/s_{sa} = \frac{RSH(kW)}{1\cdot21(t_{rm} - t_{la})°C}$$

where $\text{cfm}_{sa} =$ supply air volume measured in cubic feet per minute or $m^3/s_{sa} =$ *supply air volume measured in cubic metres per second.*

When establishing the room sensible heat allowance is made for the heat gain from the supply fan and for duct heat gains and for supply duct leakage losses, if both occur outside the conditioned space. Figure 1.5 shows the actual air condition to the space, point F, where

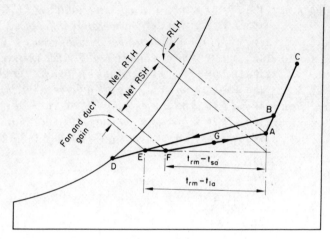

FIG. 1.5. Effect of fan and duct gains.

allowance has been made for losses. A rearrangement of the previous equation gives the supply air temperature to the room (t_{sa}) as follows:

$$t_{sa} = t_{rm} - \frac{\text{nett RSH}}{1\cdot 08 \times \text{cfm}_{sa}} \qquad t_{sa} = t_{rm} - \frac{\text{nett RSH}}{1\cdot 21 \times m^3/s_{sa}}$$

where nett RSH = RSH − (fan gain − duct losses).

The above value will be required when considering air distribution within the conditioned space.

If the supply air temperature calculated is too low to satisfy proper room air distribution, the basic psychrometrics of the cooling coil need not be changed, since room air can be by-passed around the coil to provide a higher supply air temperature (Fig. 1.6). Once the desired supply air temperature is established the air volume can be calculated.

$$\text{cfm}_{da} = \frac{\text{nett RSH(BTU/h)}}{(t_{rm} - t_{sa})1\cdot 08} \qquad m^3/s_{da} = \frac{\text{nett RSH}(kW)}{(t_{rm} - t_{sa})1\cdot 21}$$

where cfm_{da} = air volume of the delivered air to the room. The by-passed air quantity, $\text{cfm}_{ba} = \text{cfm}_{da} - \text{cfm}_{sa}$ or $m^3/s_{ba} = m^3/s_{da} - m^3/s_{sa}$.

The effect of mixing room air with supply air is shown as point G in Fig. 1.5.

Further examples of cooling coil performance are shown in Chapters 7 and 9.

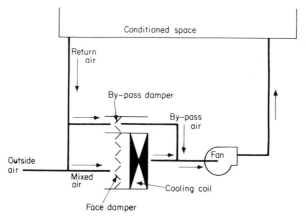

Fig. 1.6. Coil by-pass apparatus.

The psychrometric process using an air washer (adiabatic cooling) is shown in Fig. 1.7 as line AE, which follows the wet bulb line. If the air washer was 100 per cent effective then the air would leave at saturation point D. However, the saturation efficiency of such equipment varies between 80 and 99 per cent. Line AD' represents the process line when cooling is added to the spray water, whereas AD" represents the same process line when heating is added.

$$\text{saturation efficiency} = \frac{t_{edb} - t_{ldb}}{t_{edb} - t_{es}}$$

$$= \frac{W_{ea} - W_{la}}{W_{ea} - W_{es}}$$

$$= \frac{L_{ea} - L_{la}}{L_{ea} - L_{es}}$$

where t_{edb} = temperature entering: dry bulb
t_{ldb} = temperature leaving: dry bulb
t_{es} = saturation temperature
W_{ea} = moisture content: entering air
W_{la} = moisture content: leaving air
W_{es} = moisture content at saturation
L_{ea} = enthalpy of entering air
L_{la} = enthalpy of leaving air
L_{es} = enthalpy at saturation

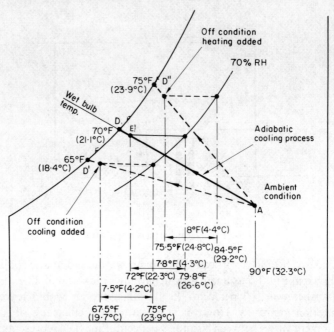

Fig. 1.7. Adiabatic cooling and humidification.

The processes shown in Fig. 1.7 are common in industrial applications, such as textile plants, where humidity control is the prime factor. The resulting room dry bulb temperatures at the desired relative humidity have a direct influence on the amount of cooling or heating added to the spray water. The examples shown are for 90 per cent Saturation Efficiency with an ambient air condition of 90°F (*32.3°C*) dry bulb 70°F (*21.1°C*) wet bulb, and a desired relative humidity of 70 per cent. The amount of cooling required can be expressed as:

$$\text{cooling load BTU/h} = \text{cfm} \times 4{\cdot}45 \times (H_{ea} - H_{la}) \text{ BTU/h}$$
$$kW = m^3/s \times 1{\cdot}19 \times (H_{ea} - H_{la}) \; kJ/kg$$

Similarly

$$\text{heating load BTU/h} = \text{cfm} \times 4{\cdot}45 \times (H_{la} - H_{ea}) \text{ BTU/h}$$
$$kW = m^3/s \times 1{\cdot}19 \times (H_{ea} - H_{la}) \; kJ/kg$$

It can also be seen that temperature differences between room air and supply air do not vary greatly, being 7·5°F (*4.2°C*) for cooling

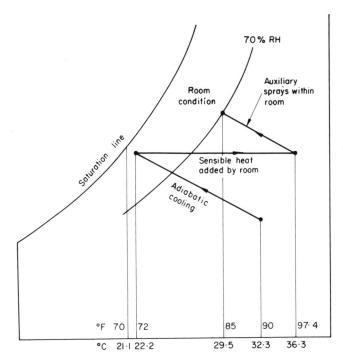

FIG. 1.8. Spray humidification in conditioned space.

application, 7·8°F *(4·3°C)* for adiabatic application and 8·0°F *(4·4°C)* for heating application, with resulting variations of only 6 per cent in air volume between the extreme conditions.

Applications requiring high relative humidity conditions require extremely high air volumes since the temperature difference between room and supply air is quite low. Figure 1.8 shows how auxiliary water sprays within the conditioned space can substantially reduce the air volume to be handled by the central apparatus. Assuming the same conditions as the previous example using adiabatic cooling only, it can be seen that the effective temperature difference between room and leaving has increased from (79·8 − 72) *(26·6 − 22·3)*, 7·8 °F *(4·3°C)* to (97·4 − 72) *(36·3 − 22·3)*, 25·4°F *(14·0°C)*, providing an air volume of some 31 per cent of the all-air system. The penalty is felt, however, in the resulting higher room dry bulb temperature 85°F *(29·5°C)* for the example chosen, or some 5·2°F *(2·7°C)* higher than the all-air system.

A further example of spray humidification is in primary plants at winter conditions. Figure 1.9 shows how air entering at condition A is first preheated to condition B where it passes through the humidifier. Sensible heat is converted into latent heat by the addition of moisture to condition C. Further sensible heat is added by a reheater as shown by the line CD. The temperature difference between D and room condition E is required to offset room heat losses.

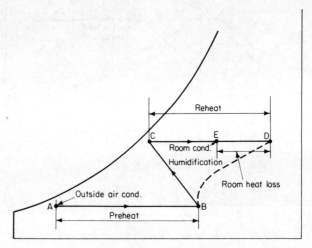

FIG. 1.9. Heating and humidification process.

Chemical dehumidification is used for applications calling for very low relative humidities and Fig. 1.10 shows the ideal process line AB. However, this process calls for the desiccant to be constantly changed such that it is able to absorb the water vapour within the air. This desiccant is heated such that the retained water vapour is evaporated and when reintroduced to the air will still retain some of this sensible heat. The process line AC will therefore result. Because the air leaving the dehumidifier is at a very low dewpoint, sensible cooling can take place as shown by line CD. The cooling coil effective surface temperature need not be at saturation, and can be analysed in the same way as a dehumidifying coil where:

$$\text{effective coil surface temp.} = t_{ea} - \frac{(t_{ea} - t_{la})}{(1 - BF)}$$

An obvious advantage of this system is that cooling surfaces need

1 PSYCHROMETRICS   17

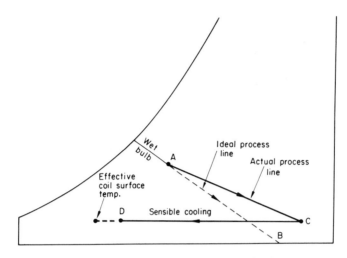

Fig. 1.10. Chemical dehumidification.

not be controlled below freezing point, with resulting coil frost-free operation.

The psychrometric chart discussed so far is for applications at sea level.

The design of an air-conditioning system for high altitude locations should take into account the decreased barometric pressure, the decreased density of the air and the increased moisture capacity which occur as the elevation above sea level increases (Table 1.2).

TABLE 1.2
ALTITUDE DENSITY TABLE FOR AIR

Altitude (ft)	Relative density (at 70°F)[a]	Barometric pressure (in Hg)	Altitude (m)	Relative density (at 21·1°C)[a]	Barometric pressure (kN/m^2)
0	1·000	29·92(**)	0	1·000	101·32(**)
1 000	0·964	28·85	300	0·965	97·78
2 000	0·930	27·82	600	0·931	94·33
3 000	0·896	26·81	900	0·898	90·09
4 000	0·864	25·84	1 200	0·866	87·75
5 000	0·832	24·89	1 500	0·835	84·61
10 000	0·687	20·57	3 000	0·692	70·12

[a] Relative density is numerically equal to the ratio of barometric pressures to standard at sea level (**).

LOAD CALCULATION

Standard room sensible heat calculations (with allowance for high altitude on solar gains) and room latent heat calculations can be made. Since this load must be absorbed by a weight of air the air volume will change at different altitudes to obtain the same effect.

DESIGN CONDITIONS

Room and outside air conditions must be adjusted to the required elevation. When DB and percentage RH are given, divide the grains per pound (grains per kilogramme) at sea level for the corresponding condition by the relative density to give the grains per pound (grains per kilogramme) content.

EFFECTIVE ROOM SENSIBLE HEAT FACTOR (See Chapter 9)

When the by-passed air load is added to the RSH and RLH respectively (use ventilation standards to obtain required cubic feet per minute (*cubic metres per second*) at sea level, and use these figures with standard air factors with ΔT and ΔG obtained from design conditions above) calculate the ESHF and adjust it as below:

$$ESHF(E) = \frac{1/(P_1)(1 - ESHF)}{(P_0)(ESHF)} + 1$$

where P_0 = barometric pressure at sea level
P_1 = barometric pressure at high level
ESHF = obtained from load estimate
ESHF(E) = equivalent ESHF referred to at sea level psychrometric charts

DESIGN AIR VOLUME

With the new ESHF find the ADP on the standard psychrometric chart. Use the formula for calculating 'dehumidified cfm', and divide by the relative density factor to give the design air volume. (The actual

1 PSYCHROMETRICS 19

TABLE 1.3
ADJUSTMENTS FOR HIGH ALTITUDE APPLICATIONS

Design calculation for	Correction applied to	Correction	Result	Used to determine
Cooling load	Factors 1·08 (*1·21*) and 0·68 (*2·98*)	1	Value at job condition	Outside air and infiltration air loads. Dehumidified air quantity
	Air moisture content	2	Value at job condition	Outside air and infiltration air loads. Dehumidified air quantity
	Solar gain	3	Value at job condition	Solar gain through glass
	Sensible heat factor	4	Equivalent SHF	ADP from sea level psychrometric chart
Dehumidifier	Air volume at job condition	1	Air volume at sea level	Apparatus size from manufacturer's data. Sea level pressure drop through cooling coil
	Sea level static pressure loss	5	Pressure loss at job	System pressure loss

TABLE 1.3—continued

Design calculation for	Correction applied to	Correction	Result	Used to determine
Condensers				
water-cooled	No correction required			
air-cooled	Air volume at job conditions	6	Power necessary at job	Apparatus size and sea level pressure drop
	Fan power from ratings			Condenser fan motor size
evaporative	Select cond. at sea level	None		
	Fan power from ratings	1	Power necessary at job	Fan motor size
Ductwork pressure loss	Pressure losses from friction chart using air volume job conditions	1	True pressure loss at job conditions	Ductwork pressure loss
Fans	Total system pressure loss	5	Sea level static pressure	Fan size from standard fan selection charts
	Table rev/min	None		
	Table power required	1	Power at job conditions	Fan motor size
Compressor	No correction necessary for selection. Correct gauge pressure for lower atmospheric pressure			
Electric motors	Check with motor manufacturer above 3000 ft (900 m)			
Cooling tower	No correction required			
Heating coil	Same as dehumidifier			
Filters	Same as dehumidifier			
Pumps	No correction for closed system. Allow for decreased atmospheric pressure on suction side of pump			

OA volume will be that required for standard air also divided by the relative density factor.)

EQUIPMENT

At high altitudes certain items of air-conditioning apparatus are affected. These are shown in Table 1.3 where the correctors are:

1. Multiply by relative density at given altitude.
2. Correct for grains of moisture at given altitude.
3. Check with appropriate system design manual.
4. Use ESHF(E) as shown.
5. Divide by relative density at given altitude.
6. Divide by the square of the relative density at given altitude.

It should be noted that psychrometric charts are available for different altitudes; when using these only the ventilation air cubic feet per minute and dehumidified air cubic feet per minute need be adjusted by the relative density factor.

Chapter 2

THE REFRIGERATION CYCLE

An understanding of the refrigeration cycle is best obtained by considering the most common fluid—water. Water exists in three states, solid (ice), liquid (water) and gas (steam) and the various states depend on a pressure–temperature relationship.

Consider ice, at 32 °F *(0 °C)* melted to water at atmospheric pressure by the application of heat. The temperature would remain the same, because the heat has been used to change the state of the ice, from solid to liquid. This is known as the latent heat of fusion, and at atmospheric pressure would amount to 144 BTU/lb *(337 kJ/kg)* of ice. To raise the temperature of the water by 1 °F would require 1 BTU/lb or *1 °C would require 4·18 kJ/kg*.

It is interesting to note at this point that, in the days before mechanical refrigeration, when the 'ice box' was used, once the ice was melted and the latent heat removed, little value was placed on the remaining water.

When water is boiled at atmospheric pressure it does so at 212 °F *(100 °C)* and turns to steam. This change in state also occurs at constant temperature, and is known as the latent heat of evaporation.

However, if the pressure at which the water is being boiled is changed, so the temperature at which boiling takes place changes, together with the amount of heat required to change this state. Water boiled on the top of a mountain in a rarefied atmosphere, or in a pressure cooker, are examples.

Any additional heat to the steam would be superheat, which would transform the dry, saturated vapour to superheated steam. The very first patent to be granted on a mechanical refrigeration machine perhaps explains best of all how the properties of a fluid can be used to

2 THE REFRIGERATION CYCLE

advantage. An extract reads: 'What I claim is an arrangement whereby I am enabled to use volatile fluids for the purpose of producing the cooling or freezing of fluids and yet at the same time constantly condensing such volatile fluids and bringing them again and again into operation without waste.'

If water was considered as the volatile fluid, the cycle could be considered as follows. Heat would be added to water to bring it to boiling temperature and then change its state from liquid to gas (water to steam) and the addition of superheat would make the gas extra dry. This would take place in an evaporator. This dry steam could then be cooked and be de-superheated, condensed from dry saturated vapour to liquid by removing the latent heat and then subcooled back to water. This would take place in a condenser.

If a compressor was introduced after the evaporator and the dry gas compressed, further superheat would be added by the act of compression and the high pressure gas would now boil or condense at a higher temperature. Therefore the act of condensing would take place at a higher temperature than the evaporation. If, after condensing, with a certain amount of subcooling to the water, this water could be expanded at no change in total heat (the evaporation of part of the fluid cooling the balance) the resulting low pressure water and vapour mixture could then be reheated in the evaporator, recompressed and so used over and over again without waste (Fig. 2.1).

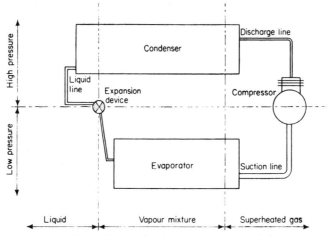

FIG. 2.1. The refrigeration cycle, showing suction and discharge lines.

Unfortunately, water is not an ideal fluid to use in a refrigeration plant, since to achieve the low temperatures required for air cooling, evaporation would have to take place at a high vacuum, requiring a large compressor to move the high-volume, low-weight, superheated steam.

There are, however, many other fluids which boil at much lower temperatures than water, which can be economically used in refrigeration plant. Table 2.1 shows commonly used refrigerants and their pressure–temperature relationship.

TABLE 2.1
COMMON REFRIGERANTS

Refrigerant	Boiling temp. at standard atm. pressure		Pressure			
	°F	°C	at 40°F sat. 4·4°C sat.		at 100°F sat. 37·8°C sat.	
			lb/in² abs.	kN/m²	lb/in² abs.	kN/m²
R.11	74·7	23·7	7·02	48·3	23·45	161·5
R.12	−21·6	−29·8	51·67	356	131·86	908
R.22	−41·4	−40·8	83·20	574	210·60	1451
R.502	−50·0	−45·6	94·90	654	229·10	1578

It can readily be seen, by comparing the two most common refrigerants used in air conditioning systems, R.12 and R.22, that the pressure associated with R.22 at 40°F ($4·4°C$), a common evaporating temperature in air conditioning systems, is much higher than R.12. What this effectively means is that with reciprocating compressor plants the weight handled (and consequently refrigeration effect) by each stroke of a compressor is greater, meaning smaller compressors per unit of refrigeration with R.22 than R.12. However, the actual power input per unit of refrigeration is about the same, provided the evaporating and condensing levels are comparable.

The common unit of refrigeration is the ton R. This unit is worth 12 000 BTU/h which is the amount of heat required to melt one ton (2 000 lb) of ice at atmospheric pressure in one day.

$$1 \text{ ton R} = \frac{2000 \text{ lb} \times 144 \text{ BTU/h (latent heat of fusion)}}{24}$$
$$= 12\,000 \text{ BTU/h}$$
$$(= 3517\,W)$$

This unit has been readily adopted by air-conditioning engineers because of its low value in terms of units, and also that at normal air-conditioning levels it requires about one horsepower of compressor work.

To examine the refrigeration cycle closely the easiest tool is the pressure–enthalpy (or P–H) chart. If a receiver of refrigerant R.22 was readily available at a temperature of 100 °F (*37·8 °C*) and was at a

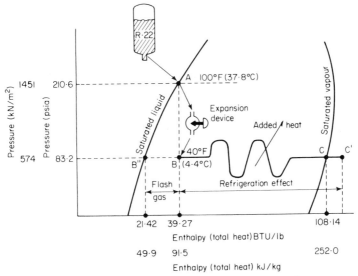

FIG. 2.2. P–H chart showing the expansion and evaporation process.

saturated state (no subcooling) it would exist at a pressure of 210·6 lb/in² absolute (*1451 kN/m²*). This liquid could be throttled through a metering device (expansion valve) and brought to a low pressure without change of total heat; this process is shown in Fig. 2.2 by the line AB. The resulting point B on the P–H chart shows that the liquid is no longer in a saturated state and a certain amount has 'flashed off' to cool the remaining liquid to (in this example) 40 °F (*4·4 °C*), corresponding to a pressure of 83·2 lb/in² absolute (*574 kN/m²*). Heat can be added to this vapour at point B at constant pressure to point C by passing the fluid through a coil or evaporator. At point C the vapour has been turned to a dry saturated vapour which is now capable of being compressed. In practice the metering device, or thermal expansion valve, is controlled from a sensor located at point C, which controls the amount of fluid handled to ensure the gas is slightly superheated, giving a little extra refrigeration effect, and ensuring that only dry gas is handled by the compressor. The actual point of entry to the compressor is shown by the point C'.

If at the pressure of 83·2 lb/in² absolute ($574 \, kN/m^2$), the fluid was saturated at point B″ a greater refrigeration effect would result.

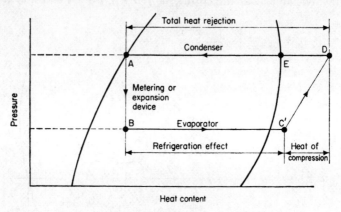

FIG. 2.3. Simplified refrigeration cycle.

However, the action of throttling produces flash gas which can be represented by the line B″B which compared to line B″C can be termed the 'per cent flash gas'.

Figure 2.3 shows the completed P–H chart for the refrigeration cycle. The gas at point C' is compressed and has heat added—the heat of compression, which is the work required by the compressor motor to elevate the pressure from 83·2 lb/in² absolute ($574 \, kN/m^2$) to return to the condensing level of 210·6 lb/in² absolute ($1451 \, kN/m^2$). The gas at the compressor outlet is at point D. Passing this gas through the condenser to point E removes the superheat, and the gas is then condensed from point E to A, where the whole process can be repeated.

It can be seen readily from Fig. 2.3 that the amount of heat to be removed by the condenser, the total heat rejection, is equal to the sum of the refrigeration effect plus the heat of compression.

Between the two saturation lines of the P–H chart, for any given refrigerant, the pressure is directly related to temperature, and for convenience in selecting refrigeration equipment the performance is always related to the saturation vapour temperatures, rather than the pressure. In the subcooled region (see Fig. 2.4), constant temperature lines can be shown as vertical lines on the P–H chart. In the superheated region constant temperature lines are shown as almost vertical lines sloping away from the saturated vapour lines. Also in the superheat region constant volume lines are shown as almost

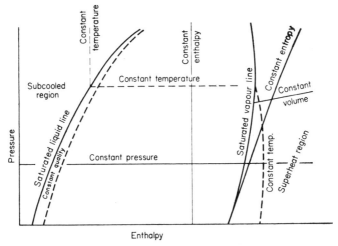

FIG. 2.4. Skeleton P–H chart.

horizontal lines sloping upwards. Constant entropy lines (lines of ideal compression) are shown as almost vertical lines moving away from the saturated vapour line.

Within the saturated liquid and vapour lines are constant quality lines which can be read as the per cent flash gas.

With this information the actual refrigeration cycle (Fig. 2.5) with allowances for the pressure drops in the suction and discharge (hot gas) lines as shown in Fig. 2.1, together with superheat subcooling effects, is readily shown.

In practice, suction and discharge lines are designed and sized to a pressure drop equivalent to a 2°F (*1·1°C*) drop in saturate at 40°F (*4·4°C*) and to condense at 100°F (*37·8°C*). Therefore, if a refrigeration system was designed to evaporate at 40°F (*4·4°C*) and condense at 100°F (*37·8°C*), the compressor would be selected at 38°F (*3·3°C*) saturated suction temperature and 102°F (*38·9°C*) saturated discharge temperature.

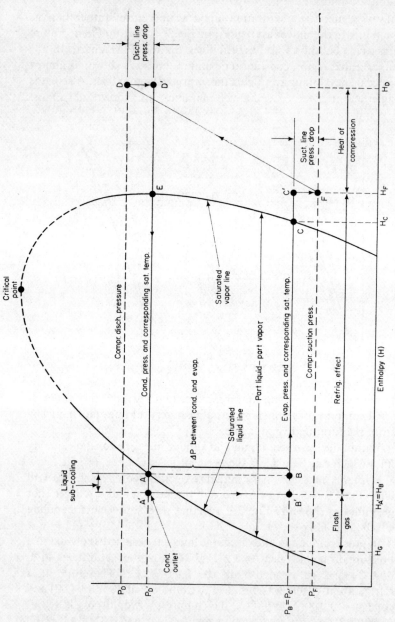

Fig. 2.5. Refrigeration cycle on idealised pressure–enthalpy diagram.

Most condensers, air, water or evaporative cooled, will provide 5 °F (*2·8 °C*) of subcooling. This, in the example shown, would be the equivalent enthalpy or total heat of saturated liquid at 100 °F − 5 °F or 95 °F (*37·8 °C − 2·8 °C or 35 °C*), thus providing a greater refrigeration effect. The addition of a subcooling coil on a condenser could further enhance the refrigeration effect. For example, if 15 °F (*8·3 °C*) subcooling was provided this would be equivalent to a saturated liquid condition of 85 °F (*29·4 °C*). The effect of subcooling would be that for every extra degree Fahrenheit of subcooling, one-half per cent (or for every extra degree Celsius subcooling about one-quarter per cent) extra refrigeration effect could be achieved.

As with subcooling the more superheat added to the gas the greater the refrigeration effect. Compressor selections have correction factors which take this into account. By installing a heat exchanger between the suction gas and liquid line, extra subcooling and superheating is achieved. However, while with R.12 systems this is permissible, care must be exercised with R.22 systems since high superheats could result in high discharge temperatures, which may exceed the manufacturer's recommended limit.

From the P–H chart the following information can readily be obtained:

$$\text{coefficient of performance (COP)} = \frac{H_{c'} - H_b}{H_d - H_f} = \frac{\text{refrigeration effect}}{\text{heat of compression}}$$

$$= \frac{\text{capacity in tons R} \times 12\,000}{\text{compressor bhp} \times 2545}$$

or

$$= \frac{\text{capacity in tons R} \times 12\,000}{\text{compressor kW} \times 3414}$$

In the heat pump cycle the COP is measured as the heat rejection as a ratio of the compressor power

$$= \frac{\text{refrigeration effect} + \text{heat of compression}}{\text{heat of compression}}$$

$$= COP + 1$$

$$\text{ideal theoretical COP (carnot COP)} = \frac{T_c}{T_{d'} - T_c}$$

where T_c = evaporator temperature in absolute units (evaporating temp. °F + 460 °F [*°C + 273 °C*]) and $T_{d'}$ = condensing temperature in absolute units (condensing temp. °F + 460°F [°C + 273°C])

$$\text{carnot efficiency} = \frac{\text{actual COP}}{\text{carnot COP}}$$

which is always less than 1.

per cent flash gas = quality of the refrigerant

$$= \frac{H_a - H_g}{H_c - H_g} = \frac{H_{b'} - H_g}{H_c - H_g}$$

$$\text{compression ratio} = \frac{\text{discharge pressure} = P_d}{\text{suction pressure} = P_f}$$

These pressures must be expressed in absolute units.

$$\text{Discharge temperature } T_d = T_{c'} \left(\frac{P_d}{P_f}\right)^{(n-1)/n}$$

where n is the compression exponent.

n varies with different refrigerants, different compressors and with the compression ratio. Therefore, every manufacturer has his own compression exponent. Since n is dependent upon the compression ratio and refrigerant for any given compressor the above formula could be rewritten as follows:

For any given compression ratio:

$$T_d = T_{c'} \times C$$

$$\text{where } C = \left(\frac{P_d}{P_f}\right)^{(n-1)/n}$$

Typical C values are listed in columns.

Compression ratio	C Value R.12	R.22	R.502
2	1·14	1·17	1·13
3	1·19	1·25	1·22
4	1·23	1·31	1·27
5	1·26	1·36	1·30
6	1·29	1·40	1·33
8	1·33	1·47	1·37
10	1·36	1·53	1·40
12	1·38	1·57	1·41

2 THE REFRIGERATION CYCLE

For example, if the previous figures were used for R.22 at 40 °F (4·4 °C) evaporating and 100 °F (37·8 °C) condensing and allowing 10 °F (5·5 °C) superheat.

P_f = Saturated pressure at 40 °F − 2 °F suction line loss (4·4 °C − 1·1 °C)
 = 80·34 lb/in² absolute (554 kN/m²)
P_d = Saturated pressure at 100 °F + 2 °F discharge line loss (37·8 °C + 1·1 °C)
 = 216·45 lb/in² absolute (1492 kN/m²)

$$\text{Compression ratio} = \frac{P_d}{P_f} = \frac{216\cdot 45}{80\cdot 34} = 2\cdot 7$$

$$(or \quad = \frac{1492}{554} = 2\cdot 7)$$

From the above table, the value for C for R.22 at 2·7 compression ratio would be 1·22

$$T_{c'} = (40\,°F + 10\,°F \text{ Superheat}) + 460\,°F = 510\,°R$$

Therefore,

$$T_d = 510 \times 1\cdot 22 = 622\,°R = 162\,°F$$

(or $T_{c'} = (4\cdot 4\,°C + 5\cdot 5\,°C$ Superheat$) + 273\,°C = 282\cdot 9\,K$
therefore $T_d = 282\cdot 9 \times 1\cdot 22 = 295\cdot 1\,K = 72\cdot 1\,°C$).

When using hermetic compressors with refrigerant cooled motor windings, in the absence of manufacturer's data, it is good practice to make an additional superheat allowance of 25 °F (14 °C) over and above that considered for the action of the thermal expansion valve. Similarly, allowance should always be made for systems using suction/liquid interchangers.

Chapter 3

SITE SURVEY

Enough emphasis cannot be placed on this very fundamental subject. From a site survey all other aspects spring: load estimation; system design and application; location of plant and services; minimal installation requirements; efficient service point location; running costs; cost estimate; company profitability and reputation. All too often such important data are presented to a design engineer 'on the back of a cigarette packet' with resultant inaccuracies.

A proper site survey will indicate:

1. The true cooling and/or heating requirements.
2. The possibilities of greatest load reduction with least cost.
3. The most economical equipment selection and location.
4. The most efficient air distribution.

To establish the above, much information is required, and this can take considerable time to evaluate. In consideration of this the sales personnel, who either gain the information or instruct an engineer to gain the information, should first consider the seriousness of an enquiry. All too often schemes requiring hours of preparatory work are presented, only to be shelved later through lack of finance. A very sensible first step before a detailed site survey is undertaken is to present the client with some idea of cost. This is only established after the experience of several jobs by each particular company who have evaluated an average cost per floor area for different types of applications. Having assured himself that the exercise is worthwhile, the engineer should set about establishing the following information.

1. Physical plans and elevations of the area to be air conditioned. Where possible, architects' drawings are naturally preferred. The

3 SITE SURVEY

drawings should indicate the usage of adjacent areas and whether they are already air conditioned and/or adequately heated.

2. The areas and construction of the various fabrics: sufficient to establish Heat Transfer Coefficients for load estimation.

3. The orientation of the building or area considered.

4. The location of adjacent buildings offering shade. The building plan alone may, for example, show a south-facing showroom window which is never in sunlight because it is located in a narrow main street.

5. Are any shading devices such as venetian blinds or outside awnings fitted to windows? If not, would they be acceptable to the client if they considerably reduced the cooling load, which decreases the capital and running costs of the plant?

6. What is the occupancy and the diversity? Several factors should be considered here, if for example it is a comfort application, the peak solar and transmission gain would occur when the building is unoccupied (i.e. a department store may have its maximum occupancy on December 24th; a restaurant may have a poor lunch trade, but a good evening trade; a showroom with east-facing glass giving a high solar gain in the morning may only do business in the afternoon).

7. What is the lighting load? Often spaces with large window areas do not have the lights turned on at times of peak solar gain, and it may be worth examining whether the lighting or solar gain provide the greatest load if they are not cumulative. It should be remembered that fluorescent lights contain a ballast which increases the heat input to the conditioned space by approximately 25 per cent above the rated output.

8. What is the machinery load and its diversity? Rarely does machinery dissipate its nameplate horsepower into a conditioned space. Probably the best example of this is a tool-room which requires 'standard' conditions. There could be several lathes which are not working when the operator is setting up, there could be machines with driving motor or the driven component outside the conditioned space. Often electric-meter readings giving average use or tariff maximum demand figures can give a much more accurate assessment of the true heat gain from machinery.

9. Is there any process load? Materials used in a process or manufacturing area can add or subtract both sensible and latent heat. For example raw material may be introduced into a manufacturing area at a high temperature and after processing may leave at

conditions near room temperature, or the raw material may be hygroscopic and absorb or adsorb moisture. In such cases, the quantity, quality entering and leaving the conditioned space should be evaluated together with the specific heat of the product.

10. What is the conditioned space to be used for? Often the engineer can advise the client as to what temperature and/or humidity conditions should be maintained in the room together with the economical control tolerances. Too often the client has his own fixed and uninitiated ideas of what conditions he requires, which, without him being aware of it, may radically increase the installation cost of the plant. Particular emphasis should be placed on 'comfort' installations which do not require such close tolerances as many 'process' or 'manufacturing' installations.

11. Are there any special filtration requirements? The process within the conditioned space will most often dictate its own specialities. Particular attention should be paid to any filtration methods already adopted by the client who because of satisfaction and/or standardisation may specify a certain type.

12. Where can the plant be located? Particular attention should be paid to the location of and access to equipment rooms, with emphasis on the following:

(a) Is the plant room close to the conditioned space so that air and/or water services may be installed without serious disruption of existing or planned walls, furnishings, or other services?
(b) Can equipment be lifted or positioned into the area allocated?
(c) Can drains from evaporators or humidifiers be run to a convenient point and proper trapping be ensured?
(d) Can fresh air for ventilation requirements be introduced?
(e) Are the locations of external areas for condenser or cooling tower installations sufficiently adjacent to avoid the need for extensive pipework runs?
(f) Is there a local water supply for humidification purposes? If so at what pressure?
(g) Is there an electrical supply available in the area for the machinery needed? Particular note should be made as to power available from existing mains since the provision of an additional supply can be both expensive and inconvenient. It should also be remembered that refrigeration plants over 2

3 SITE SURVEY

tons R (6 kW) capacity practically always require a three-phase supply, which is rarely available in residential or small commercial establishments.

13. Is there an existing heating medium? In most existing buildings heating will have already been catered for and it would more than likely prove economical to tap this heating source, if of sufficient capacity, to provide the heat required for the air-conditioning system. On new installations procedure 12 must be carried out to cater for a boiler-room or additional electrical requirements.

14. Is free-blow equipment or ducting required? Obviously free-blow is more economical but it could impair room distribution and it requires equipment servicing within the conditioned space.

15. Whose responsibility are the following, which are often excluded by the air conditioning contractor but must be considered by the client in establishing his final cost?

(a) Builders' work in cutting away holes and access points, making or preparing foundations, etc.
(b) Electrical wiring from the main isolation point to the equipment and its controls.
(c) The supply of electrical switch-gear and isolators. Often a client may have standards which must be adhered to.
(d) Plumbers' work running drains from evaporators, humidifiers, and cooling towers and water supplies to humidifiers and cooling towers.
(e) The provision of the heating medium from source to equipment.
(f) Off-loading and positioning the equipment into its final position. Attention should be paid to the requirements of any special lifting gear or cranes.

16. Is there any other special requirement other than those above which the client wants taken into consideration?

With the above information a load estimate and system design can be prepared. However, all too often the person making the load survey is not the design engineer preparing the scheme. Despite all the information established, on existing buildings it is very difficult to present it in such a way that the design engineer can get a proper 'feel' for the job. For example, a dimensional drawing may not reveal a Grecian freize or Adam ceiling which the design engineer may have

DETAILS OF SITE SURVEY

Client .... Contact .... Survey by ....
Address .... Site .... Date ....
.... Proposal reqd. ....
.... Inquiry No. ....
.... Design by ....

1. Space is used for ....
2. Required room conditions .... °F (°C) ± .... °F (°C) .... %RH ± .... % Filtration
3. Floor Area .... × .... = ... ft² (m^2) × Height = ... ft³ (m^3) room volume
   See overleaf/architects' drawings for details of fabric areas and construction, and space orientation
   See overleaf/architects' drawings of adjacent building location and heights
4. Occupancy .... People at .... a.m./p.m. to .... a.m./p.m. Activity ....
5. Shading devices Type .... Window orientation ....
6. Lighting load .... Watts (nameplate), Fluorescent/incandescent. In use during solar gain ....
7. Machinery load .... kW/h Diversity ....
8. Process load details ....
9. Existing heating services .... Where ....
10. Water services .... Where ....
11. Builder's work by ....
12. Electrical wiring by ....

13. Isolators and switchgear by ....
14. Plumbing work by ....
15. Off loading by ....
16. Remarks or special instructions ....
....
17. Approximate price .... ft² (m²) ×
18. Approximate load .... ft² (m²) × BTU/h/ft² (W/m²) .... BTU/h (W) .... Tons refrigeration
19. Condensing medium .... Location ....
20. Tentative selections
    Equipment   A   B   C

    A. Filter removal
    B. Motor service
    C. Shaft removal
    D. Drains
    E. Heat exchange removal
    F. Condenser air flow
    G.

Notes:
1. Show probable plant room location, equipment position and ductwork run on drawing overleaf.
2. Show best refrigeration, water, heating medium, and drain pipework runs on drawing overleaf.

adopted as his support for sheet-metal trunking to the dismay of the client. Here both sales and design engineer should appreciate what the probable equipment and associated services would be before a detailed appraisal is made. Very approximate check figures can be used to establish probable equipment sizes on the following basis.

30 BTU/h/ft^2 ($95\ W/m^2$) floor area for external areas with average occupancy.

40 BTU/h/ft^2 ($125\ W/m^2$) floor area for internal areas with average occupancy and exposure, or internal areas with high occupancy.

50 BTU/h/ft^2 ($160\ W/m^2$) floor area for external areas with large glass exposures or high occupancy, or for process work.

While the above figures are very approximate they do make possible a tentative equipment selection which can be visualised at the site survey. There are three basic advantages in making this approximate assessment. Firstly, it could confirm or be the basis of the initial costing to establish the client's interest. Secondly, it enables the sales engineer, if he knows what equipment is available, to establish whether the area suggested for a plant room is adequate and to gain an idea of the cross-section and probable run of sheet-metal ductwork. This information can be shown on the sketch plans so that the engineer has a start in appreciating the job. Thirdly, it gives the client some idea of what he is getting, giving him an interest and an opportunity to raise objections to duct or pipework runs before they are detailed.

Practically every problem has more than just one answer in terms of an air-conditioning system and the survey could show immediately two probable solutions, although the best answer may only be forthcoming after a detailed economic appraisal. Nevertheless, from the alternative systems which could suit the job the following information can be established.

1. What forms of condensing? (Air or water cooled.)
2. Is there sufficient space to remove filters for cleaning or disposal?
3. Can the shaft on any centrifugal fan be withdrawn? (This would also cover sufficient space to remove wheels or fan scrolls on packaged equipment.)
4. Are fan motors in a serviceable and ventilated position?
5. Are the refrigeration controls and automatic controls in an accessible position?

3 SITE SURVEY

6. Can drains be run with proper trapping? (This has already been mentioned, but since it is the most common error made it is felt that a second reference would be of no harm.)

7. What is the refrigeration pipework, condenser water pipework and/or heating pipework run? Will they impair aesthetic appearance of the building? Are they too far for proper or economical installation? Are there too many walls or floors to go through?

8. Is there sufficient space for the removal of heat exchangers for service or cleaning? (i.e. cooling or heating coils, water cooled condenser or direct expansion evaporator tubes.)

9. Is there sufficient free space to allow proper air flow to and from air cooled condensers or cooling towers without causing starvation or short-circuiting of the air?

10. Because of adjacent or adjoining property is there a noise problem with condensers or cooling towers that are located outside?

11. Is local code approval required for site equipment located outside which may be unsightly because of its bulk and/or location?

Having established this considerable amount of information a fair appreciation is gained of the problems involved, and a proper load estimate can be prepared which will result in economical equipment selection, proper air distribution and an accurate estimate of cost. From such a firm foundation satisfactory and profitable installations can be made.

To assist in the preparation of a site survey a detailed proposal sheet on which pertinent data can be recorded is shown on pages 36–37 and is reproduced by courtesy of 'Refrigeration and Air Conditioning' Magazine.

Chapter 4

LOAD ESTIMATING

Once a true site survey has been made an accurate assessment of the heat gain to the area must be established. The gain to an area can be split into four categories:

(a) Solar Gain through glazing and building fabric.
(b) Transmission Gain through glazing and building fabric.
(c) Internal Gain from occupants, lights, machinery, etc.
(d) Outside Air required for ventilation purposes.

All the above values depend on the time of day and time of year for which the estimate is made, which of course should be the maximum which can be expected.

Solar gain through glazing will most often dictate the time of day and time of year for which an estimate is made, always assuming of course that occupancy is constant. Examples from the tabular data for solar gain through glass show that east and west exposures reach a maximum at 10 a.m. and 5 p.m. respectively in June, whereas south exposures reach a maximum at noon to 2 p.m. in February or October. It may be necessary therefore to consider more than one load estimate to establish a maximum.

Before commencement of the load estimate, ambient design conditions and required room conditions must be established or assumed. Details of ambient design conditions have been established by meteorological establishments throughout the world and a comprehensive account of these can be found in 'ASHRAE Fundamentals' 1967. An extract from these data for the United Kingdom is shown in Tables 4.1 and 4.1a. The figures presented are for a design day which would occur at 3 p.m. in July or August.

4 LOAD ESTIMATING

Corrections to design conditions for other times can be found in Tables 4.2 and 4.3.

Internal conditions will depend on a project application in consideration of the need for human comfort or for process control. Recommended inside design conditions for various applications are presented in Table 4.4. It should be noted that commercial practice allows for a temperature swing during the day. This is highly

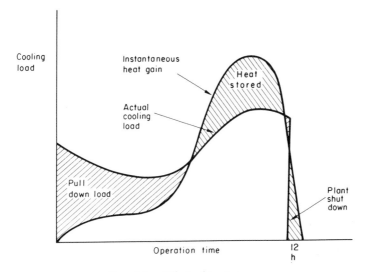

FIG. 4.1. Effect of heat storage.

recommended since it reduces the installed capacity of the plant, allowing it to run more continually, and in addition will allow the temperature difference between room and outside to be more tolerable.

The tabular data within this chapter consider the full effect of the storage which can be obtained from the building fabric during periods of maximum solar heat gain. Figure 4.1 illustrates how the actual cooling load lags behind the instantaneous solar gain for a west exposure and how it is substantially lower in value. The values assume a 12-hour plant operation, and it can clearly be seen that on start-up the plant requires to reduce that load which had been retained within the building fabric subsequent to the plant having previously been shut down.

Figure 4.2 shows how the installed plant can be further reduced by allowing a temperature swing at the time of maximum gain. The result is that the cooling capacity supplied to the space is lower than the cooling load, which means the temperature will rise. As the space temperature increases, less heat is allowed to enter the room from the building structure, and a greater solar gain is stored in the structure. It should be noted that allowance for temperature swing is made for a

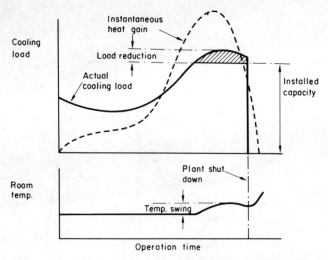

FIG. 4.2. Load reduction due to temperature swing.

design day, and under most frequent operation no swing in design temperature would occur. Tables 4.5 and 4.5a show the storage factors to be applied for room temperature swing.

Solar heat gains through glass are shown in Tables 4.6 to 4.14 with corrections in Table 4.15. The tables are presented in two sections, those for bare glass where solar gain to the internal surfaces is felt more quickly than the second section, and those for applications using venetian blinds. All values are for applications at 50 °N latitude, 12-hour plant operation and for building constructions equivalent to 100 lb/ft^2 ($488 kg/m^2$) of conditioned space floor area.

The cooling load from solar gain through glass is calculated as follows:

Glass area × solar gain × correction factors.
Less storage factor due to room temperature swing.

4 LOAD ESTIMATING

The above figures include only the solar gain through glass and do not consider transmission gain, which is discussed below.

Shading provided by adjacent buildings and/or overhangs and/or reveals must be considered when evaluating the solar gain through glass. The degree of shading for a given time of day and time of year will depend upon the solar altitude and azimuth angles. Table 4.16 lists these angles for 50°N from February to October. Figure 4.3 shows plans and elevations of air-conditioned buildings in relation to adjacent buildings providing shade. From the example it can clearly be seen that solar azimuth angles from 225° to 270° will offer some degree of shade as will solar altitude angles from 0° to 45°. Figure 4.4 shows the plan and elevation of a window having a reveal and an overhang.

The amount of shading from above is shown as x = tan (altitude angle) y.

The amount of shading from the side for various exposures can be shown as:

$$NE, x = \tan(45° - \text{azimuth angle}) y$$
$$E, x = \tan(90° - \text{azimuth angle}) y$$
$$SE, x = \tan(135° - \text{azimuth angle}) y$$
$$S, x = \tan(180° - \text{azimuth angle}) y$$
$$SW, x = \tan(225° - \text{azimuth angle}) y$$
$$W, x = \tan(270° - \text{azimuth angle}) y$$
$$NW, x = \tan(315° - \text{azimuth angle}) y$$

Other considerations to solar gain must be made to walls and roofs; this gain, however, must take into account the time lag for the solar energy to penetrate within the building. Very often the full effect of this gain, particularly with west-facing walls or south walls of heavy construction, is not felt until very late in the day, well after the peak solar gain through glass and often after the plant has shut down. It is convenient to show the solar gain through walls and roof, including the transmission gain, that is due to the temperature difference between room and ambient. Table 4.17 shows the equivalent temperature differences for sunlit and shaded walls and roofs and considers an ambient condition of 82°F (27·8°C) DB coincident with a room temperature of 72°F (22·2°C), a 16°F (8·9°C) Daily Range and for July at 50°N latitude. Corrections to these equivalent temperatures for different room and ambient conditions with various daily range temperatures are shown in Tables 4.18 and 4.18a.

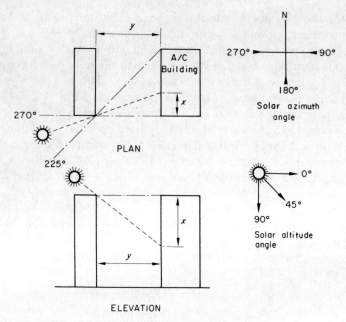

Fig. 4.3. Shading from adjacent buildings.

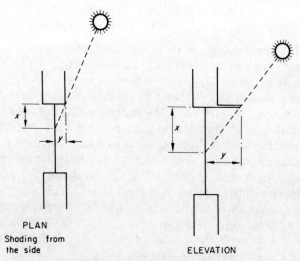

Fig. 4.4. Shading from overhangs and reveals.

4 LOAD ESTIMATING

The tables refer to dark coloured surfaces which are most capable of absorbing the radiant solar energy. The values assume an absorbivity of 0·9 whereas medium and light coloured walls have absorbivities of 0·7 and 0·5 respectively. The transmission quotient of the equivalent temperature is independent of colour and therefore only a partial correction should be made. The difference between the equivalent temperature at a given time for a given exposure and the same value for a wall in shade is the value which must be corrected. The following corrections can therefore be made.

For medium colours

eff. temp. = eff. shade temp. $+ \dfrac{0\cdot 7}{0\cdot 9}$(eff. solar temp. − eff. shade temp.)

For light colours:

eff. temp. = eff. shade temp. $+ \dfrac{0\cdot 5}{0\cdot 9}$(eff. solar temp. − eff. shade temp.)

The basic heat transfer equation for all surfaces is:

$$Q = UAt$$

where Q = heat flow BTU/h (W)
U = transmission coefficient BTU/h/ft^2/°F ($W/m^2/°C$)
A = area ft^2 (m^2)
t = equivalent temperature difference

For transmission gain through glass, the equivalent temperature difference can be considered as the design ambient temperature for the time of day considered, less the desired room condition, since there is little mass to withhold heat flow. Partition walls, ceilings and floors must take into consideration the temperature of the adjacent area, assuming that it is not air conditioned. (No consideration should be made for heat gain from adjacent areas which are air conditioned.) It is normal practice to assume that adjacent areas will attain ambient conditions, and the temperature difference can be taken as that for glass. However, where specific information is available true values must be taken. The designer must consider, for example, walls adjoining boiler rooms, where the temperature could rise 20 or 30°F (*11 or 17°C*) above ambient, or floors above shaded basements where the temperature could be 5°F (*3°C*) lower than ambient. When ground floors are considered allowance must be made for the effect of stabilised ground temperatures, which would approximate to 5°F (*3°C*) below the ambient temperature.

Tables 4.19 and 4.19a present U values—transmission coefficients, for commonly used building fabrics.

The transmission coefficient is the reciprocal of the summation of thermal resistances of the various fabrics constituting the wall, floor, ceiling or roof construction. This is shown by the equation:

$$U = \frac{1}{\Sigma R}$$

where $\Sigma R = r_1 + r_2 + r_3 + \ldots r_n$.

It can clearly be seen from Tables 4.20, 4.20a, 4.21 and 4.21a that the resistance to heat flow caused by air films, or air gaps can constitute a major factor.

Internal gains can account for a major proportion of the cooling load, and close attention should be paid to any diversity which can occur. Table 4.22 considers this diversity for various applications.

The heat generated from people is dissipated in three ways: by radiation in the form of sensible heat; by convection in the form of sensible heat; and by evaporation of moisture in the form of latent heat. The higher the dry bulb temperature of the conditioned space, the more the body depends on evaporation to provide its cooling, and so the greater the latent gain. The degree of activity also plays a major role in the split of sensible and latent heat dissipation and the total metabolic rate. Tables 4.23 and 4.23a show the heat gain that can be expected from occupants in various applications.

As the demands for increased lighting intensities grow so this heat gain is becoming more and more the dominant factor in cooling load estimates. Tables 4.24 and 4.24a show the heat gain in BTU/h (*W*) per nameplate watt for various applications and types of lighting. It should be stressed that the choke or ballast of fluorescent lights increases the heat output against nameplate watts by 25 per cent.

Heat gain from electric motors must be considered, i.e. whether the motor and driven machine are within the conditioned space, or just the motor, or just the driven machine, and also the size of the motor, since efficiencies differ. Tables 4.25 and 4.25a show the heat gain from motors, but close attention must be given to service factors and diversity, wherever possible a meter check should be made to assess the true loading for a given area, since machines in industrial applications are seldom used continually.

4 LOAD ESTIMATING

Infiltration of ambient air into the conditioned space can impose substantial latent and to a lesser degree sensible gains to the room. In most applications using proper ventilation standards (see Table 4.30) this infiltration can be offset by conditioned and filtered air from the air-conditioning apparatus. An allowance of 20 per cent should be added to the calculated infiltration rate to provide the minimum amount of ventilation air. Systems using all recirculated air, i.e. without proper ventilation, must consider infiltration as a room gain. The amount of infiltration to any conditioned area cannot easily be defined and several methods are available to the designer. The simplest method is to allow 1 to $1\frac{1}{2}$ air changes per hour, i.e. room volume × air change rate ÷ 60 = cfm infiltration (*room volume* (m^3) × *air change rate* ÷ *3600* = m^3/s *infiltration*). This arbitrary method is in common use and can be used as a check figure against a calculated method of, for example, considering the infiltration created by loosely fitted windows, and the effect of opening and shutting doors. Applications with fixed windows, particularly with double glazing, can consider infiltration at an absolute minimum, and air change rates of 0·25 to 0·5 are applicable.

Table 4.26 shows the infiltration rate for windows per linear of crack, and Table 4.27 the infiltration through doors in common usage. The addition to room gains can be expressed as:

Room sensible heat (BTU/h)
= 1·08 × cfm infiltration × temp. (outside − inside)°F
Room sensible heat (kW)
= *1·21 × m^3/s infiltration × temp. (outside − inside)°C*
Room latent heat (BTU/h)
= 0·68 × cfm infiltration × grs/lb (outside − inside)
Room latent heat (kW)
= *2·98 × m^3/s infiltration × grs/kg (outside − inside)*

At this stage both nett room sensible heat and room latent heat can be established. In addition to both these loads a percentage should be added for Duct Leakage Loss. This value can be ignored where all ducting is within the conditioned space, but should be as high as 10 per cent for high-velocity systems and 5 per cent for low-velocity systems.

Further additions to the room sensible heat are supply duct heat gain (Tables 4.28 and 4.28a) and fan horsepower gain (Table 4.29).

The requirement for proper ventilation standards imposes a

demand on the air-conditioning apparatus, but does not necessarily play a part in the room loads since the heat is taken out before this air enters the conditioned space. Table 4.30 shows the recommended ventilation rates per person for various applications. The heat gain can be calculated as:

Sensible cooling (BTU/h)
$= 1 \cdot 08 \times$ cfm ventilation $\times$ temp. (outside $-$ inside)°F
Sensible cooling (kW)
$= 1 \cdot 21 \times m^3/s \text{ ventilation} \times \text{temp. (outside} - \text{inside)}°C$
Latent cooling (BTU/h)
$= 0 \cdot 68 \times$ cfm ventilation $\times$ grs/lb (outside $-$ inside)
Latent cooling (kW)
$= 2 \cdot 98 \times m^3/s \text{ ventilation} \times grs/kg \text{ (outside} - \text{inside)}$

Alternatively:

Total cooling (BTU/h)
$= 4 \cdot 45 \times$ cfm ventilation $\times (H_{oa} - H_{ra})$ BTU/lb
Total cooling (kW)
$= 1 \cdot 19 \times m^3/s \text{ ventilation } (H_{oa} - H_{ra}) \, kJ/kg$

where H_{oa} = enthalpy outside air
H_{ra} = enthalpy room air.

4 LOAD ESTIMATING

UK WEATHER DATA
(Reprinted courtesy of ASHRAE, Fundamentals)

(Col. 1) Country and station	(Col. 2) Latitude and longitude	(Col. 3) Elevation (ft)	Winter (Col. 4) Mean of annual extremes (°F)	99% (°F)	97½% (°F)	Summer (Col. 5) Design dry bulb (°F) 1%	2½%	5%	(Col. 6) Outdoor daily range (°F)	(Col. 7) Design wet bulb (°F) 1%	2½%	5%
Belfast	54°36′N/5°55′W	24	19	23	26	74	72	69	16	65	64	62
Birmingham	52°29′N/1°56′W	535	21	24	27	79	76	73	15	66	64	63
Cardiff	51°28′N/3°10′W	203	21	24	27	79	76	73	14	64	63	62
Edinburgh	55°55′N/3°11′W	441	22	25	28	73	70	68	13	64	62	61
Glasgow	55°52′N/4°17′W	85	17	21	24	74	71	68	13	64	63	61
London	51°29′N/00°00′	149	20	24	26	82	79	76	16	68	66	65

TABLE 4.1a
UK WEATHER DATA

(Col. 1) Country and station	(Col. 2) Latitude and longitude	(Col. 3) Elevation (m)	Winter (Col. 4) Mean of annual extremes (°C)	99% (°C)	97½% (°C)	Summer (Col. 5) Design dry bulb (°C) 1%	2½%	5%	(Col. 6) Outdoor daily range (°C)	(Col. 7) Design wet bulb (°C) 1%	2½%	5%
Belfast	54°36′N/5°55′W	7	−7·2	−5·0	−3·3	23·3	22·2	20·6	8·9	18·3	17·7	16·7
Birmingham	52°29′N/1°56′W	163	−6·1	−4·5	−2·8	26·1	24·5	22·8	8·3	18·9	17·7	17·2
Cardiff	51°28′N/3°10′W	62	−6·1	−4·5	−2·8	26·1	24·5	22·8	7·7	17·7	17·2	16·7
Edinburgh	55°55′N/3°11′W	135	−5·5	−3·9	−2·2	22·8	21·1	20·0	7·2	17·7	16·7	16·1
Glasgow	55°52′N/4°17′W	26	−8·3	−6·1	−4·5	23·3	21·7	20·0	7·2	17·7	17·7	16·1
London	51°29′N/00°00′	45	−6·7	−4·5	−3·3	27·8	26·1	24·5	8·9	20·0	18·9	18·3

TABLE 4.2
CORRECTION IN AMBIENT DRY BULB TEMPERATURE FOR TIME OF DAY

Daily range temperature	Time										
	8 a.m.	9 a.m.	10 a.m.	11 a.m.	noon	1 p.m.	2 p.m.	3 p.m.	4 p.m.	5 p.m.	6 p.m.
10°F	−9	−8	−7	−6	−5	−3	−1	0	−1	−1	−2
5°C	−4.5	−4.0	−3.5	−3.0	−2.5	−1.5	−0.5	0	−0.5	−0.5	−1.0
15°F	−12	−11	−9	−7	−5	−3	−1	0	−1	−1	−2
7.5°C	−6.0	−5.5	−4.5	−3.5	−2.5	−1.5	−0.5	0	−0.5	−0.5	−1.0
20°F	−14	−12	−10	−8	−5	−3	−1	0	−1	−2	−3
10°C	−7.0	−6.0	−5.0	−4.0	−2.5	−1.5	−0.5	0	−0.5	−1.0	−1.5
25°F	−16	−14	−10	−8	−5	−3	−1	0	−1	−2	−3
12.5°C	−8.0	−7.0	−5.0	−4.0	−2.5	−1.5	−0.5	0	−1	−2	−4
30°F	−18	−16	−12	−9	−6	−3	−1	0	−1	−2	−4
15°C	−8.5	−7.5	−5.5	−4.5	−3.0	−1.5	−0.5	0	−0.5	−1.0	−2.0

Notes:
1. To find the design dry bulb temperature for a given time of day apply the above corrections to the maximum design temperature.
2. During a design day the ambient dewpoint would remain constant. From the psychrometric chart the wet bulb temperature can be established at design dewpoint coincident with corrected dry bulb temperature.

4 LOAD ESTIMATING

TABLE 4.3
CORRECTIONS IN AMBIENT DESIGN TEMPERATURES FOR TIME OF YEAR

Yearly range temperature		Temp.	Time of year												
			March		April		May		June		September		October		November	
°F	°C		°F	°C	°F	°C	°F	°C	°F	°C	°F	°C	°F	°C	°F	°C
50	28	DB	−5	−3	−4	−2	−3	−1·5	−1	−0·5	−2	−1	−4	−2	−7	−4
		WB	−3	−1·5	−2	−1	−1	−0·5	0	0	−1	−0·5	−2	−1	−3	−1·5
60	33	DB	−9	−5	−7	−4	−3	−1·5	−1	−0·5	−2	−1	−5	−3·0	−10	−5·5
		WB	−4	−2	−3	−1·5	−2	−1	0	0	−1	−0·5	−3	−1·5	−5	−3·0
70	39	DB	−13	−7·5	−9	−5	−4	−2	−1	−0·5	−2	−1	−7	−4	−14	−8
		WB	−6	−3·5	−4	−2	−2	−1	0	0	−1	−0·5	−4	−2	−6	−3·5
80	44	DB	−24	−13·5	−16	−9	−8	−4·5	−3	−1·5	−4	−2	−12	−6·5	−20	−11
		WB	−13	−7·5	−9	−5	−4	−2·0	−2	−1·0	−2	−1	−6	−3·5	−11	−6
90	50	DB	−29	−16	−19	−11	−10	−5·5	−3	−1·5	−6	−3·5	−16	−9	−26	−14
		WB	−14	−8	−10	−5·5	−5	−3·0	−2	−1·0	−3	−1·5	−8	−4·5	−14	−8
100	55	DB	−29	−16	−19	−11	−10	−5·5	−3	−1·5	−6	−3·5	−16	−9	−27	−15
		WB	−14	−8	−10	−5·5	−5	−3·0	−2	−1·0	−3	−1·5	−8	−4·5	−14	−8

Notes:
1. Yearly range temperature is the difference between design summer and winter dry bulb temperatures.
2. The above corrections should be applied to the design summer dry bulb and wet bulb temperatures.

TABLE 4.4
RECOMMENDED INSIDE DESIGN CONDITIONS (NORTHERN EUROPE)

Application	Summer			Winter	
	Dry bulb	RH	Temp. swing	Dry bulb	RH
High latent load (restaurants, theatres, etc.)	70–73 °F 21–23 °C	60–50	1–2 °F 0·5–1 °C	68–70 °F 20–21 °C	40–35
Retail shops	72–74 °F 22–23·5 °C	55–50	2–4 °F 1–2 °C	66–68 °F 19–20 °C	35–30
General offices	72–74 °F 22–23·5 °C	50–45	2–4 °F 1–2 °C	68–70 °F 20–21 °C	40–35
De luxe	72 °F 22 °C	50	none	70 °F 21 °C	40
Factory	73–76 °F 22·5–24·5 °C	55–45	3–6 °F 1·5–3·5 °C	65–68 °F 18–20 °C	35–30

4   LOAD ESTIMATING

TABLE 4.5
EFFECT OF TEMPERATURE SWING, BTU/H/DEGREE FAHRENHEIT SWING/SQUARE FOOT FLOOR AREA

% Glass area to exposed wall	24-h operation (temp. swing °F)			16-h operation (temp. swing °F)			12-h operation (temp. swing °F)		
	1–2	3–4	5–6	1–2	3–4	5–6	1–2	3–4	5–6
25	1·35	1·25	1·20	1·25	1·00	0·90	1·20	0·95	0·70
50	1·50	1·40	1·30	1·35	1·30	1·20	1·30	1·25	1·10
75	1·70	1·60	1·45	1·50	1·45	1·35	1·40	1·35	1·30

Notes:
1. **Reduction in peak load** = (floor area) × (temp. swing) × (factor).
2. The above figures do not apply to north or shaded windows.

TABLE 4.5a
EFFECT OF TEMPERATURE SWING, W/DEGREE CELSIUS SWING/SQUARE METRE FLOOR AREA

% Glass area to exposed wall	24-h operation (temp. swing °C)			16-h operation (temp. swing °C)			12-h operation (temp. swing °C)		
	0·5–1	1·5–2	2·5–3·5	0·5–1	1·5–2	2·5–3·5	0·5–1	1·5–2	2·5–3·5
25	7·8	7·1	6·8	7·1	5·7	5·1	6·8	5·4	4·0
50	8·5	8·0	7·4	7·7	7·4	6·8	7·4	7·1	6·2
75	9·6	9·1	8·2	8·5	9·1	7·7	8·0	7·7	7·4

Notes:
1. **Reduction in peak load** = (floor area) × (temp. swing) × (factor).
2. The above figures do not apply to north or shaded windows.

TABLE 4.6
EFFECTIVE SOLAR HEAT GAIN THROUGH GLASS
(NORTH AND SHADE EXPOSURES, 50°N LATITUDE, 12-H OPERATION)

(A) BARE SINGLE GLAZING

Time of day (BST)	June BTU/h/ft²	June W/m²	July–May BTU/h/ft²	July–May W/m²	August–April BTU/h/ft²	August–April W/m²	September–March BTU/h/ft²	September–March W/m²	October–February BTU/h/ft²	October–February W/m²
7 a.m.	15	47	13	41	10	31	5	16	5	16
8 a.m.	16	49	13	42	10	31	5	16	5	16
9 a.m.	16	50	14	43	10	31	5	16	5	16
10 a.m.	17	53	14	44	11	33	5	16	5	16
11 a.m.	17	54	15	46	11	34	5	16	5	16
Noon	18	55	15	47	11	35	6	19	6	19
1 p.m.	18	56	15	47	11	35	6	19	6	19
2 p.m.	18	57	15	47	11	35	6	19	6	19
3 p.m.	18	57	15	47	11	35	6	19	6	19
4 p.m.	18	57	15	47	11	35	6	19	6	19
5 p.m.	18	57	15	47	11	35	6	19	6	19
6 p.m.	18	57	15	47	11	35	6	19	6	19

(B) SINGLE GLAZING WITH LIGHT COLOURED VENETIAN BLINDS

Time of day (BST)	June BTU/h/ft²	June W/m²	July–May BTU/h/ft²	July–May W/m²	August–April BTU/h/ft²	August–April W/m²	September–March BTU/h/ft²	September–March W/m²	October–February BTU/h/ft²	October–February W/m²
7 a.m.	11	35	9	30	7	22	6	19	4	13
8 a.m.	11	35	9	30	7	22	6	19	4	13
9 a.m.	11	35	9	30	7	22	6	19	4	13
10 a.m.	11	35	9	30	7	22	6	19	4	13
11 a.m.	11	35	9	30	7	22	6	19	4	13
Noon	11	35	9	30	7	22	6	19	4	13
1 p.m.	11	35	9	30	7	22	6	19	4	13
2 p.m.	11	35	9	30	7	22	6	19	4	13
3 p.m.	11	35	9	30	7	22	6	19	4	13
4 p.m.	11	35	9	30	7	22	6	19	4	13
5 p.m.	11	35	9	30	7	22	6	19	4	13
6 p.m.	11	35	9	30	7	22	6	19	4	13

4 LOAD ESTIMATING

EFFECTIVE SOLAR HEAT GAIN THROUGH GLASS
(NORTH EAST EXPOSURE, 50° N LATITUDE, 12-H OPERATION)

(A) BARE SINGLE GLAZING

Time of day (BST)	June BTU/h/ft²	June W/m²	July–May BTU/h/ft²	July–May W/m²	August–April BTU/h/ft²	August–April W/m²	September–March BTU/h/ft²	September–March W/m²	October–February BTU/h/ft²	October–February W/m²
7 a.m.	52	163	48	151	39	122	24	75	12	38
8 a.m.	67	211	62	195	50	157	31	98	15	47
9 a.m.	74	233	69	217	55	173	34	107	17	53
10 a.m.	73	230	67	211	54	170	33	104	17	53
11 a.m.	67	211	62	195	50	157	31	98	15	47
Noon	62	195	58	182	46	145	29	91	14	44
1 p.m.	50	157	47	148	37	116	23	72	12	38
2 p.m.	44	138	41	129	33	104	20	63	10	31
3 p.m.	40	126	37	116	30	94	18	57	9	28
4 p.m.	38	119	36	113	29	91	18	56	9	28
5 p.m.	34	107	32	101	25	79	16	50	8	25
6 p.m.	30	94	27	85	22	69	14	44	7	22

(B) SINGLE GLAZING WITH LIGHT COLOURED VENETIAN BLINDS

Time of day (BST)	June BTU/h/ft²	June W/m²	July–May BTU/h/ft²	July–May W/m²	August–April BTU/h/ft²	August–April W/m²	September–March BTU/h/ft²	September–March W/m²	October–February BTU/h/ft²	October–February W/m²
7 a.m.	50	157	45	142	36	113	22	69	11	34
8 a.m.	57	179	52	163	41	129	26	82	13	41
9 a.m.	54	170	49	154	39	122	24	76	12	37
10 a.m.	44	138	40	126	32	101	20	63	10	31
11 a.m.	29	91	27	85	21	66	13	41	7	22
Noon	24	75	22	69	18	56	11	34	7	22
1 p.m.	20	63	18	56	15	47	9	28	6	19
2 p.m.	19	60	17	53	14	44	9	28	4	13
3 p.m.	17	53	15	47	12	38	8	25	4	13
4 p.m.	16	50	14	44	12	38	7	22	4	13
5 p.m.	14	44	13	41	10	31	6	19	3	10
6 p.m.	13	41	11	34	9	28	6	19	3	10

See Table 4.15 for correction factors.

TABLE 4.8
EFFECTIVE SOLAR HEAT GAIN THROUGH GLASS
(EAST EXPOSURE, 50°N LATITUDE, 12-H OPERATION)

(A) BARE SINGLE GLAZING

Time of day (BST)	June BTU/h/ft²	June W/m²	July–May BTU/h/ft²	July–May W/m²	August–April BTU/h/ft²	August–April W/m²	September–March BTU/h/ft²	September–March W/m²	October–February BTU/h/ft²	October–February W/m²
7 a.m.	66	207	65	204	63	198	55	173	42	132
8 a.m.	85	267	84	264	82	258	71	223	55	173
9 a.m.	104	327	104	327	100	314	87	273	67	210
10 a.m.	112	352	111	349	108	340	94	296	72	226
11 a.m.	110	346	109	343	106	336	92	290	71	223
Noon	98	308	98	308	95	299	83	261	63	198
1 p.m.	85	267	84	264	82	258	71	223	55	173
2 p.m.	75	235	75	235	73	229	63	198	48	151
3 p.m.	66	207	65	204	63	198	55	173	42	132
4 p.m.	60	188	60	188	58	182	50	157	38	119
5 p.m.	54	170	54	170	52	163	45	141	35	110
6 p.m.	46	145	46	145	45	141	39	123	30	94

(B) SINGLE GLAZING WITH LIGHT COLOURED VENETIAN BLINDS

Time of day (BST)	June BTU/h/ft²	June W/m²	July–May BTU/h/ft²	July–May W/m²	August–April BTU/h/ft²	August–April W/m²	September–March BTU/h/ft²	September–March W/m²	October–February BTU/h/ft²	October–February W/m²
7 a.m.	57	179	56	176	55	173	47	148	36	113
8 a.m.	74	232	72	226	70	220	60	188	47	148
9 a.m.	80	252	79	253	77	247	66	207	51	160
10 a.m.	77	247	76	244	74	232	63	198	49	154
11 a.m.	68	214	63	198	61	192	52	163	41	129
Noon	44	138	43	135	42	132	36	113	28	88
1 p.m.	32	101	31	98	30	94	26	82	20	63
2 p.m.	29	91	28	88	27	85	24	76	18	57
3 p.m.	26	82	26	82	25	79	22	69	17	54
4 p.m.	23	72	23	72	22	69	19	60	15	47
5 p.m.	21	66	21	66	20	63	17	54	13	41
6 p.m.	18	57	17	54	17	54	14	44	11	34

4 LOAD ESTIMATING

(SOUTH EAST EXPOSURE, 50° N LATITUDE, 12-H OPERATION)

(A) BARE SINGLE GLAZING

Time of day (BST)	June BTU/h/ft²	W/m²	July–May BTU/h/ft²	W/m²	August–April BTU/h/ft²	W/m²	September–March BTU/h/ft²	W/m²	October–February BTU/h/ft²	W/m²
7 a.m.	46	145	49	154	54	170	56	176	54	170
8 a.m.	52	163	55	173	61	192	63	198	61	192
9 a.m.	65	204	69	217	76	238	79	249	76	238
10 a.m.	81	255	86	270	94	296	98	308	94	296
11 a.m.	92	290	97	305	107	337	111	349	107	337
Noon	97	305	102	320	113	355	117	368	113	355
1 p.m.	97	305	102	320	113	355	117	368	113	355
2 p.m.	89	279	94	296	104	326	108	340	104	326
3 p.m.	78	244	82	258	91	287	94	296	91	287
4 p.m.	70	220	74	232	81	255	84	264	81	255
5 p.m.	59	185	62	195	68	214	71	223	68	214
6 p.m.	52	163	55	173	61	192	63	198	61	192

(B) SINGLE GLAZING WITH LIGHT COLOURED VENETIAN BLINDS

Time of day (BST)	June BTU/h/ft²	W/m²	July–May BTU/h/ft²	W/m²	August–April BTU/h/ft²	W/m²	September–March BTU/h/ft²	W/m²	October–February BTU/h/ft²	W/m²
7 a.m.	16	50	17	54	19	60	19	60	19	60
8 a.m.	36	113	38	119	42	132	43	135	42	132
9 a.m.	51	160	54	170	59	185	61	192	59	185
10 a.m.	62	195	66	207	73	229	75	235	73	229
11 a.m.	67	210	71	223	78	244	80	252	78	244
Noon	64	201	68	214	75	235	77	241	75	235
1 p.m.	56	176	59	185	66	207	67	210	66	207
2 p.m.	44	138	46	145	51	160	52	163	51	160
3 p.m.	30	94	32	101	35	110	36	113	35	110
4 p.m.	25	79	26	82	29	91	30	94	29	91
5 p.m.	22	69	24	76	26	82	27	85	26	82
6 p.m.	19	60	20	63	22	69	22	69	22	69

See Table 4.15 for correction factors.

TABLE 4.10
EFFECTIVE SOLAR HEAT GAIN THROUGH GLASS
(SOUTH EXPOSURE, 50°N LATITUDE, 12-H OPERATION)

(A) BARE SINGLE GLAZING

Time of day (BST)	June BTU/h/ft²	June W/m²	July–May BTU/h/ft²	July–May W/m²	August–April BTU/h/ft²	August–April W/m²	September–March BTU/h/ft²	September–March W/m²	October–February BTU/h/ft²	October–February W/m²
7 a.m.	48	151	55	173	71	223	82	258	86	270
8 a.m.	40	126	46	145	60	188	69	217	73	229
9 a.m.	43	135	49	154	63	198	73	229	76	238
10 a.m.	47	148	54	170	70	220	80	252	84	264
11 a.m.	55	173	63	198	81	255	93	293	98	308
Noon	62	195	71	223	92	290	106	332	112	352
1 p.m.	70	220	80	252	104	326	119	373	125	393
2 p.m.	74	232	85	267	110	346	126	396	133	417
3 p.m.	76	238	88	276	113	355	130	408	137	430
4 p.m.	74	232	85	267	110	346	126	396	133	417
5 p.m.	69	217	79	249	102	320	117	367	123	387
6 p.m.	58	182	66	207	86	270	99	311	104	326

(B) SINGLE GLAZING WITH LIGHT COLOURED VENETIAN BLINDS

Time of day (BST)	June BTU/h/ft²	June W/m²	July–May BTU/h/ft²	July–May W/m²	August–April BTU/h/ft²	August–April W/m²	September–March BTU/h/ft²	September–March W/m²	October–February BTU/h/ft²	October–February W/m²
7 a.m.	16	50	18	57	23	72	27	85	29	91
8 a.m.	13	41	15	47	20	63	23	72	24	76
9 a.m.	23	72	26	82	34	107	39	123	42	132
10 a.m.	31	98	35	110	46	145	53	167	56	176
11 a.m.	39	123	44	138	58	182	66	207	70	220
Noon	45	141	50	157	66	207	75	235	80	252
1 p.m.	48	151	55	173	71	223	81	255	87	273
2 p.m.	48	151	55	173	71	223	81	255	87	273
3 p.m.	46	145	53	167	69	217	79	249	85	267
4 p.m.	40	126	45	141	59	185	67	210	72	226
5 p.m.	31	98	35	110	46	145	53	167	56	176
6 p.m.	19	60	21	66	28	88	32	101	34	107

4 LOAD ESTIMATING

(SOUTH WEST EXPOSURE, 50°N LATITUDE, 12-H OPERATION)

(A) BARE SINGLE GLAZING

Time of day (BST)	June BTU/h/ft²	June W/m²	July–May BTU/h/ft²	July–May W/m²	August–April BTU/h/ft²	August–April W/m²	September–March BTU/h/ft²	September–March W/m²	October–February BTU/h/ft²	October–February W/m²
7 a.m.	84	264	89	279	98	308	102	320	98	308
8 a.m.	70	220	74	232	81	255	84	264	81	255
9 a.m.	59	185	62	195	68	214	71	223	68	214
10 a.m.	56	176	59	185	65	204	67	211	65	204
11 a.m.	49	154	52	163	57	179	60	188	57	179
Noon	52	163	55	173	61	192	63	198	61	192
1 p.m.	62	195	66	207	72	226	75	235	72	226
2 p.m.	73	229	77	242	85	267	88	276	85	267
3 p.m.	87	271	92	290	102	320	106	334	102	320
4 p.m.	99	311	104	326	115	361	119	373	115	361
5 p.m.	102	320	108	340	118	370	123	387	118	370
6 p.m.	95	299	101	323	111	349	115	361	111	349

(B) SINGLE GLAZING WITH LIGHT COLOURED VENETIAN BLINDS

Time of day (BST)	June BTU/h/ft²	June W/m²	July–May BTU/h/ft²	July–May W/m²	August–April BTU/h/ft²	August–April W/m²	September–March BTU/h/ft²	September–March W/m²	October–February BTU/h/ft²	October–February W/m²
7 a.m.	29	91	31	98	34	107	35	110	34	107
8 a.m.	25	79	26	82	29	91	30	94	29	91
9 a.m.	22	69	24	76	26	82	27	85	26	82
10 a.m.	20	63	22	69	24	76	25	79	24	76
11 a.m.	20	63	22	69	24	76	25	79	24	76
Noon	31	98	33	104	36	113	37	116	36	113
1 p.m.	45	141	47	147	52	163	54	170	52	163
2 p.m.	57	179	60	188	67	210	68	213	67	210
3 p.m.	66	207	70	220	77	242	79	249	77	242
4 p.m.	69	216	72	226	80	252	82	258	80	252
5 p.m.	62	195	66	207	73	229	75	235	73	229
6 p.m.	45	141	47	148	52	163	54	170	52	163

See Table 4.15 for correction factors.

TABLE 4.12
EFFECTIVE SOLAR HEAT GAIN THROUGH GLASS
(WEST EXPOSURE, 50°N LATITUDE, 12-H OPERATION)

(A) BARE SINGLE GLAZING

Time of day (BST)	June BTU/h/ft²	June W/m²	July–May BTU/h/ft²	July–May W/m²	August–April BTU/h/ft²	August–April W/m²	September–March BTU/h/ft²	September–March W/m²	October–February BTU/h/ft²	October–February W/m²
7 a.m.	116	364	115	361	111	346	97	209	74	232
8 a.m.	100	314	100	314	97	309	84	264	64	201
9 a.m.	85	267	85	267	82	258	71	223	55	173
10 a.m.	75	235	75	235	73	229	63	198	48	151
11 a.m.	66	207	65	204	63	198	55	173	42	132
Noon	60	188	60	188	58	182	50	157	38	119
1 p.m.	56	176	56	176	54	170	47	148	36	113
2 p.m.	54	170	54	170	52	163	45	142	35	110
3 p.m.	64	201	63	198	61	192	53	167	41	129
4 p.m.	83	261	83	261	80	252	67	210	53	167
5 p.m.	98	308	98	308	95	203	83	261	63	198
6 p.m.	110	346	109	343	106	334	92	290	71	223

(B) SINGLE GLAZING WITH LIGHT COLOURED VENETIAN BLINDS

Time of day (BST)	June BTU/h/ft²	June W/m²	July–May BTU/h/ft²	July–May W/m²	August–April BTU/h/ft²	August–April W/m²	September–March BTU/h/ft²	September–March W/m²	October–February BTU/h/ft²	October–February W/m²
7 a.m.	74	232	72	226	70	220	60	188	47	148
8 a.m.	36	113	36	113	35	110	30	94	23	72
9 a.m.	31	98	30	94	29	91	25	79	20	63
10 a.m.	29	91	28	88	27	85	23	73	18	57
11 a.m.	26	82	26	82	25	79	22	69	17	54
Noon	24	76	24	76	23	73	20	63	15	47
1 p.m.	22	69	22	69	21	66	18	57	14	44
2 p.m.	31	98	30	94	29	91	25	79	20	63
3 p.m.	48	151	48	151	46	145	40	126	31	98
4 p.m.	67	210	66	207	64	201	55	173	43	135
5 p.m.	79	249	78	246	76	239	65	204	50	157
6 p.m.	80	252	79	249	77	242	66	207	51	160

4 LOAD ESTIMATING

EFFECTIVE SOLAR HEAT GAIN THROUGH GLASS
(NORTH WEST EXPOSURE, 50°N LATITUDE, 12-H OPERATION)

(A) BARE SINGLE GLAZING

Time of day (BST)	June BTU/h/ft²	June W/m²	July–May BTU/h/ft²	July–May W/m²	August–April BTU/h/ft²	August–April W/m²	September–March BTU/h/ft²	September–March W/m²	October–February BTU/h/ft²	October–February W/m²
7 a.m.	80	252	74	232	59	185	37	116	18	57
8 a.m.	73	229	67	210	54	170	33	104	17	54
9 a.m.	61	192	56	176	45	141	28	88	14	44
10 a.m.	52	163	48	151	39	123	24	76	12	37
11 a.m.	46	145	42	132	34	107	21	66	11	34
Noon	41	129	38	119	31	98	19	60	10	31
1 p.m.	37	116	34	107	28	88	17	54	9	28
2 p.m.	34	107	32	101	25	79	16	50	8	25
3 p.m.	36	113	33	104	26	82	16	50	8	25
4 p.m.	44	138	41	129	33	104	20	63	10	31
5 p.m.	58	182	53	167	43	135	27	85	13	41
6 p.m.	71	223	66	207	53	167	33	104	16	50

(B) SINGLE GLAZING WITH LIGHT COLOURED VENETIAN BLINDS

Time of day (BST)	June BTU/h/ft²	June W/m²	July–May BTU/h/ft²	July–May W/m²	August–April BTU/h/ft²	August–April W/m²	September–March BTU/h/ft²	September–March W/m²	October–February BTU/h/ft²	October–February W/m²
7 a.m.	60	188	54	170	43	135	27	85	13	41
8 a.m.	26	82	24	76	19	60	12	37	6	19
9 a.m.	23	72	21	66	16	50	10	31	5	16
10 a.m.	20	63	18	57	15	47	9	28	5	16
11 a.m.	18	57	17	54	13	41	8	25	4	13
Noon	18	57	16	50	13	41	8	25	4	13
1 p.m.	16	50	14	44	12	38	7	22	4	13
2 p.m.	15	47	14	44	11	34	7	22	3	10
3 p.m.	19	60	17	54	14	44	9	28	4	13
4 p.m.	34	107	30	94	24	76	15	47	8	25
5 p.m.	49	154	44	138	35	110	22	69	11	34
6 p.m.	59	185	53	167	43	135	27	85	13	41

See Table 4.15 for correction factors.

TABLE 4.14
EFFECTIVE SOLAR HEAT GAIN THROUGH GLASS
(HORIZONTAL, 50°N LATITUDE, 12-H OPERATION)

(A) BARE SINGLE GLAZING

Time of day (BST)	June BTU/h/ft²	June W/m²	July–May BTU/h/ft²	July–May W/m²	August–April BTU/h/ft²	August–April W/m²	September–March BTU/h/ft²	September–March W/m²	October–February BTU/h/ft²	October–February W/m²
7 a.m.	114	369	110	348	96	301	77	242	48	151
8 a.m.	96	301	92	289	81	254	64	201	41	129
9 a.m.	101	318	97	304	85	267	68	214	43	135
10 a.m.	111	351	107	339	94	296	75	236	47	148
11 a.m.	130	409	125	393	109	343	87	274	55	173
Noon	148	465	142	447	124	390	99	311	63	198
1 p.m.	166	522	159	501	140	440	111	351	70	220
2 p.m.	176	553	169	533	148	465	118	372	75	236
3 p.m.	181	569	174	547	153	481	122	384	77	242
4 p.m.	176	553	169	533	148	465	118	372	75	236
5 p.m.	163	514	157	495	137	431	110	348	69	217
6 p.m.	137	431	132	415	116	366	92	290	58	182

(B) SINGLE GLAZING WITH LIGHT COLOURED VENETIAN BLINDS

Time of day (BST)	June BTU/h/ft²	June W/m²	July–May BTU/h/ft²	July–May W/m²	August–April BTU/h/ft²	August–April W/m²	September–March BTU/h/ft²	September–March W/m²	October–February BTU/h/ft²	October–February W/m²
7 a.m.	38	119	36	113	31	98	25	79	16	50
8 a.m.	32	101	31	98	27	85	22	69	14	44
9 a.m.	55	173	53	167	46	145	37	116	24	76
10 a.m.	74	233	71	223	62	195	50	157	32	101
11 a.m.	93	293	89	280	77	242	63	198	40	126
Noon	106	336	101	318	88	276	72	227	45	141
1 p.m.	115	363	110	348	96	301	77	242	49	154
2 p.m.	115	363	110	348	96	301	77	242	49	154
3 p.m.	112	354	107	339	93	293	75	236	48	151
4 p.m.	94	296	90	282	79	248	64	201	40	126
5 p.m.	74	233	71	224	62	195	50	157	32	101
6 p.m.	45	141	43	135	38	119	30	94	19	60

4 LOAD ESTIMATING

CORRECTION FACTORS TO TABLES 4.6 TO 4.14

Type of glazing	Tables 4.6(A) to 4.14(A)			Tables 4.6(B) to 4.14(B)	
	No shading	Outside ven. blind	Outside awning	Med. colour ven. blind	Dark colour ven. blind
Ordinary glass	1·00	0·15	0·20	1·16	1·32
Plate glass	0·94	0·14	0·19	1·16	1·31
Double glazing	0·90	0·14	0·18	1·09	1·19
Double plate	0·80	0·12	0·16	1·05	1·16
Triple glazing	0·83	0·12	0·16	1·00	1·15
Triple plate	0·69	0·10	0·14	0·93	1·02

Notes:
1. Cooling load = (window area) × (solar heat gain) × (correction factor).
2. All values refer to metal framed windows. For wooden frames deduct 15 per cent from above values.

TABLE 4.16
SOLAR ALTITUDE AND AZIMUTH ANGLES, LATITUDE 50°N

Sun time	Feb. 20–Oct. 23		March 22–Sept. 22		April 20–Aug. 24		May 21–July 23		June 21	
	Alt.	Az.	Alt.	Az.	Alt.	Az.	Alt.	Az.	Alt.	Az.
6 a.m.					9	83	15	77	18	74
7 a.m.			10	101	18	94	25	88	27	85
8 a.m.	10	121	19	114	28	106	34	100	37	97
9 a.m.	17	134	27	127	37	120	44	114	46	110
10 a.m.	23	148	34	143	44	137	52	131	55	128
11 a.m.	27	164	39	160	49	157	58	152	61	151
Noon	29	180	40	180	51	180	60	180	63	180
1 p.m.	27	196	39	200	49	203	58	208	61	209
2 p.m.	23	212	34	217	44	223	52	229	55	232
3 p.m.	17	226	27	233	37	240	44	246	46	250
4 p.m.	10	121	19	246	28	254	34	260	37	263
5 p.m.			10	259	18	266	25	272	27	275
6 p.m.					9	277	15	283	18	286

TABLE 4.17
EQUIVALENT TEMPERATURE DIFFERENCES, SOLAR AND TRANSMISSION GAIN THROUGH WALLS AND ROOFS
(50°N LATITUDE, 82°F MAX. AMBIENT DRY BULB, 72°F ROOM)

Exposure	Wt of fabric (lb/ft²)	7 a.m	8 a.m	9 a.m	10 a.m	11 a.m	12 Noon	1 p.m.	2 p.m.	3 p.m.	4 p.m.	5 p.m.	6 p.m.	7 p.m.
NE wall	100	1	1	1	6	12	11	10	8	7	8	8	8	8
E wall	100	3	5	11	17	21	22	21	17	15	13	11	11	11
SE wall	100	4	4	4	9	15	16	17	18	17	14	12	11	10
S wall	100	-2	-2	-2	2	3	8	14	18	19	22	22	17	14
SW wall	100	7	5	5	6	7	8	10	15	15	21	25	26	26
W wall	100	4	4	5	4	4	8	6	8	10	15	18	23	26
NW wall	100	2	2	2	2	2	5	2	5	6	9	11	16	17
N or shade wall	100	-2	-2	-2	-2	-2	-1	0	-1	2	3	3	3	5
Exposed roof	20	-4	0	0	6	13	20	26	32	37	38	38	35	31
Shaded roof	20	-6	-4	-2	0	4	7	10	11	12	11	10	8	6
Exposed roof	40	-4	-3	0	7	13	19	24	29	34	36	36	34	31
Shaded roof	40	-6	-5	-4	-2	0	3	6	8	10	11	10	9	8
Exposed roof	60	3	4	5	12	13	18	23	27	31	34	35	34	32
Shaded roof	60	-4	-4	-4	-3	-2	0	2	4	6	7	8	8	8

TABLE 4.18
CORRECTIONS TO EQUIVALENT TEMPERATURES.
SOLAR AND TRANSMISSION GAIN THROUGH WALLS AND ROOF

Daily range (°F)	Design outdoor temp. (°F) at 3 p.m. − room temp. (°F)					
	5	10	15	20	25	30
10	-2	3	8	13	18	23
15	-5	0	5	10	15	20
20	-7	-2	3	8	13	18
25	-10	-5	0	5	10	15
30	-12	-7	-2	3	7	12

4   LOAD ESTIMATING

EQUIVALENT TEMPERATURE DIFFERENCES, SOLAR AND TRANSMISSION GAIN THROUGH WALLS AND ROOFS
(50°N LATITUDE, 27·8°C MAX. AMBIENT DRY BULB, 22·2°C ROOM)

Exposure	Wt of fabric (kg/m²)	Time of day (BST)											
		7 a.m.	8 a.m.	9 a.m.	10 a.m.	11 a.m.	12 Noon	1 p.m.	2 p.m.	3 p.m.	4 p.m.	5 p.m.	6 p.m.	7 p.m.
NE wall	488	0·5	0·5	0·6	3·3	6·7	6·2	5·6	4·5	3·9	4·5	4·5	4·5	4·5
E wall	488	1·7	2·8	6·2	9·5	11·7	12·2	11·7	9·5	8·3	7·2	6·2	6·2	6·2
SE wall	488	2·2	2·2	2·2	5·0	8·3	8·9	9·5	10·0	9·5	7·8	6·7	6·7	5·6
S wall	488	−1·1	−1·1	−1·1	1·1	1·7	4·5	7·8	10·0	10·5	12·8	12·2	9·5	7·8
SW wall	488	3·9	2·8	2·8	3·3	3·9	4·5	5·6	8·3	8·3	12·2	13·9	14·5	14·5
W wall	488	2·2	2·2	2·2	2·2	2·2	2·8	3·3	4·5	5·6	8·3	10·0	12·8	14·5
NW wall	488	1·1	1·1	1·1	1·1	1·1	1·1	1·1	2·8	3·3	5·0	6·2	8·9	9·5
N or shade wall	488	−1·1	−1·1	−1·1	−1·1	−1·1	−0·5	0	0·5	1·1	1·7	1·7	1·7	2·8
Exposed roof	98	−2·2	0	0	3·3	7·2	11·1	14·5	17·8	20·6	21·1	21·1	19·5	17·2
Shaded roof	98	−3·3	−2·2	−1·1	0	2·2	3·9	5·6	6·2	6·7	6·2	5·6	4·5	3·3
Exposed roof	196	−2·2	−1·7	0	3·9	7·2	10·5	13·3	16·1	18·9	20·0	20·0	18·9	17·2
Shaded roof	196	−3·3	−2·8	−2·2	−1·1	0	1·7	3·3	4·5	5·6	6·2	5·6	5·0	4·5
Exposed roof	292	1·7	2·2	−2·8	6·7	7·2	10·0	12·8	15·0	17·2	18·9	19·5	18·9	17·8
Shaded roof	292	−2·2	−2·2	−2·2	−1·6	−1·1	0	1·1	2·2	3·3	3·9	4·5	4·5	4·5

TABLE 4.18a
CORRECTIONS TO EQUIVALENT TEMPERATURES.
SOLAR AND TRANSMISSION GAIN THROUGH WALLS AND ROOF

Daily range (°C)	Design outdoor temp. (°C) at 3 p.m. − room temp. (°C)					
	2·5	5	7·5	10·0	12·5	15
5·0	−1·1	1·4	3·9	6·4	8·9	11·4
7·5	−2·5	0	2·5	5·0	7·5	10·0
10·0	−3·9	−1·4	1·1	3·6	6·1	8·6
12·5	−5·3	−2·8	−0·3	2·2	4·7	7·2
15·0	−6·7	−4·2	−1·7	0·8	3·3	5·8

TABLE 4.19
THERMAL TRANSMISSION COEFFICIENTS (U-VALUES)

1. EXTERNAL WALLS

External surface	Thickness (in)	Weight (lb/ft²)	None	Internal finish ⅝ in plaster (6)	⅜ in plaster board (2)	⅜ in plaster board on battens (3)	Insulating boards on battens ½ in thick (2)	Insulating boards on battens 1 in thick (4)
Solid brickwork	4½	(45)	0·54	0·51	0·40	0·29	0·22	0·16
	9	(90)	0·37	0·35	0·29	0·23	0·18	0·14
	13½	(135)	0·28	0·27	0·23	0·19	0·16	0·12
Cavity brickwork	11	(88)	0·27	0·26	0·23	0·19	0·15	0·12
Poured concrete	6	(40)	0·48	0·45	0·36	0·26	0·20	0·15
	8	(53)	0·40	0·38	0·31	0·24	0·18	0·14
	10	(66)	0·34	0·33	0·28	0·22	0·16	0·13
Hollow concrete blocks	4	(19)	0·61	0·57	0·43	0·31	0·23	0·16
	8	(38)	0·49	0·46	0·38	0·27	0·21	0·15
4½ in brick with cinder block	7½	(58)	0·45	0·42	0·35	0·26	0·20	0·15
	8½	(63)	0·40	0·38	0·31	0·24	0·19	0·14
4½ in brick with cavity cinder block	9½	(58)	0·31	0·30	0·26	0·21	0·17	0·13
	10½	(63)	0·29	0·28	0·24	0·19	0·16	0·12
Corrugated asbestos on studding		(4)	0·70	0·64	0·48	0·33	0·24	0·17
Corrugated asbestos with ½ in insulating board on studding		(5)	0·31	0·29	0·25	0·20	0·16	0·13

Note:
The above values based on 7½ mph wind.
For still air:

$$\text{U-value} = \frac{1}{(1/\text{Table U-value}) + 0\cdot 43}$$

For 15 mph wind (winter):

$$\text{U-value} = \frac{1}{(1/\text{Table U-value}) - 0\cdot 08}$$

4   LOAD ESTIMATING

THERMAL TRANSMISSION COEFFICIENTS (U-VALUES)

1. EXTERNAL WALLS

External surface	Thickness (mm)	Weight (kg/m^2)	None	16 mm plaster (29)	9 mm plaster board (10)	9 mm plaster board on battens (15)	Insulating boards on battens 13 mm thick (10)	Insulating boards on battens 25 mm thick (19)
Solid brickwork	100	(218)	2·06	2·90	2·28	1·65	1·25	0·91
	200	(436)	2·10	1·99	1·65	1·31	1·02	0·80
	300	(654)	1·59	1·53	1·31	1·08	0·91	0·68
Cavity brickwork	280	(425)	1·53	1·48	1·31	1·08	0·85	0·68
Poured concrete	150	(193)	2·74	2·56	2·04	1·48	1·14	0·85
	200	(256)	2·28	2·16	1·76	1·37	1·02	0·80
	250	(319)	1·93	1·88	1·59	1·25	0·91	0·74
Hollow concrete blocks	100	(92)	3·46	3·24	2·25	1·76	1·31	0·91
	200	(184)	2·79	2·62	2·16	1·53	1·20	0·85
100 mm brick with cinder block	190	(280)	2·56	2·40	1·99	1·48	1·14	0·85
100 mm brick with cavity cinder block	220	(304)	2·28	2·16	1·76	1·37	1·08	0·80
	240	(280)	1·76	1·70	1·48	1·20	0·97	0·74
Corrugated asbestos on studding	270	(304)	1·65	1·59	1·37	1·08	0·91	0·68
		(19)	3·98	3·64	2·74	1·88	1·37	0·97
Corrugated asbestos with 12 mm insulating board on studding		(24)	1·76	1·65	1·42	1·14	0·91	0·74

Note:
The above values based on 3·35 m/s wind.
For still air:

$$U\text{-value} = \frac{1}{(1/\text{Table } U\text{-value}) + 0·076}$$

For 6·7 m/s wind:

$$U\text{-value} = \frac{1}{(1/\text{Table } U\text{-value}) - 0·014}$$

TABLE 4.19
THERMAL TRANSMISSION COEFFICIENTS (U-VALUES)

2. FLAT ROOFS (HEAT FLOW DOWN—SUMMER)

External surface	Thickness (in)	Insulation on top (in)	Weight (lb/ft²)	Ceiling None	Plaster (6)	Plaster board on battens (3)	Suspended acoustic tile (1)
Concrete	4	None	(27)	0·51	0·48	0·29	0·21
	4	1	(31)	0·21	0·20	0·16	0·14
	4	2	(35)	0·13	0·13	0·11	0·10
	6	None	(40)	0·42	0·40	0·26	0·20
	6	1	(44)	0·20	0·19	0·15	0·13
	6	2	(48)	0·13	0·12	0·11	0·09
	8	None	(53)	0·36	0·35	0·24	0·18
	8	1	(57)	0·18	0·18	0·14	0·12
	8	2	(61)	0·12	0·12	0·10	0·09
Hollow cinder block	5	None	(23)	0·35	0·34	0·23	0·18
	5	1	(27)	0·18	0·17	0·14	0·12
	5	2	(31)	0·12	0·12	0·10	0·09
Flat metal with external roofing felt		None	(2)	0·78	0·71	0·36	0·25
		1	(6)	0·25	0·24	0·19	0·15
		2	(10)	0·15	0·15	0·12	0·11
Preformed wood fibre slabs with cement binder	2	None	(3)	0·21	0·20	0·16	0·14
	2	1	(7)	0·13	0·13	0·12	0·11
	2	2	(11)	0·10	0·10	0·09	0·08
Wood with external roofing felt	1	None	(4)	0·40	—	0·25	0·19
	1	1	(8)	0·19	—	0·15	0·13
	1	2	(12)	0·12	—	0·11	0·09

Note:
The above values based on 7½ mph wind.
For 15 mph wind:

$$\text{U-value} = \frac{1}{(1/\text{Table U-value}) - 0·08}$$

For still air heat flow down:

2. FLAT ROOFS (HEAT FLOW DOWN—SUMMER)

THERMAL TRANSMISSION COEFFICIENTS (U-VALUES)

External surface	Thickness (mm)	Insulation on top (mm)	Weight (kg/m²)	Ceiling None	Ceiling Plaster (29)	Ceiling Plaster board on battens (15)	Ceiling Suspended acoustic tile (5)
Concrete	100	None	(130)	2·90	2·74	1·65	1·20
		25	(150)	1·20	1·14	0·91	0·80
		50	(169)	0·74	0·74	0·62	0·57
	150	None	(193)	2·40	2·28	1·48	1·14
		25	(212)	1·14	1·08	0·85	0·74
		50	(232)	0·74	0·68	0·62	0·51
	200	None	(261)	2·04	1·99	1·37	1·02
		25	(275)	1·02	1·02	0·80	0·68
		50	(295)	0·68	0·68	0·57	0·51
Hollow cinder block	125	None	(111)	1·99	1·93	1·31	1·02
		25	(130)	1·02	0·97	0·80	0·68
		50	(150)	0·68	0·68	0·57	0·51
Flat metal with external roofing felt		None	(10)	4·44	4·03	2·04	1·42
		25	(29)	1·42	1·37	1·08	0·85
		50	(48)	0·85	0·85	0·68	0·62
Preformed wood fibre slabs with cement binder	50	None	(15)	1·20	1·14	0·91	0·80
		25	(34)	0·74	0·74	0·68	0·62
		50	(53)	0·57	0·57	0·51	0·46
Wood with external roofing felt	25	None	(19)	2·28	—	1·42	1·08
		25	(39)	1·08	—	0·85	0·74
		50	(58)	0·68	—	0·62	0·51

Note:
The above values based on 3·35 m/s wind.
For 6·7 m/s wind:

$$U\text{-value} = \frac{1}{(1/\text{Table } U\text{-value}) - 0.014}$$

For still air heat flow down:

$$U\text{-value} = \frac{1}{(1/\text{Table } U\text{-value}) + 0.118}$$

TABLE 4.19
THERMAL TRANSMISSION COEFFICIENTS (U-VALUES)

3. FLAT ROOFS (HEAT FLOW UP—WINTER)

External surface	Thickness (in)	Insulation on top (in)	Weight (lb/ft²)	None	Ceiling Plaster (6)	Plaster board on battens (3)	Suspended acoustic tile (1)
Concrete	4	None	(27)	0·60	0·56	0·34	0·23
	4	1	(31)	0·23	0·22	0·17	0·15
	4	2	(35)	0·14	0·14	0·12	0·11
	6	None	(40)	0·49	0·46	0·30	0·23
	6	1	(44)	0·21	0·20	0·16	0·14
	6	2	(48)	0·13	0·13	0·11	0·10
	8	None	(53)	0·41	0·39	0·27	0·21
	8	1	(57)	0·19	0·18	0·15	0·13
	8	2	(61)	0·13	0·12	0·11	0·10
Hollow cinder block	5	None	(23)	0·39	0·37	0·26	0·20
	5	1	(27)	0·18	0·18	0·15	0·13
	5	2	(31)	0·12	0·12	0·11	0·10
Flat metal with external roofing felt		None	(2)	1·02	0·91	0·44	0·30
		1	(6)	0·28	0·27	0·20	0·17
		2	(10)	0·16	0·15	0·13	0·12
Preformed wood fibre slabs with cement binder	2	None	(3)	0·23	0·22	0·18	0·15
	2	1	(7)	0·14	0·12	0·12	0·10
	2	2	(11)	0·10	0·10	0·09	0·08
Wood with external roofing felt	1	None	(4)	0·45	—	0·28	0·22
	1	1	(8)	0·20	—	0·16	0·14
	1	2	(12)	0·13	—	0·11	0·10

Note:
The above values based on 7½ mph wind.
For 15 mph wind:

$$\text{U-value} = \frac{1}{(1/\text{Table U-value}) - 0.08}$$

For still air heat flow up:

$$\frac{1}{}$$

3. FLAT ROOFS (HEAT FLOW UP—WINTER)

				Ceiling			
External surface	Thickness (mm)	Insulation on top (mm)	Weight (kg/m²)	None	Plaster (29)	Plaster board on battens (15)	Suspended acoustic tile (5)
Concrete	100	None	(130)	3·40	3·18	1·93	1·31
		25	(150)	1·31	1·25	0·97	0·85
		50	(169)	0·80	0·80	0·68	0·62
	150	None	(193)	2·79	2·62	1·70	1·31
		25	(212)	1·20	1·14	0·91	0·80
		50	(232)	0·74	0·74	0·62	0·57
	200	None	(261)	2·34	2·22	1·53	1·20
		25	(275)	1·08	1·02	0·85	0·74
		50	(295)	0·74	0·68	0·62	0·57
Hollow cinder block	125	None	(111)	2·22	2·10	1·48	1·14
		25	(130)	1·02	1·02	0·85	0·74
		50	(150)	0·68	0·68	0·62	0·57
Flat metal with external roofing felt		None	(10)	5·79	5·16	2·50	1·70
		25	(29)	1·59	1·53	1·14	0·97
		50	(48)	0·91	0·85	0·74	0·68
Preformed wood fibre slabs with cement binder	50	None	(15)	1·31	1·25	1·02	0·85
		25	(34)	0·80	0·68	0·68	0·57
		50	(53)	0·57	0·57	0·51	0·46
Wood with external roofing felt	25	None	(19)	2·36	—	1·59	1·14
		25	(39)	1·14	—	0·91	1·80
		50	(58)	0·74	—	0·62	0·57

Note:
The above values based on 3·35 m/s wind.
For 6·7 m/s wind:

$$U\text{-value} = \frac{1}{(1/\text{Table } U\text{-value}) - 0·014}$$

For still air heat flow up:

$$U\text{-value} = \frac{1}{(1/\text{Table } U\text{-value}) + 0·063}$$

TABLE 4.19
THERMAL TRANSMISSION COEFFICIENTS (U-VALUES)

4. PITCHED ROOFS (HEAT FLOW DOWN—SUMMER)

External surface	Sheathing	Weight (lb/ft²)	Internal cladding								Flat ceiling			
			None	Wood panel	Plaster board	Insulating board				Insulating board				Acoustic tile	Plaster board
						½ in thick	1 in thick	½ in thick	1 in thick			
			(2)	(2)	(2)	(2)	(4)	(2)	(4)	(1)	(2)
Slates (8), tile (14) or sheet metal (2)	Plywood 1 in wood	(2) (4)	0·55 0·41	0·28 0·23	0·30 0·25	0·21 0·19	0·15 0·14	0·18 0·16	0·14 0·12	0·20 0·17	0·24 0·21
Corrugated asbestos	Plywood 1 in wood	(2) (4)	0·47 0·38	0·26 0·22	0·27 0·24	0·20 0·18	0·15 0·14	0·18 0·16	0·13 0·12	0·18 0·16	0·22 0·20

Note:
The above values based on 7½ mph wind.
For 15 mph wind (winter):

$$\text{U-value} = \frac{1}{(1/\text{Table U-value}) - 0·08}$$

For still air heat flow down:

$$\text{U-value} = \frac{1}{(1/\text{Table U-value}) + 0·51}$$

4 LOAD ESTIMATING

TABLE 4.19a
THERMAL TRANSMISSION COEFFICIENTS (U-VALUES)
4. PITCHED ROOFS (HEAT FLOW DOWN—SUMMER)

External surface	Sheathing	Weight (kg/m²)	Internal cladding						Flat ceiling				
			None	Wood panel	Plaster board	Insulating board			Insulating board		Acoustic tile	Plaster board	
						13 mm thick	25 mm thick		13 mm thick	25 mm thick			
				(10)	(10)	(10)	(19)		(10)	(19)	(5)	(10)	
Slates, tile or sheet metal	Plywood	(10)	3·12	1·59	1·70	1·20	0·85		1·02	0·80	1·14	1·37	
	25 mm wood	(19)	2·34	1·31	1·42	1·08	0·80		0·91	0·68	0·97	1·20	
Corrugated asbestos	Plywood	(10)	2·68	1·48	1·53	1·14	0·85		1·02	0·74	1·02	1·25	
	25 mm wood	(19)	2·16	1·25	1·37	1·02	0·80		0·91	0·68	0·91	1·14	

Note:
The above values based on 3·35 m/s wind.
For 6·7 m/s wind (winter):

$$U\text{-}value = \frac{1}{(1/Table\ U\text{-}value) - 0\cdot014}$$

For still air heat flow down:

$$U\text{-}value = \frac{1}{(1/Table\ U\text{-}value) + 0\cdot090}$$

TABLE 4.19
THERMAL TRANSMISSION COEFFICIENTS (U-VALUES)

5. PITCHED ROOFS (HEAT FLOW UP—WINTER)

External surface	Sheathing	Weight (lb/ft²)	Internal cladding					Flat ceiling			
			None	Wood panel (2)	Plaster board (2)	Insulating board ½ in thick (2)	Insulating board 1 in thick (4)	Insulating board ½ in thick (2)	Insulating board 1 in thick (4)	Acoustic tile (1)	Plaster board (2)
Slates (8), tile (14) or sheet metal (2)	Plywood 1 in wood	(2) (4)	0·59 0·43	0·40 0·32	0·32 0·26	0·22 0·19	0·15 0·14	0·21 0·19	0·15 0·14	0·23 0·20	0·30 0·25
Corrugated asbestos (3)	Plywood 1 in wood	(2) (4)	0·50 0·38	0·36 0·28	0·29 0·25	0·21 0·18	0·15 0·13	0·20 0·18	0·15 0·13	0·21 0·19	0·27 0·23

Note:
The above values based on 7½ mph wind.
For 15 mph wind (winter):

$$\text{U-value} = \frac{1}{(1/\text{Table U-value}) - 0\cdot08}$$

For still air heat flow up:

$$\text{U-value} = \frac{1}{(1/\text{Table U-value}) + 0\cdot37}$$

TABLE 4.19a
THERMAL TRANSMISSION COEFFICIENTS (U-VALUES)

5. PITCHED ROOFS (HEAT FLOW UP—WINTER)

External surface	Sheathing	Weight (kg/m²)	Internal cladding					Flat ceiling				
			None	Wood panel	Plaster board	Insulating board		Insulating board		Acoustic tile	Plaster board
						13 mm thick	25 mm thick	13 mm thick	25 mm thick		
				(10)	(10)	(10)	(19)	(10)	(19)	(5)	(10)
Slates, tile or sheet metal	Plywood	(10)	3·35	2·28	1·82	1·25	0·85	1·20	0·85	1·31	1·70
	25 mm wood	(19)	2·45	1·82	1·48	1·08	0·80	1·08	0·80	1·14	1·42
Corrugated asbestos	Plywood	(10)	2·84	2·05	1·65	1·20	0·85	1·14	0·85	1·20	1·53
	25 mm wood	(19)	2·16	1·59	1·42	1·02	0·74	1·02	0·74	1·08	1·31

Note:
The above values based on 3·35 m/s wind.
For 6·7 m/s wind (winter):

$$U\text{-value} = \frac{1}{(1/\text{Table U-value}) - 0.014}$$

For still air heat flow up:

$$U\text{-value} = \frac{1}{(1/\text{Table U-value}) + 0.065}$$

TABLE 4.19
THERMAL TRANSMISSION COEFFICIENTS (U-VALUES)

6. FLOORS AND CEILINGS (HEAT FLOW UP)

Flooring	Sub floor	Weight (lb/ft^2)	Ceiling[a]			Suspended ceiling[b]			Ground[c]
			None	Plaster board (2)	Acoustic tile (1)	Plaster board (2)	Acoustic tile (1)	½ in Insulation board (2)	
⅛ in tile or linoleum	4 in concrete	(51)	0·48	0·39	0·28	0·29	0·23	0·21	0·67
	6 in concrete	(76)	0·40	0·34	0·25	0·26	0·21	0·19	0·53
	8 in concrete	(101)	0·34	0·30	0·23	0·24	0·19	0·18	0·44
	1 in wood on joists	(4)	0·39	0·26	0·21	0·26	0·21	0·19	0·51
	½ in insulation board on 1 in wood on joists	(9)	0·30	0·18	0·15	0·18	0·15	0·14	0·36
⅛ in tile or linoleum on ⅝ in wood on 2 in sleepers	4 in concrete	(53)	0·27	0·24	0·19	0·24	0·19	0·18	0·32
	6 in concrete	(78)	0·24	0·22	0·18	0·22	0·18	0·17	0·28
	8 in concrete	(103)	0·22	0·20	0·17	0·20	0·17	0·16	0·25
	None	(3)	0·48	0·30	0·23	0·30	0·23	0·21	0·43
Wood block	4 in concrete	(52)	0·40	0·34	0·25	0·27	0·21	0·19	0·53
	6 in concrete	(77)	0·35	0·30	0·23	0·24	0·19	0·18	0·44
	8 in concrete	(102)	0·31	0·27	0·21	0·22	0·18	0·17	0·38

[a] For heat flow down:

$$\text{U-value} = \frac{1}{(1/\text{Table U-value}) + 0\cdot 62}$$

[b] For heat flow down:

$$\text{U-value} = \frac{1}{(1/\text{Table U-value}) + 1\cdot 00}$$

[c] For heat flow down:

$$\text{U-value} = \frac{1}{(1/\text{Table U-value}) + 0\cdot 31}$$

TABLE 4.19a
THERMAL TRANSMISSION COEFFICIENTS (U-VALUES)
6. FLOORS AND CEILINGS (HEAT FLOW UP)

Flooring	Sub floor	Weight (kg/m²)	Ceiling[a]				Suspended ceiling[b]			Ground[c]
			None	Plaster board (10)	Acoustic tile (5)	Plaster board (10)	Acoustic tile (5)	13 mm Insulation board (10)	
3 mm tile or linoleum	100 mm concrete	(246)	2·74	2·22	1·59	1·65	1·31	1·20	3·81
	150 mm concrete	(369)	2·28	1·93	1·42	1·48	1·20	1·08	2·01
	200 mm concrete	(492)	1·93	1·70	1·31	1·37	1·08	1·02	2·50
	25 mm wood on joists	(19)	2·22	1·48	1·20	1·48	1·20	1·08	2·90
	13 mm insulation board on 25 mm wood on joists	(43)	1·70	1·02	0·85	1·02	0·85	0·80	2·05
3 mm tile or linoleum on 15 mm wood on 50 mm sleepers	100 mm concrete	(256)	1·53	1·37	1·08	1·37	1·08	1·02	1·82
	150 mm concrete	(379)	1·37	1·25	1·02	1·25	1·02	0·97	1·59
	200 mm concrete	(502)	1·25	1·14	0·97	1·14	0·97	0·91	1·42
	None	(14)	2·74	1·70	1·31	1·70	1·31	1·20	2·45
Wood block	100 mm concrete	(251)	2·28	1·93	1·42	1·53	1·20	1·08	3·01
	150 mm concrete	(374)	1·99	1·70	1·31	1·37	1·08	1·02	2·50
	200 mm concrete	(497)	1·76	1·53	1·20	1·25	1·02	0·97	2·16

[a] For heat flow down:
$$U\text{-value} = \frac{1}{(1/\text{Table } U\text{-value}) + 0.109}$$

[b] For heat flow down:
$$U\text{-value} = \frac{1}{(1/\text{Table } U\text{-value}) + 0.176}$$

[c] For heat flow down:
$$U\text{-value} = \frac{1}{(1/\text{Table } U\text{-value}) + 0.055}$$

TABLE 4.19
THERMAL TRANSMISSION COEFFICIENTS (U-VALUES)

7. PARTITION WALLS (STILL AIR BOTH SIDES)

Construction	Weight (lb/ft²)	Finish								
		One side					Both sides			
		None	Gypsum plaster	Plaster board on battens	Insulating board on battens		Gypsum plaster	Plaster board on battens	Insulating board on battens	
					½ in thick	1 in thick			½ in thick	1 in thick
			(6)	(3)	(2)	(4)	(6)	(3)	(2)	(4)
4½ in brick	(45)	0·44	0·38	0·27	0·20	0·15	0·33	0·20	0·13	0·09
9 in brick	(90)	0·32	0·28	0·22	0·17	0·13	0·25	0·17	0·12	0·08
4 in cinder block or 4 in hollow concrete block	(20)	0·33	0·29	0·22	0·17	0·13	0·26	0·17	0·12	0·08
8 in hollow concrete block	(37)	0·29	0·26	0·21	0·16	0·12	0·24	0·16	0·11	0·08
None								0·31	0·17	0·10
2 in paper honeycomb with lining	(1)	0·32								
Glass, wood 2 in thick	(1)	0·11								
or mineral fibre 3 in thick	(2)	0·08								
Fibreboard 2 in thick	(3)	0·19								
3 in thick	(4)	0·14								

TABLE 4.19a
THERMAL TRANSMISSION COEFFICIENTS (U-VALUES)

7. PARTITION WALLS (STILL AIR BOTH SIDES)

Construction		Weight (kg/m²)	Finish							
			One side				Both sides			
			None	Gypsum plaster	Plaster board on battens	Insulating board on battens		Gypsum plaster	Plaster board on battens	Insulating board on battens	
						13 mm thick	25 mm thick			13 mm thick	25 mm thick
				(29)	(15)	(10)	(19)	(29)	(15)	(10)	(19)
115 mm brick		(218)	2·50	2·16	1·53	1·14	0·85	1·88	1·14	0·74	0·51
225 mm brick		(436)	1·82	1·59	1·25	0·97	0·74	1·42	0·97	0·68	0·46
100 mm cinder block or 100 mm hollow concrete block		(97)	1·88	1·65	1·25	0·97	0·74	1·48	0·97	0·68	0·46
200 mm hollow concrete block		(179)	1·65	1·48	1·20	0·91	0·68	1·37	0·91	0·62	0·46
None											
50 mm paper honeycomb with lining		(5)	1·82						1·76	0·97	0·57
Glass, wood or mineral fibre	50 mm thick	(5)	0·62								
	75 mm thick	(10)	0·46								
Fibreboard	50 mm thick	(15)	1·08								
	75 mm thick	(19)	0·80								

TABLE 4.19
THERMAL TRANSMISSION COEFFICIENTS (U-VALUES)
8. WINDOWS

Glazing	Air space thickness (in)	Vertical summer and winter	Horizontal Summer	Horizontal Winter
Single	—	1·13	0·86	1·40
Double	1/4	0·61	0·50	0·70
	1/2	0·55		
	4	0·53		
Triple	1/4	0·41		
	1/2	0·36		
	4	0·34		

TABLE 4.19a
THERMAL TRANSMISSION COEFFICIENTS (U-VALUES)
8. WINDOWS

Glazing	Air space thickness (mm)	Vertical summer and winter	Horizontal Summer	Horizontal Winter
Single	—	6·42	4·89	7·95
Double	6	3·46	2·84	3·98
	13	3·12		
	100	3·01		
Triple	6	2·34		
	13	2·05		
	100	1·93		

TABLE 4.19a
THERMAL TRANSMISSION COEFFICIENTS (U-VALUES)
9. DOORS

Type	U-value
25 mm wood	3·92
38 mm wood	2·96
50 mm wood	5·96

TABLE 4.19
THERMAL TRANSMISSION COEFFICIENTS (U-VALUES)
9. DOORS

Type	U-value
1 in wood	0·69
1½ in wood	0·52

TABLE 4.20
THERMAL RESISTANCES (R-VALUES)

Material	Description		Density (lb/ft³)	Weight (lb/ft²)	Resistance R Per in thickness (i/k)	Resistance R For stated thickness (i/c)
Masonry	Common brick	(4½ in)	120	45	0.20	0.90
	Facing brick	(4½ in)	130	49	0.11	0.50
	Hollow concrete block	(4 in)	70	23	—	0.70
	Hollow concrete block	(8 in)	64	43	—	1.10
	Cinder block	(3 in)	80	20	—	0.40
	Cinder block	(4 in)	70	23	—	0.71
	Concrete—cement mortar		116	—	0.20	—
	Concrete—sand and gravel		140	—	0.11	—
	Lightweight aggregate	(3 in)	60	15	—	0.56
	Lightweight aggregate	(6 in)	46	23	—	1.50
	Lightweight aggregate	(12 in)	43	24	—	0.89
Plastering	Cement and sand plaster		116	—	0.20	—
	Gypsum plaster		45	—	0.64	—
	Sand aggregate		105	—	1.08	—
Building board	Plaster board	(½ in)	50	2	0.90	0.45
	Asbestos–cement board	(⅛ in)	120	1.2	0.25	0.03
Wood	Plywood	(½ in)	34	1.5	1.25	0.62
	Wood fibre board		26	—	2.38	—
	Hardboard	(¼ in)	65	1.4	0.72	0.18
	Softwoods		32	—	1.25	—
	Hardwoods		45	—	0.91	—

TABLE 4.20—continued
THERMAL RESISTANCES (R-VALUES)

Material	Description		Density (lb/ft³)	Weight (lb/ft²)	Resistance R Per in thickness (i/k)	Resistance R For stated thickness (i/c)
Flooring materials	Plastic tile	(⅛ in)	110	1·15	—	0·05
	Linoleum	(⅛ in)	80	0·83	—	0·08
	Asphalt tile	(⅛ in)	120	1·25	—	0·04
	Carpet and rubber pad		—	—	—	1·23
	Carpet and fibrous pad		—	—	—	2·08
	Cork tile	(⅛ in)	25	0·26	2·22	0·28
	Plywood subfloor	(½ in)	34	1·4	1·24	0·62
	Terrazzo	(1 in)	140	11·7	0·08	0·08
	Hardwood block	(¾ in)	45	2·81	1·01	0·68
Roofing	Asbestos–cement shingles		120	—	—	0·21
	Asphalt shingles		70	—	—	0·44
	Asphalt roll		70	—	—	0·15
	Hollow cinder block	(5 in)	—	—	—	1·30
	Slate		201	24·00	0·11	—
	Corrugated asbestos		—	—	—	0·50
	Wood shingles		40	—	—	0·94
Insulation	Roof insulation		15·6	—	2·78	—
	Cotton fibre		3·4	—	3·85	—
	Wood fibre		9·5	—	4·00	—
	Glass fibre		—	—	4·00	—
	Mineral wool		22·4	—	3·70	—
	Acoustic tile		1·62	—	2·38	—
	Foamed plastic		7·00	—	3·45	—
	Cork board		7·00	—	3·70	—
	Expanded vermiculite			—	2·08	—

TABLE 4.20a
THERMAL RESISTANCES (R-VALUES)

Material	Description		Density (kg/m³)	Weight (kg/m²)	Resistance R Per m thickness (1/k)	Resistance R For stated thickness (1/c)
Masonry	Common brick	(115 mm)	1920	220	1·38	0·159
	Facing brick	(115 mm)	2080	240	0·75	0·087
	Hollow concrete block	(100 mm)	1120	112	—	0·123
	Hollow concrete block	(200 mm)	1025	205	—	0·194
	Cinder block	(75 mm)	1280	96	—	0·071
	Cinder block	(100 mm)	1120	112	—	0·125
	Concrete–cement mortar		1860	—	1·38	—
	Concrete–sand and gravel		2240	—	0·76	—
	Lightweight aggregate	(75 mm)	960	72	—	0·099
	Lightweight aggregate	(150 mm)	735	110	—	0·264
	Lightweight aggregate	(300 mm)	690	207	—	0·157
Plastering	Cement and sand plaster		1860	—	1·38	—
	Gypsum plaster		720	—	4·40	—
	Sand aggregate		1680	—	7·50	—
Building board	Plaster board	(13 mm)	800	10·4	6·20	0·080
	Asbestos–cement board	(3 mm)	1920	5·8	1·70	0·005
Wood	Plywood	(13 mm)	545	7·1	8·60	0·109
	Wood fibre board		415	—	17·80	—
	Hardboard	(6 mm)	1040	6·2	5·00	0·032
	Softwoods		510	—	8·60	—
	Hardwoods		720	—	6·30	—

TABLE 4.20a—continued
THERMAL RESISTANCES (R-VALUES)

Material	Description		Density (kg/m³)	Weight (kg/m²)	Resistance R Per m thickness (i/k)	Resistance R For stated thickness (i/c)
Flooring materials	Plastic tile	(3 mm)	1760	5·3	—	0·009
	Linoleum	(3 mm)	1280	3·9	—	0·014
	Asphalt tile	(3 mm)	1920	5·8	—	0·007
	Carpet and rubber pad		—	—	—	0·217
	Carpet and fibrous pad		—	—	—	0·367
	Cork tile	(3 mm)	400	1·2	15·30	0·050
	Plywood subfloor	(13 mm)	545	7·1	8·55	0·109
	Terrazzo	(25 mm)	2240	56	0·54	0·014
	Hardwood block	(19 mm)	720	13·8	7·00	0·120
Roofing	Asbestos–cement shingles		1920	—	—	0·037
	Asphalt shingles		1120	—	—	0·078
	Asphalt roll		1120	—	—	0·026
	Hollow cinder block	(125 mm)	—	—	—	0·230
	Slate		3220	115	0·76	—
	Corrugated asbestos		—	—	—	0·088
	Wood shingles		640	—	—	0·166
Insulation	Roof insulation		250	—	19·20	—
	Cotton fibre		54	—	26·60	—
	Wood fibre		152	—	27·60	—
	Glass fibre		—	—	27·60	—
	Mineral wool		—	—	25·50	—
	Acoustic tile		360	—	16·40	—
	Foamed plastic		26	—	23·80	—
	Cork board		112	—	25·50	—
	Expanded vermiculite		112	—	14·30	—

4 LOAD ESTIMATING

AIR FILM COEFFICIENTS (R)

Wind speed	Facing	Heat flow	Thickness (in)	R
Still air	Horizontal	Up	—	0·61
	Horizontal	Down	—	0·92
	Vertical	Horizontal	—	0·68
	Sloping 45°	Up	—	0·62
	Sloping 45°	Down	—	0·76
7½ mph (summer)	Any position	Any direction	—	0·25
15 mph (winter)	Any position	Any direction	—	0·17
Air space	Horizontal	Up	¾ to 4	0·85
		Down	¾	1·02
		Down	4	1·23
	Vertical	Horizontal	¾ to 4	0·97
	Sloping 45°	Up	¾ to 4	0·90
	Sloping 45°	Down	¾ to 4	1·03

TABLE 4.21a
THERMAL RESISTANCES (R-VALUES)

AIR FILM COEFFICIENTS (R)

Wind speed	Facing	Heat flow	Thickness (mm)	R
Still air	*Horizontal*	*Up*	—	*0·107*
	Horizontal	*Down*	—	*0·162*
	Vertical	*Horizontal*	—	*0·120*
	Sloping 45°	*Up*	—	*0·109*
	Sloping 45°	*Down*	—	*0·134*
3·35 m/s (summer)	*Any position*	*Any direction*	—	*0·044*
6·7 m/s (winter)	*Any position*	*Any direction*	—	*0·030*
Air space	*Horizontal*	*Up*	*18 to 100*	*0·150*
		Down	*18*	*0·180*
		Down	*100*	*0·216*
	Vertical	*Horizontal*	*18 to 100*	*0·171*
	Sloping 45°	*Up*	*18 to 100*	*0·159*
	Sloping 45°	*Down*	*18 to 100*	*0·182*

TABLE 4.22
DIVERSITY OF INTERNAL LOADS

Application	Diversity factor	
	People	Lights
Industrial	0·85 to 0·95	0·85 to 0·90
Offices	0·75 to 0·90	0·75 to 0·85
Hotels, residences	0·40 to 0·60	0·30 to 0·50
Stores	0·80 to 0·90	0·90 to 1·0

TABLE 4.23
METABOLIC RATES OF PEOPLE, BTU/H

Degree of activity	80°F Sensible	80°F Latent	78°F Sensible	78°F Latent	76°F Sensible	76°F Latent	74°F Sensible	74°F Latent	72°F Sensible	72°F Latent	70°F Sensible	70°F Latent
Seated at rest	195	155	210	140	225	125	235	115	245	105	260	90
Seated light work	195	205	215	185	230	170	245	155	260	140	275	125
Office worker	200	250	215	235	230	220	250	200	270	180	285	165
Standing, walking slowly	200	300	220	280	240	260	255	245	270	230	290	210
Sedentary	220	330	240	310	260	290	280	270	300	250	320	230
Light bench work	220	530	245	505	275	475	305	445	335	415	365	385
Walking	300	700	330	670	360	640	395	605	430	570	460	540

TABLE 4.23a
METABOLIC RATES OF PEOPLE, W

Degree of activity	26°C Sensible	26°C Latent	25°C Sensible	25°C Latent	24°C Sensible	24°C Latent	23°C Sensible	23°C Latent	22°C Sensible	22°C Latent	21°C Sensible	21°C Latent
Seated at rest	57	46	61	42	65	38	69	34	73	30	77	26
Seated light work	60	57	64	53	68	49	72	45	76	41	80	37
Office worker	64	68	68	64	72	60	76	56	80	52	84	48
Standing, walking slowly	65	82	69	78	73	74	77	70	81	64	85	62
Sedentary	69	93	74	88	79	83	84	78	89	73	94	68
Light bench work	69	151	77	143	84	136	92	128	99	121	107	113
Walking	95	199	103	191	111	183	119	175	127	167	135	159

TABLE 4.24
HEAT GAIN FROM LIGHTS, BTU/H/NAMEPLATE WATTS

Application	Heat gain to room	Heat gain to return air
Exposed incandescent	3·4	—
Exposed fluorescent	4·25	—
Incandescent recessed in suspended return air plenum	3·0	0·4
Fluorescent recessed in suspended return air plenum	3·0	1·25
Fluorescent in patent return air extract fittings	1·7	2·55

TABLE 4.24a
HEAT GAIN FROM LIGHTS, W/NAMEPLATE WATTS

Application	Heat gain to room	Heat gain to return air
Exposed incandescent	1·0	—
Exposed fluorescent	1·25	—
Incandescent recessed in suspended return air plenum	0·90	0·1
Fluorescent recessed in suspended return air plenum	0·90	0·35
Fluorescent in patent return air extract fittings	0·50	0·75

TABLE 4.25
HEAT GAIN FROM ELECTRIC MOTORS, BTU/H PER ABSORBED BRAKE HORSEPOWER

Nameplate horsepower	Motor and driven in conditioned space	Motor out, driven in conditioned space	Motor in, driven out of conditioned space
$1/20-\frac{1}{4}$	4 260	2 540	1 810
$\frac{1}{3}-\frac{3}{4}$	3 640	2 540	1 100
1–5	3 150	2 540	610
$7\frac{1}{2}-20$	2 970	2 540	430
25–60	2 880	2 540	340
75 and over	2 790	2 540	250

Note:
The above values are for absorbed horsepower and not nameplate.
Where absorbed horsepower is not known assume: 85 per cent nameplate horsepower Class 'E' windings; 100 per cent nameplate horsepower Class 'A' windings.

TABLE 4.25a
HEAT GAIN FROM ELECTRIC MOTORS, W PER ABSORBED BRAKE HORSEPOWER

Nameplate horsepower	Motor and driven in conditioned space	Motor out, driven in conditioned space	Motor in, driven out of conditioned space
$1/20-\frac{1}{4}$	1 250	746	504
$\frac{1}{3}-\frac{3}{4}$	1 066	746	320
1–5	922	746	176
$7\frac{1}{2}-20$	871	746	125
25–60	846	746	100
75 and over	818	746	72

Note:
The above values are for absorbed horsepower and not nameplate.
Where absorbed horsepower is not known assume: 85 per cent nameplate horsepower Class 'E' windings; 100 per cent nameplate horsepower Class 'A' windings.

TABLE 4.26
INFILTRATION THROUGH WINDOWS

Type of window	cfm per linear foot of crack	Litres/s per linear metre of crack
Wooden sash (good condition)	0·25	0·36
Wooden sash (poor condition)	0·70	1·00
Metal: sash	0·50	0·72
Metal: industrially pivoted	1·20	1·70
Metal: casement	0·35	0·50
Metal: vertically pivoted	1·00	1·44

Note:
The above values are based on $7\frac{1}{2}$ mph (3·35 m/s) wind velocity in summer. For other velocities use ratio of wind speed to $7\frac{1}{2}$ mph (3·35 m/s).

TABLE 4.27
INFILTRATION THROUGH DOORS

Type of door		cfm infiltration			Litres/s infiltration		
		Closed	Average use	Open	Closed	Average use	Open
Revolving doors		28	180	1200	13	85	565
Glass door	(3 ft 0 in × 7 ft 0 in) (0·9 m × 2·13 m)	95	350	700	45	165	330
Wooden door	(2 ft 6 in × 6 ft 6 in) (0·76 m × 2·0 m)	10	75	500	5	35	236
Wooden door	(3 ft 0 in × 7 ft 0 in) (0·92 m × 2·13 m)	20	135	700	9	64	330
Wooden door	(6 ft 0 in × 7 ft 0 in) (1·84 m × 2·13 m)	80	180	1400	38	85	660

Note:
The above values are based on $7\frac{1}{2}$ mph (3·35 m/s) wind velocity in summer. For other velocities use ratio of wind speed to $7\frac{1}{2}$ mph (3·35 m/s).

TABLE 4.28
SUPPLY DUCT HEAT GAIN, PER CENT ROOM SENSIBLE HEAT PER 10 FT OF DUCT IN UNCONDITIONED SPACE

Room sensible heat BTU/h		50 000	100 000	200 000	400 000 and upwards
Uninsulated	(% per 10 ft)	3·2	2·4	1·6	1·2
1 in insulation	(% per 10 ft)	1·45	1·1	0·75	0·6
2 in insulation	(% per 10 ft)	0·95	0·75	0·50	0·40

Notes:
Figures based on 30 °F temperature difference between unconditioned space and supply air temperature.
For other temperature differences use ratio to 30 °F.

TABLE 4.28a
SUPPLY DUCT HEAT GAIN, PER CENT ROOM SENSIBLE HEAT PER METRE OF DUCT IN UNCONDITIONED SPACE

Room sensible heat W		*15 000*	*30 000*	*45 000*	*60 000 and upwards*
Uninsulated	*(% per m)*	*1·05*	*0·80*	*0·53*	*0·40*
25 mm insulation	*(% per m)*	*0·47*	*0·36*	*0·25*	*0·20*
50 mm insulation	*(% per m)*	*0·31*	*0·25*	*0·16*	*0·13*

Note:
Figures based on 16 °C temperature difference between space and supply air temperature. For other temperature differences use ratio to 16 °C.

TABLE 4.29
FAN HEAT GAIN, PER CENT ROOM SENSIBLE HEAT

Fan total pressure (N/m^2)	(in. w.g.)	Packaged system: motor in conditioned air	Central system: motor out of conditioned air
125	0·50	1·1	0·6
250	1·00	2·4	1·4
375	1·50	3·5	2·3
500	2·00	5·0	3·1
625	2·50	6·5	4·1
750	3·00	8·5	5·2
1000	4·00	—	7·0
1250	5·00	—	9·6
1500	6·00	—	12·2
1750	7·00	—	15·0
2000	8·00	—	19·0

Note:
The above values are based on a room–supply air temperature of 20°F (*11·1°C*). For other temperature differences use inverse ratio to 20°F (*11·1°C*) (e.g. for central system with 10°F (*5·55°C*) temperature difference at 3·0 in (*750 N/m²*) total pressure gain = 5·2 × 20/10 = 10·4 per cent (*or 5·2 × 11·1/5·55 = 10·4 per cent*)).

TABLE 4.30
VENTILATION REQUIREMENTS, CFM (LITRES/S) PER PERSON

Application	Smoking	Recommendation cfm	Litres/s
Small stores or shops	None	10–15	5–7
	Some	30	14
Department stores	None	7·5–10	3·6–5
Hotel rooms	Heavy	25–30	12–14
Residential	Some	20–30	9–14
General office	Some	10–15	5–7
Private office	None	15–25	7–12
	Heavy	25–30	12–14
Meeting or board rooms	Heavy	30–50	14–24
Restaurants or bars	Some	10–15	5–7
Theatres	None	5–7·5	2·4–3·6
	Some	10–15	5–7
Hospital wards	None	20–30	9–14
Laboratories	Some	15–20	7–9
Factories	None	7·5–10	3·6–5

Notes:
1. The above recommendations must be considered in the light of local regulations.
2. Some applications may be governed by exhaust systems.

Chapter 5

ROOM AIR DISTRIBUTION

No matter how accurately a room-heat-gain estimate is calculated, no matter how well equipment has been selected and applied to meet the job requirements, it is all wasted unless the conditioned air is introduced properly. Too much stress cannot be placed on this subject, which may form the least costly sector of an air-conditioning system.

The first principle which should be observed is that it is the distribution ductwork which is designed or selected to provide air to the outlets, not the outlets selected for a given ductwork layout.

Most complaints arising from a faulty air distribution system can be put into three classifications, draughts, stuffiness, or noise. Noise problems can be avoided by using manufacturers' recommendations of maximum diffuser discharge velocities (the largest contributor to noise), and ensuring that the application falls within such limits.

Draughts cause problems when sensitive parts of the body, such as the neck or ankles, are exposed to air which is moving too fast or is too cold. A combination of the two will accelerate the cooling of the body and cause discomfort, for example air moving at 50 fpm ($0.254\,m/s$) should be 3–4°F (1.7–$2.2°C$) higher in temperature than still air at a given temperature to provide the same comfort conditions. It is difficult to satisfy everybody in an air-conditioned space, but occupied area room velocities between 15 fpm ($0.076\,m/s$) (minimum) and 50 fpm ($0.254\,m/s$) (maximum) are desirable.

When room velocities fall below 15 fpm ($0.076\,m/s$) a feeling of stuffiness would occur because body heat is not removed fast enough, or insufficient ventilation or lack of fresh air make-up may provide the same discomfort. A poorly designed distribution system may starve

5 ROOM AIR DISTRIBUTION

certain areas, which will both decrease air motion and increase room temperature thus enhancing the feeling of stuffiness.

The two most sensitive areas of the body are the neck and ankles. Ideally, air should be directed towards the face, but never at the back of the neck or ankles.

Draughts and stuffiness are not entirely dependent on air motion, but are also a function of temperature, humidity and direction. All too frequently, bad air distribution is caused through lack of

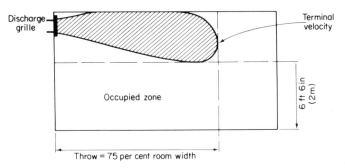

FIG. 5.1. Room air distribution from sidewall grille.

understanding of these four points, which when designed properly will provide little or no temperature stratification in the occupied zone, will avoid fluctuating gusts of air, and hot or cold spots, and local high velocities. In understanding the selection of supply grilles or diffusers which are used to provide proper air distribution the following terminology should be considered.

Throw: The horizontal or vertical axial distance that an air stream travels on leaving an outlet. Usually measured where the terminal velocity reaches 50 fpm (*0·254 m/s*) at a height of 6 ft 6 in (*2 m*).

Drop: The vertical distance the lower edge of the air stream drops between leaving the outlet and reaching the end of its throw, associated with cooling systems.

Rise: The converse of drop—associated with heating systems.

Grille: A functional or decorative covering for an outlet or intake.

Diffuser: An outlet grille designed to guide the direction of air.

Register: A grille provided with a damper.

Free area: The nett area of the openings in a grille.

Gross or core area: Total area of the grille opening.

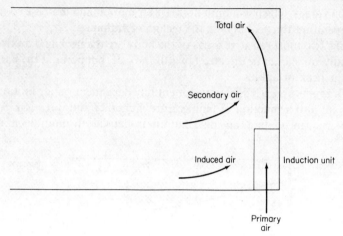

Fig. 5.2. Under-the-window air distribution.

Primary air: Air coming from an outlet (Fig. 5.2).
Secondary air: The room air picked up by the primary air by entrainment.
Total air: Summation of primary and secondary airs.
Entrainment ratio: Total air divided by the primary air.
Induction: Air drawn from the room into an outlet or air conditioning unit by the primary air stream.
Induction ratio: Induced air plus primary air divided by the primary air.
Outlet velocity: The average air velocity emerging from the outlet.
Terminal velocity: Average air stream velocity at the end of the throw (usually taken at 50 fpm (*0·254 m/s*)).
Spread: The divergence of the air stream in the horizontal or vertical plane after it leaves the outlet.
Temperature differential: The temperature difference between primary and room air.
Diffusion area: The useful room area or coverage served by the outlet.
Ceiling effect: The air stream moving along the ceiling creates a low pressure area between the air stream and the ceiling. This results in part of the discharge air remaining in contact with the ceiling throughout the length of the throw (Fig. 5.3).

The type of outlet to be used is governed essentially by its location and can be categorised into three: (1) Grilles and registers for sidewall

5 ROOM AIR DISTRIBUTION

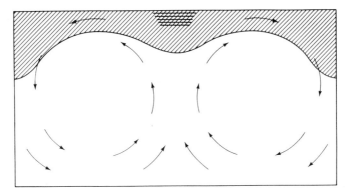

Fig. 5.3. Ceiling effect.

application; (2) Slot diffusers for some ceiling, sidewall, baseboard or sill applications; and (3) Ceiling diffusers.

Supply outlets are not designed to correct unreasonable conditions of air flow supplied to them. An outlet mounted directly on the side of the duct will not discharge air in a normal direction because of the effect of the air velocity in the duct. This is shown in Fig. 5.4 where the static pressure in the duct causes the air to pass normally into the room and the velocity pressure will cause it to travel in its original direction. The nett result would be the velocity and direction as shown by resultant vector V_R.

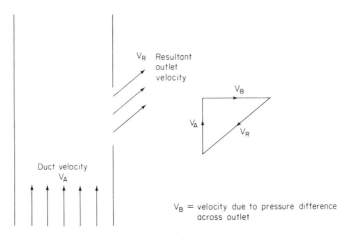

Fig. 5.4. Outlet velocity without vanes.

To overcome this a short collar with vanes should be fitted so that the air discharge is perpendicular to the ducting air flow with resultant improvement of diffuser performance (see Fig. 5.5). Whenever the outlet velocity of the diffuser is less than the ducting velocity, straightening vanes should always be used.

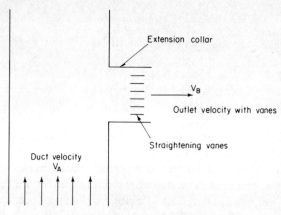

FIG. 5.5. Outlet velocity with vanes.

Grilles and registers have relatively long throws, with the lowest entrainment rate and lowest temperature diffusion. Ceiling diffusers have the advantage of a very high entrainment ratio, because the ceiling effect allows high temperature differentials. Table 5.1 shows the application of the various outlet categories. Where low temperature differentials are considered special attention should be paid to room air velocity against air volume and temperature differential; it may be easier to introduce half the air volume with a 20°F (*11°C*) differential than that required for a 10°F (*5·5°C*) differential and still maintain better comfort conditions with the space.

TABLE 5.1

Outlet type	Temperature differential (°F)	(*°C*)	Floor area (cfm/ft^2)	*Floor area (litres/s/m^2)*
Ceiling diffusers	20–35	*11–20*	3–3·5	*1·5–17·5*
Slot diffusers	20–25	*11–14*	1·5–2·5	*7·5–12·5*
Grilles or registers	16–25	*9–14*	1–2	*5–10*

5 ROOM AIR DISTRIBUTION

TABLE 5.2
RECOMMENDED AIR VELOCITIES IN OCCUPIED ZONES

Room air velocity (fpm)	Application	Comments	Room air velocity (m/s)
0–15	None	Stagnant air	0–0.076
25	All commercial	Ideal design	0.127
25–50	Commercial	50 fpm (0.25 m/s) is max. tolerable velocity for seated persons	0.127–0.254
65	Retail or department	Light papers blow off desk	0.330
75	Stores, banks	Upper limit for people moving slowly	0.382
700–300	Some factory installations	High velocities, used for slot cooling	0.356–1.52

Table 5.2 shows recommendations of room air velocity against various applications. Particular attention must be paid to ceiling height when considering any outlet selection. Catalogue data may prove adequate for a given application, but because of spread or drop if the high velocity air leaving the outlet comes into the occupied zone before it reaches its terminal velocity what may have appeared to be an acceptable room velocity may prove too high.

If straightening vanes are fitted prior to a register then the air approaching the register neck is perpendicular. When the guide vanes of the diffuser are placed parallel there will be a small amount of divergence or spread because of the entrainment of secondary air into the perimeter of the primary air stream. The included angle of the spread with straight vanes is 19°, whether directed perpendicular to the duct air flow or at an angle in order to avoid some obstruction. This divergence occurs in both the horizontal and vertical planes, and therefore for cooling applications attention should be paid to the divergence into the occupied zone where long air throws are required.

Grilles with vanes set at $22\frac{1}{2}°$ have a divergence angle of 30° and those set at 45° an angle of 60°. The spread from an outlet can be expressed in terms of the throw for each vane angle setting as follows:

$$\text{Spread at } 0° \text{ vane setting} = \text{throw} \times 0.3$$
$$\text{Spread at } 22\tfrac{1}{2}° \text{ vane setting} = \text{throw} \times 0.5$$
$$\text{Spread at } 45° \text{ vane setting} = \text{throw} \times 1.2$$

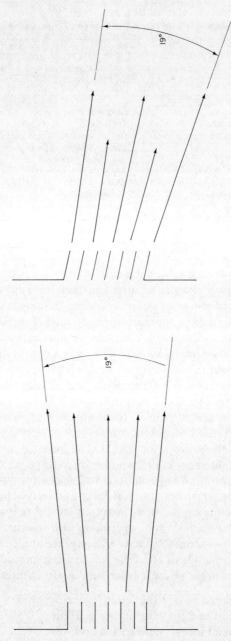

FIG. 5.6. Spread with straight turning vanes.

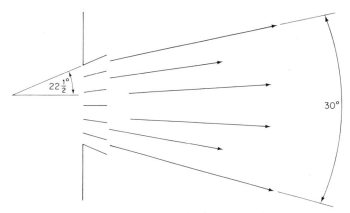

FIG. 5.7. Spread with $22\frac{1}{2}°$ vane setting.

The above information will assist in the selection of the quantity of grilles required for a given application. For example a room 72 ft (*23·6 m*) long requiring a total air volume of 2400 cfm (*1·32 m³/s*), and a sidewall register throw of 20 ft (*6·5 m*) would require 12 grilles at 0° vane setting each at 200 cfm (*0·094 m³/s*) [spread = 20 × 0·3 = 6 ft (*6·5 × 0·3 = 1·95 m*)] or 7 grilles at $22\frac{1}{2}°$ vane setting each at 343 cfm (*0·162 m³/s*) [spread 20 × 0·5 = 10 ft (*6·5 × 0·5 = 3·25 m*)] or 3

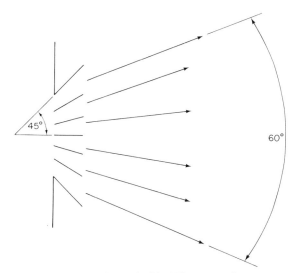

FIG. 5.8. Spread with 45° vane setting.

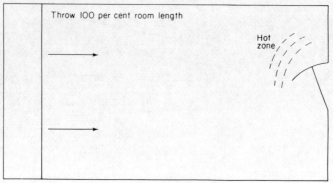

FIGURE 5.9.

grilles at 45° vane setting each at 800 cfm ($0.378\,m^3/s$) [spread = 20 × 1·2 = 24 ft ($6.5 \times 1.2 = 7.8\,m$)].

Examination of manufacturers' data will show recommended grille sizes for the three applications, one of which will best suit the problem. It is assumed that the conditions set out meet the job requirements; examination of manufacturers' data may show that none of the selections are suitable and it may be necessary to relocate the grilles, say to the centre of the room, to allow 1200 cfm ($0.566\,m^3/s$) per side with only 10 ft ($3.3\,m$) throw.

When selecting grilles or registers the throw should be equal to 75 per cent of the total distance from the grille face to the facing wall or to the point where opposing air streams meet. Overthrow will cause draughts and discomfort for anybody sitting adjacent to the wall, and underthrow will cause stuffiness in the same position.

Figures 5.9 to 5.14 show various applications of distribution ducting and grille locations for a shop with a door opening to the street.

Figure 5.9 shows air supplied from the rear of the store; if this grille location is used the throw must be sized for the total length of the room to alleviate the possibility of a hot zone near the door due to outdoor infiltration. This location is not recommended for long rooms because in order to obtain 100 per cent throw, high air velocities are required and excessive room air turbulence occurs.

Figure 5.10 shows the air supplied through grilles above the door, this location has the disadvantage of causing excessive outdoor air infiltration through the door due to the induction effect caused by the supply grilles.

5 ROOM AIR DISTRIBUTION

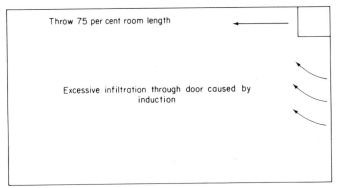

FIGURE 5.10.

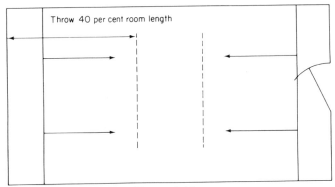

FIGURE 5.11.

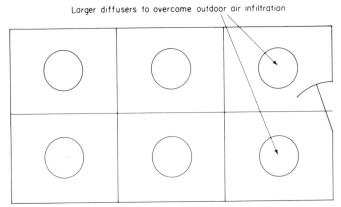

FIGURE 5.12.

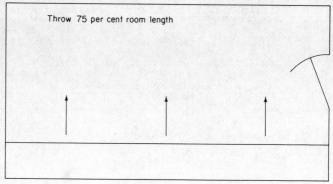

FIGURE 5.13.

The above methods may be combined as shown in Figure 5.11 so that air is supplied from the front and rear of the store blowing towards the centre; if this method is used each outlet should be sized to give a throw of approximately 40 per cent of the total room length.

By far the best method of air distribution for this type of store is obtained by using ceiling diffusers as shown in Fig. 5.12 with the two outlets near the door that is larger than the others in order to combat outdoor air infiltration; this system uses more ductwork and is generally more expensive to install than wall outlets.

Two other satisfactory methods of distribution are shown in Figs. 5.13 and 5.14, but extreme care must be taken in their use to ensure that overblow does not occur.

Let us now look at a room with a reasonable area of exposed glass such as a hotel bedroom or modern office building. The major

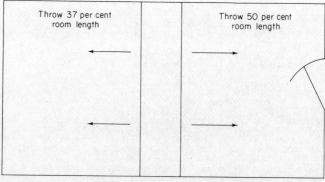

FIGURE 5.14.

5 ROOM AIR DISTRIBUTION

problem with exposed glass occurs in low ambient temperatures. The cold glass cools the layer of air next to it causing a downward motion of air; this is called downdraught and can reach velocities of up to 200 fpm (*1·02 m/s*) usually resulting in complaints of cold feet from anyone near the exposed glass area. The temperature at low level near an exposed glass area may be as much as 10 °F (*5·5 °C*) lower than that of the remainder of the room.

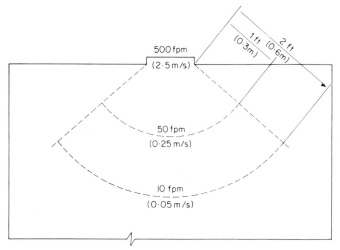

FIG. 5.15. Velocity fall off away from return air grille.

The only satisfactory solution to this is to supply air from beneath the window blowing vertically up the window; this will combat any downdraught effect and give excellent room air distribution and therefore excellent temperature conditions throughout the room.

The location of return air or exhaust outlets is much less important than the positioning of inlet grilles or registers. Figure 5.15 shows the velocity fall off away from the grille. It can be clearly seen that providing occupants keep more than 2 ft (*0·61 m*) from the outlet no discomfort from draughts will occur, nor will noise be a problem providing the outlet is sized to deal with between 400 and 800 fpm (*2·03 and 4·06 m/s*) across the gross area.

Providing that grilles are not located such that they discharge directly towards the return air grilles no short circuiting of conditioned air should occur.

A wall return at low level proves best for both heating and cooling applications. Stratification due to poor room mixing in winter is offset by a low return since cool air at floor level is returned first and is replaced by warm air from above, also short circuiting is completely eliminated.

TABLE 5.3
RECOMMENDED OUTLET VELOCITIES FOR GRILLES AND REGISTERS

Outlet velocity (fpm)	Applications	Outlet velocity (m/s)
300–500	Broadcasting studios	1·5–2·5
500–750	Residences, hotel bedrooms, private offices	2·5–3·75
500–1 000	Theatres, restaurants	2·5–5·0
1 000–1 250	General office, small shops	5·0–6·25
1 000–1 500	Department stores	5·0–7·5
1 000–2 000	Factories	5·0–10·0

Applications calling for both heating and cooling should avoid ceiling return locations, since difficulty may be experienced during the season when the heating is in operation and room circulation, due to low induction, is insufficient to cause the warm air to flow to the floor. A ceiling return may also prove to be a short circuit, particularly with ceiling diffusers and high level registers or grilles which depend upon ceiling effect. When it is necessary to provide a ceiling return, diffusers should be arranged such that they are directed away from the outlet. In ceiling diffusers this may be achieved by using blanking plates on one side or on one quadrant; and with registers by adjusting the vane settings.

Floor returns, whilst ideal in terms of function, are best avoided because they become a catch-all for dirt, which may impose heavy demands on the system filters and thereby increase running costs. Also, floor grilles have to be substantially made in order that they can withstand the weight associated with the job application.

Chapter 6

DUCT DESIGN

The object of a good air distribution system is to provide conditioned air at the required quantities at given outlets with the minimum of balancing, minimum of material, minimum of friction loss, minimum of leakage and at a velocity that is acoustically suitable to the application.

Balancing a badly designed system could take many hours, if not days, of a commissioning engineer's time, and the precaution of having a well designed job should always prove more economical. In a badly designed job every outlet must be provided with a damper so that balancing can be achieved, whereas arranging every outlet to have the correct air volume at the correct pressure could well eliminate many dampers, with a saving in both capital and running costs that would be needed to create the pressure to overcome their resistance.

Correct sizing of ductwork will provide the minimum of material at the most economical friction loss. An oversized section of ductwork will require throttling by a damper; thus expensive ductwork requires expensive fittings together with the possible addition of higher running costs and the generation of noise.

Leakage should be kept to an absolute minimum, and particular care must be taken with regard to choosing the correct type of manufacture for the application. It is common practice, for systems with their distribution systems outside the conditioned space, to allow 5 to 10 per cent for leakage loss. Care must be taken to ensure that thermal losses are also kept to a minimum and, if necessary, ducts outside the conditioned space should be insulated.

The velocities in ductwork should be such that noise generation is kept to an acceptable level. Table 6.1 shows typical velocities for various applications. Where it is necessary because of physical limitations to increase the velocity then sound attenuation should be considered.

The total pressure of air in a ducted system is split into two components, velocity head and static head, both of which are

TABLE 6.1
RECOMMENDED MAXIMUM DUCT VELOCITIES—ft/min (m/s)
(LOW VELOCITY SYSTEMS)

Application	Controlling factor				
	Noise generation	Duct friction			
		Main duct		Take offs or branches	
		Supply	Return	Supply	Return
Residences	600 (3)	800 (4)	600 (3)	600 (3)	600 (3)
Theatres	800 (4)	1 200 (6)	1 000 (5)	800 (4)	600 (3)
Hotel bedrooms, hospitals, apartments	1 000 (5)	1 400 (7)	1 100 (5·5)	1 000 (5)	800 (4)
Private offices, libraries	1 200 (6)	1 500 (7·5)	1 200 (6)	1 200 (6)	1 000 (5)
General offices, banks, shops, restaurants	1 500 (7·5)	1 800 (9)	1 400 (7)	1 500 (7·5)	1 200 (6)
Department stores, cafeterias	1 800 (9)	2 000 (10)	1 500 (7·5)	1 600 (8)	1 200 (6)
Industrial	2 500 (12·5)	3 000 (15)	1 800 (9)	2 200 (11)	1 500 (7·5)

produced by a fan or air moving device. The pressure or head due to fluid velocity is expressed as

$$v = (2gh)^{\frac{1}{2}}$$

where the velocity v is expressed in feet per second, g in feet per second per second and h the head of fluid in feet. In air-conditioning systems the common units of head are inches of water and velocity in feet per minute, thus the equation for standard air at a density of $0\cdot075\,\text{lb/ft}^3$ ($13\cdot4\,\text{ft}^3/\text{lb}$ sp. vol.) can be expressed as

$$v = 4005(h_v)^{\frac{1}{2}}$$

where h_v is the velocity head, or more commonly as

$$h_v = \left(\frac{v}{4005}\right)^2$$

Thus air moving at a velocity of 4005 ft/min is said to have a velocity head equal to 1 inch water gauge.

In SI units the velocity pressure h_v is measured in N/m^2, which considers absolute units of force. This differs from the gravitational units of force used in Imperial units, in that the Newton, N, induces unit acceleration in unit mass, i.e. $N = kgm/s^2$.

The original formula, $v = (2gh)^{\frac{1}{2}}$, in substituted form becomes:

$$h_v = \frac{v^2}{2g}$$

where: h_v is measured in m of air, v is measured in m/s, and g is measured in m/s^2.
Substituting units:

$$h_v(N/m^2) = \frac{v^2(m^2/s^2)}{2\,m/s^2} \times \frac{kg/m/s^2}{m}$$

or

$$h_v(N/m^2) = \frac{v^2(m^2/s^2)}{2\,m/s^2} \times \frac{kg}{m^3} \times m/s^2$$

$$= \frac{v^2(m^2/s^2)}{2} \times \frac{kg}{m^3}$$

Using the specific volume of air as $0{\cdot}832\,m^3/kg$,

$$h_v = 0{\cdot}6\,v^2.$$

Thus air moving at 10 m/s would have a velocity pressure of 60 N/m^2.

Static pressure and velocity pressure can exist together and are convertible from one to another in a ductwork system. For example, if air leaving a fan outlet has a certain velocity which can be expressed as a velocity head, and in addition has some static pressure head which is used to overcome the friction in the duct, the two pressures summated are referred to as the total pressure. Figure 6.1 shows a typical system with air leaving a fan and passing along a duct. It shows a sudden expansion and then a sudden contraction with the air patterns for

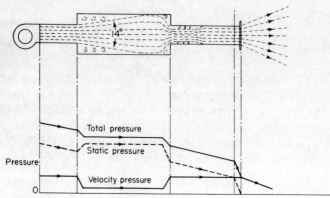

FIG. 6.1. Relationship between static and velocity head.

Section	Condition			L/D ratio	
90° elbow	R/D = 1·5			9	
90° mitred elbow	R/D = 1·5				
	3 piece			24	
	5 piece			12	
Rectangular radius elbow	W/D	R/D = 0·5	R/D = 0·75	R/D = 1·0	R/D = 1·5
	0·5	34	15	10	4
	1·0	45	18	11	5
	3·0	80	30	18	6
Radius take off piece	6·0	125	40	14	7
Rectangular vaned elbow or vaned take off piece	1 vane	18	10	9	7
	2 vanes	12	9	8	7
	3 vanes	10	8	7	6
Rectangular squared elbow	No vanes			60	
	Single thickness vanes			15	
Squared take off piece	Double thickness vanes			10	
Angle	Vaned or unvaned elbow			Equivalent 90° elbow x A/90	

FIG. 6.2. Equivalent lengths of common fittings [L = (L/D)D].

each. Plotted against the ductwork run are the total, static and velocity pressures. It can clearly be seen that whilst the velocity in the ductwork after the abrupt expansion is the same, the turbulence in static and total pressure is varied, despite at expansion the velocity head turning to static head, and at re-entry to the duct static head turning to velocity head.

The losses in a system are due to friction and dynamic losses and both for a given duct size and flow reflect in a reduction in static pressure, since the air velocity must remain constant. The pressure created by the air-moving device must therefore be able to overcome the resistance to air motion caused by duct friction and the fittings and outlets associated with such systems, and it is necessary that the

Section	Condition			n		
90° tee or 180° cross (No loss)	$v_2/v_1 = 0.2$			4.0		
	$v_2/v_1 = 0.5$			2.0		
	$v_2/v_1 = 1.0$			1.75		
	$v_2/v_1 = 5.0$			1.6		
Conical tee or cross (No loss)	$v_2/v_1 = 0.5$			0.2		
	$v_2/v_1 = 1.0$			0.5		
	$v_2/v_1 = 2.0$			1.0		
	$v_2/v_1 = 5.0$			1.2		
Transformation piece (15° or less)	$v_2 = v_1$ Loss $= n\,hv_1$			0.15		
Expansion piece	v_2/v_1	$a=5°$	$a=10°$	$a=15°$	$a=20°$	$a=30°$	$a=40°$
	0.20	0.83	0.73	0.65	0.60	0.50	0.45
	0.40	0.90	0.83	0.78	0.73	0.69	0.64
	0.60	0.95	0.92	0.89	0.87	0.85	0.83
	Pressure regain $= n\,(hv_1 - hv_2)$						
Contraction	a	30°		45°		60°	
	n	1.02		1.04		1.07	
	Pressure loss $= n\,(hv_2 - hv_1)$						
Abrupt expansion	v_2/v_1	0.20		0.40		0.60	0.80
	n	0.32		0.48		0.48	0.32
	Pressure loss $= n\,hv_1$						
Abrupt contraction	v_1/v_2	0		0.25		0.50	0.75
	n	0.35		0.30		0.20	0.10
	Pressure loss $= n\,hv_2$						

FIG. 6.3. Values of n for common fittings [$L = n(h_v \times 100/h_f)$ or $L = n(h_v/h_f)$].

engineer should be able to calculate this resistance so that the correct air handling apparatus can be selected.

Friction rate is expressed in terms of inches water gauge per 100 ft (*or* N/m^3). It is convenient therefore that friction and dynamic losses of fittings are expressed in equivalent length of duct run, where the addition of fittings and straight ducting can be used to calculate the system pressure drop.

Figures 6.2 and 6.3 show the equivalent length of various fittings commonly used in sheet metal ducting, and the pressure losses for each are shown in the two alternative forms. Firstly, the L/D ratio is the factor which multiplied by the dimension D in feet (*or metres*) will give the equivalent length of straight ducting. The second factor n, when multiplied by the velocity head (h_v) or loss in velocity head ($h_{v_1} - h_{v_2}$), whichever is referred to in the table, will give the pressure

TABLE 6.2
VELOCITY PRESSURE CONVERSION TABLE

Air velocity ft/min	Velocity head in.w.g.	Air velocity ft/min	Velocity head in.w.g.
400	0·010	3 000	0·562
500	0·016	3 100	0·601
600	0·022	3 200	0·640
700	0·031	3 300	0·681
800	0·040	3 400	0·723
900	0·051	3 500	0·766
1 000	0·062	3 600	0·810
1 100	0·076	3 700	0·856
1 200	0·090	3 800	0·903
1 300	0·106	3 900	0·951
1 400	0·122	4 000	1·000
1 500	0·141	4 100	1·051
1 600	0·160	4 200	1·103
1 700	0·181	4 300	1·156
1 800	0·202	4 400	1·210
1 900	0·226	4 500	1·266
2 000	0·250	4 600	1·323
2 100	0·276	4 700	1·381
2 200	0·302	4 800	1·440
2 300	0·331	4 900	1·501
2 400	0·360	5 000	1·562
2 500	0·391	5 500	1·891
2 600	0·422	6 000	2·250
2 700	0·456	6 500	2·641
2 800	0·490	7 000	3·063
2 900	0·526	7 500	3·516

6 DUCT DESIGN

loss in inches water gauge (N/m^3). This figure can be converted to equivalent length by dividing the pressure drop by the friction rate per 100 ft for the ducting considered.
Therefore

$$L(\text{ft}) = \frac{h_v \times 100}{h_f} \times n$$

where: h_v is in. w.g. and h_f is in. w.g.,

or,

$$L(m) = \frac{h_v}{h_f} n$$

where: h_v is N/m^2 and h_f is N/m^3.

Velocity heads are shown in Tables 6.2 and 6.2a for both SI and Imperial units.

TABLE 6.2a
VELOCITY PRESSURE CONVERSION TABLE

Air velocity m/s	Velocity head N/m^2	Air velocity m/s	Velocity head N/m^2
2	2·4	15	135·0
2·5	3·8	15·5	144·2
3	5·4	16	153·6
3·5	7·4	16·5	163·4
4	9·6	17	173·4
4·5	12·2	17·5	183·8
5	15·0	18	194·4
5·5	18·2	18·5	205·4
6	21·6	19	216·6
6·5	23·4	19·5	228·2
7	29·4	20	240·0
7·5	33·8	20·5	252·2
8	38·4	21	264·6
8·5	43·4	21·5	277·4
9	48·6	22	290·4
9·5	54·2	22·5	303·8
10	60·0	23	317·4
10·5	66·2	23·5	331·4
11	72·6	24	345·6
11·5	79·4	24·5	360·2
12	86·4	25	375·0
12·5	93·8	26	405·6
13	101·4	27	437·4
13·5	109·4	28	470·4
14	117·6	29	504·6
14·5	126·2	30	540·0

When sizing ducting it is normal practice to consider the equivalent duct diameter for a given air flow against a determined friction rate or duct velocity whichever the criteria. Figure 6.4 is used to determine the equivalent duct diameters.

Figure 6.5 is used to equate these figures into the more common cross-sections of rectangular ducts. Before using the duct-sizing chart, several considerations must be known, firstly the air volume to be handled, secondly the maximum velocity to suit the application. These two values will give a friction rate which for most applications is kept constant throughout the exercise. Maintaining the friction rate as a pivot the equivalent duct sizes and duct velocities can be read for the various air volumes used in the system.

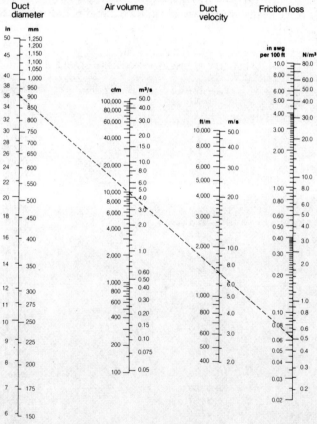

FIG. 6.4. Duct-sizing chart.

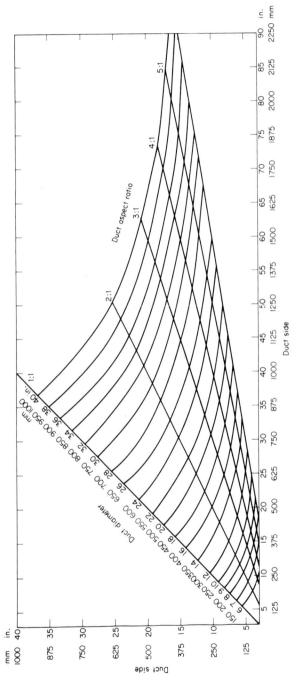

FIG. 6.5. Equivalent rectangular ductwork cross-sections.

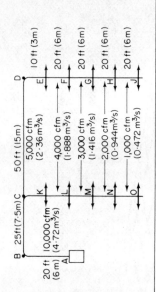

Section	Air volume		Air velocity		Duct diameter		Equivalent rect.		Ductwork length		Equivalent ductwork length		Description
	ft³/min	(m³/s)	ft/min	(m/s)	in	(mm)	in	(mm)	ft	(m)	ft	(m)	
AC	10 000	(4·72)	1500	(7·5)	35·8	(910)	46 × 20	(1150 × 500)	45	(13·5)	43	(13)	Radius elbow B, R/D = 0·75, W/D = 2·3, L/D = 26, L = 20/12 × 26 = 43 ft (or L = 0·5 × 26 = 13 m).
CE	5000	(2·36)	1200	(6·0)	27·6	(700)	32 × 20	(800 × 500)	60	(18)	37	(11)	Radius elbow C, R/D = 0·75, W/D = 1·6, L/D = 22, L = 20/12 × 22 = 37 ft (or L = 0·5 × 22 = 11 m).
EF	4000	(1·888)	1150	(5·75)	24·8	(630)	24 × 20	(600 × 500)	20	(6)	—	—	
FG	3000	(1·416)	1100	(5·5)	22·3	(570)	20 × 20	(500 × 500)	20	(6)	—	—	

Section	Flow (cfm)	(m³/s)	Velocity (fpm)	(m/s)		Size (in)	(mm)	Length (ft)	(m)	Friction		Notes
GH	2000	(0.944)	1000	(5.0)	19.2 (490)	20 × 14	(500 × 350)	20	(6)	—	—	(21) Take off J, no vanes, R/D = 0.75, L/D = 60, L = 14/12 × 60 = 70 ft (or L = 0.35 × 60 = 21 m).
HJ	1000	(0.472)	810	(4.05)	15.0 (380)	14 × 14	(350 × 350)	20	(6)	70		
							Sub total	185	(55.5)	150	(45)	Total length = 335 ft (100.5 m).
CK	5000	(2.36)	1480	(7.4)	25.2 (640)	26 × 20	(650 × 500)	10	(3)	33	(10)	Radius take off, C, R/D = 0.75, W/D = 1.3, L/D = 20, L = 20/12 × 20 = 33 ft (or L = 0.5 × 20 = 10 m).
KL	4000	(1.888)	1370	(6.85)	23.0 (590)	20 × 20	(500 × 500)	20	(6)	—		
LM	3000	(1.416)	1290	(6.46)	20.5 (520)	20 × 16	(500 × 400)	20	(6)	—		
MN	2000	(0.944)	1170	(5.85)	17.7 (450)	14 × 16	(350 × 400)	20	(6)	—		
NO	1000	(0.472)	1000	(5.0)	14.0 (360)	14 × 11	(350 × 275)	20	(6)	55	(17)	Take off O, no vanes, R/D = 0.75, L/D = 60, L = 11/12 × 60 = 55 ft (or L = 0.275 × 60 = 17 m).
							Sub total	20	(27)	88	(27)	Total length = 178 ft (54 m).

Friction loss A to J = 335 ft × 0.065/100 = 0.218 in w.g. (or 100.5 × 0.54 = 54.3 N/m²)
Friction loss at J =   0.230 in w.g.   (= 57.2 N/m²)
supply duct resistance   = 0.448 in w.g.   (= 111.5 N/m²)

FIG. 6.6. Example of duct sizing.

An example of a typical ductwork layout is shown in Fig. 6.6. The air volume to be handled by the fan is 10 000 cfm ($4{\cdot}72\,m^3/s$), and there are 20 outlets each handling 500 cfm ($0{\cdot}236\,m^3/s$). The application is a supermarket and the recommended maximum duct velocity is 1500 fpm ($7{\cdot}5\,m/s$). From the duct-sizing chart the friction rate for 10 000 cfm ($4{\cdot}72\,m^3/s$) at 1500 fpm ($7{\cdot}5\,m/s$) velocity is 0·065 in water per 100 ft ($0{\cdot}54\,N/m^3$) and the equivalent duct size is 35·8 in ($0{\cdot}91\,m$) diameter. For this example it can be assumed that the maximum depth of the ducting is 20 in ($500\,mm$) because of height restriction. For economy of duct manufacture it would be feasible to keep one size constant at transformation so that only one side is reduced.

Each outlet has to handle 500 cfm ($0{\cdot}236\,m^3/s$) and from side-wall diffuser-selection charts the recommended grille size is 20 in × 8 in, ($500\,mm \times 200\,mm$), with blades set at 45° angle. The pressure drop required at the grille is 0·23 in w.g. ($57{\cdot}2\,N/m^2$). At each transformation piece the take off to each grille should be sized as the outlet grille, i.e. 20 in × 8 in ($500\,mm \times 200\,mm$).

Below the sketch of the ductwork layout is a table showing the various ducting sizes for the different air volumes assuming the constant friction rate of 0·065 in per 100 ft ($0{\cdot}54\,N/m^2$) together with the respective velocities. It is quite clear from this table how the velocity is being reduced as the air volume is dropping, even though the friction rate is constant. The reason for this is the ratio of surface area causing friction to cross-sectional area which allows the air to pass. Although with circular ducting the ratio of cross-section to perimeter is constant it should be remembered that the air touching the perimeter is almost stationary and forms an air film, which is thicker the slower the velocity. Therefore, for a fixed velocity the smaller the cross-section, the greater the friction rate.

With rectangular ductwork, the aspect ratio between duct height and width should be kept at a maximum of 6 to 1 to keep the system economical from a standpoint of both material cost and friction rate. For example an equivalent duct diameter of 32 in ($800\,mm$) would require a duct 30 in × 30 in ($750\,mm \times 750\,mm$) in cross-section [120 in ($3\,m$) perimeter] or a duct 80 in × 14 in (188 in perimeter) [$2000\,mm \times 350\,mm$ ($4{\cdot}7\,m\ perimeter$)]. Not only is the material content higher in the high aspect ratio alternative, but also the cost of stiffening a duct so wide would be expensive and the thickness of the metal would probably be greater.

For the example considered the index run is said to be from points

6 DUCT DESIGN

A to J. This is the longest, or the run with the greatest friction drop to the final outlet. Although the run-outs DJ and CO are identical the index run has the additional friction loss due to ducting CD. There are two alternatives which can be taken, the first and most common alternative, since it takes the designer less time, is to make the run-outs the size dimensions and install a damper between CK so that a restriction can be made to flow equal to the friction loss in ductwork CD. Secondly, the friction loss between CJ and CO can be made equal.

The fittings for each take off are the same, so a direct comparison between duct lengths can be made. The length CJ is 140 ft ($42\,m$) and CO 90 ft ($27\,m$). Therefore, the tolerable friction rate in the branch CO is $140/90 \times 0.065 = 0.101$ in per 100 ft ($42/27 \times 0.54 = 0.84\ N/m^3$). At 5000 cfm ($2.36\,m^3/s$) the duct velocity is just below 1500 fpm ($7.5\,m/s$) the recommended maximum. Therefore this branch can now be sized at the new friction rate, with consequent savings in duct sizes.

From the table of duct sizes calculated for this exercise the summation of straight ducting is 185 ft ($55.5\,m$) and the fitting loss 150 ft ($45\,m$), making a total loss of 335 equivalent ft ($100.5\,m$). The pressure drop is therefore equal to 0.065×335 ft or 0.218 in w.g. ($0.54 \times 100.5\,m$ or $54.3\ N/m^2$). This value added to the outlet grille pressure drop of 0.23 in w.g. ($57.2\ N/m^2$) exercise, plus the apparatus and return air pressure drop, will give the total static pressure against which the fan must operate.

Chapter 7

PARTIAL LOAD AND ZONING

Zoning of a building or conditioned space is necessary when the cooling or heating requirements of one space vary significantly from another. The orientation of areas or frequency of occupation are major factors influencing the need for zoning and will affect the type of air-conditioning system to be installed.

The need for zoning can best be illustrated by a simple example. If an intermediate floor of an office block is considered as having east and west exposures with a central corridor and a conditioned space above and below as in Fig. 7.1 a simplified load estimate can be made as in Table 7.1.

From the load estimate it can be clearly seen that the maximum gains for the east exposure occur at 10 a.m. whereas those for the west occur at 4 p.m. To examine the methods of applying various systems to this problem it has been assumed that the outside air condition at 3 p.m. design is 82°F (*27·8°C*) DB, 68°F (*20·0°C*) WB; 500 cfm (*236 litres/s*) of ventilation air is required per exposure. The room temperature to be controlled is 72°F (*22·2°C*) DB, 50 per cent RH, and the temperature difference between supply and room air is 20°F (*11·1°C*).

The first consideration of equipment for these areas could be a single-zone air-conditioning unit having supply ducting down the centre corridor discharging into the east and west exposures. If the plant was designed to meet the maximum gains to the whole area, i.e. at 4 p.m. the air volume required for the east exposure would be 2120 cfm (*1000 litres/s*) and for the west 2950 cfm (*1400 litres/s*). At these circumstances, assuming the plant was allowed to run without

7 PARTIAL LOAD AND ZONING 121

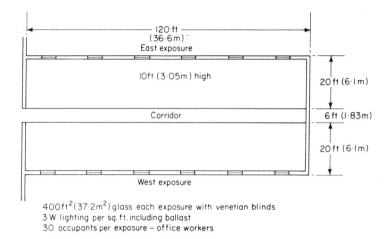

400 ft² (37·2m²) glass each exposure with venetian blinds
3 W lighting per sq. ft. including ballast
30 occupants per exposure – office workers

FIG. 7.1. Plan of office used in example.

thermostatic control, the following room temperature would probably result:

	10 a.m.	Noon	2 p.m.	4 p.m.
East exposure	75°F (23·9°C)	74°F (23·3°C)	73°F (22·8°C)	72°F (22·2°C)
West exposure	64°F (17·8°C)	64°F (17·8°C)	68°F (20·0°C)	72°F (22·2°C)

These conditions would be grossly exaggerated if a thermostat was located in a particular exposure. For example, if the east exposure was kept under control at 72°F (22·2°C), the west at its maximum cooling requirement at 4 p.m. would increase to a temperature of about 77°F (25°C). Or if the thermostat was located in the west zone, the east zone at 10 a.m. would increase to 79°F (26·2°C). These areas therefore are very much in need of zoning.

One very simple way of overcoming this problem would be to install room air conditioners in each outside wall as shown in Fig. 7.2. This system would overcome the problem of wide fluctuations in room temperature, give individual control for relatively small areas; probably give the cheapest equipment capital cost; and would provide a degree of stand-by in the event of a unit failure.

There are, however, several disadvantages to this system. The installed capacity would have to meet the maximum for each area, i.e.

TABLE 7.1
LOAD ESTIMATE FOR EXAMPLE OFFICE BLOCK

East exposure	BTU/h				W			
	10 a.m.	noon	2 p.m.	4 p.m.	10 a.m.	noon	2 p.m.	4 p.m.
Glass solar gain	21 000	10 500	8 700	6 900	6 150	3 075	2 545	2 020
Glass trans gain	340	1 700	3 050	3 050	100	500	895	895
Wall trans gain	2 600	4 600	4 600	3 200	760	1 350	1 350	935
Lighting	24 500	24 500	24 500	24 500	7 180	7 180	7 180	7 180
Occupants	8 100	8 100	8 100	8 100	2 375	2 375	2 375	2 375
Room sensible heat =	56 540	49 400	48 950	45 750	16 565	14 480	14 345	13 405
Occupants:								
Room latent heat =	5 400	5 400	5 400	5 400	1 580	1 580	1 580	1 580
Room total heat =	61 940	54 800	54 350	51 150	18 145	16 060	15 925	14 985
Fresh air load =	8 100	10 600	12 800	12 800	2 375	3 105	3 750	3 750
Grand total heat =	70 040	65 400	67 150	63 950	20 520	19 165	19 675	18 735

West exposure	10 a.m.	noon	2 p.m.	4 p.m.	10 a.m.	noon	2 p.m.	4 p.m.
Glass solar gain	8 700	7 300	15 900	26 000	2 550	2 140	4 660	7 610
Glass trans gain	340	1 700	3 050	3 050	100	500	895	895
Wall trans gain	1 000	1 000	1 400	2 200	290	290	410	645
Lighting	24 500	24 500	24 500	24 500	7 180	7 180	7 180	7 180
Occupants	8 100	8 100	8 100	8 100	2 375	2 375	2 375	2 375
Room sensible heat =	42 640	42 600	52 950	63 950	12 495	12 485	15 520	18 705
Occupants:								
Room latent heat =	5 400	5 400	5 400	5 400	1 580	1 580	1 580	1 580
Room total heat =	48 040	48 000	58 350	69 350	14 075	14 065	17 100	20 285
Fresh air load =	8 100	10 600	12 800	12 800	2 375	3 105	3 750	3 750
Grand total heat =	56 140	58 600	71 150	82 150	16 450	17 170	20 850	24 035

East and West exposures	10 a.m.	noon	2 p.m.	4 p.m.	10 a.m.	noon	2 p.m.	4 p.m.
Room sensible heat	99 180	92 000	101 900	109 700	29 060	26 965	29 865	33 110
Room total heat	109 980	102 800	112 700	120 500	32 220	30 125	33 025	35 270
Grand total heat	126 180	124 000	138 300	146 100	36 970	36 335	40 525	42 770

7 PARTIAL LOAD AND ZONING 123

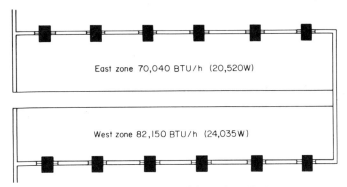

Fig. 7.2. Room air conditioner installation.

70040 BTU/h (20 520W) in the east zone and 82 150 BTU/h (24 035 W) in the west zone; the equipment must be serviced (including the changing of filters) in the conditioned space: the noise created by numerous units might prove unacceptable; the life of such units cannot be expected to be as long as central station plant; the external appearance of many units through the window might be unsightly; the running cost of many small machines operating at relatively low efficiencies would be quite high; and the cost of many power points might be expensive.

An alternative to this method, Fig. 7.3, is the use of packaged units for each area. The shape of the building rules out single 'free-blow' units for each area, therefore sheet-metal ductwork would have to be installed. As with the previous example each unit would have to be sized for the maximum requirement of each zone. Unless a plant

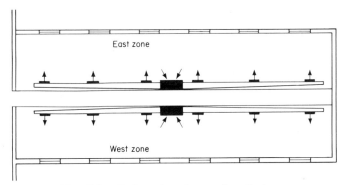

Fig. 7.3. Packaged equipment installation.

room was used to house the equipment it would have to be serviced from within the conditioned space. This system would prove relatively economical and would provide reasonable temperature control from an 'on–off' thermostat.

There are some disadvantages to this system, which can only sense the room temperature and not the RH. If the east zone was considered at 4 p.m. on a design day, but without the sun shining, the room sensible heat would be reduced to 38 850 BTU/h (*11 385 W*), whereas the plant has been designed for operation at a room sensible heat of 56 540 BTU/h (*16 565 W*) with an air volume of 2650 cfm (*1250 litres/s*). The load therefore would be at 69 per cent, and controlled from a room thermostat the plant could be expected to run for a similar time. However, the latent gain from the occupants would remain constant, and therefore the plant would only remove 69 per cent of the latent gains, less the moisture which re-evaporates from the cooling coil when it is switched off, and less the unconditioned fresh air that is not dehumidified during the off cycle. This method of control could well increase the room relative humidity by 10 per cent above design; in this particular example the problem being increased with high occupancy applications and decreased in areas of low latent heat gains.

Obviously 'on–off' control is attractive because it is the least expensive, but it may prove unacceptable unless additional control is incorporated. For example larger packaged units may have split coils such that, at partial load, one or more sections are isolated. This results in part of the air always being dehumidified, and the lowest coil split is usually of smaller area than other splits such that the same amount of refrigeration effect is provided at each coil. The result is that the lowest coil becomes a dehumidification coil and upper coils are used for sensible heat extraction. With a multi-stage thermostat or step-controller, this can prove an easy way of overcoming high room relative humidities at partial loads.

A third alternative is shown in Fig. 7.4 which diagrammatically shows a variable volume central station plant. This system can overcome some of the above difficulties inasmuch as each zone only calls for the air volume needed to cater for the actual room sensible heat gains. Because the refrigeration plant (sized for 4 p.m., i.e. 146 100 BTU/h (*42 770 W*)) is running continually, constant dehumidification takes place and the lower air volumes across the coil will increase the latent heat removal per pound of air handled and so

7 PARTIAL LOAD AND ZONING 125

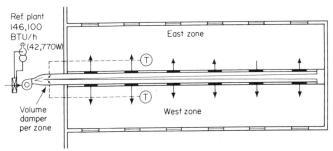

FIG. 7.4. Variable volume system.

provide better partial load conditions. However, this system has the problem that should the air volume decrease by more than 25 per cent, room air distribution suffers when using conventional distribution outlets (see Chapter 14, for Variable Air Volume Terminal Systems). The lower air volume will cause the air to drop, and areas farthest away from the outlets will not receive conditioned air.

If partial load conditions can be guaranteed not to create air volumes below 75 per cent of design, this would prove a simple system and be reasonably economical, since the total design air volume of the plant need only be sized to cater for the room sensible heat of 109 700 BTU/h (*33 110 W*) occurring at 4 p.m. in the example considered. Figure 7.5 shows probably the best way to maintain close control of room temperature and humidity at partial load. By maintaining a constant coil temperature a constant dew point can be held. Variations in room sensible heat are picked up with a reheater battery per zone. This provides excellent results but is expensive in terms of running cost. This method of control is adopted where close control is

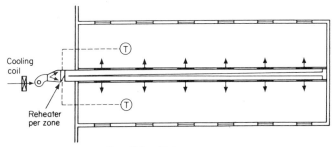

FIG. 7.5. Reheat system.

required for laboratories or test rooms, but is often frowned upon for comfort or process applications because of the waste of energy.

A common method of controlling single zones is the use of face and by-pass control of return air (see Fig. 1.6, Chapter 1 and Fig. 13.13, Chapter 13). The use of this method can be shown best by use of a psychrometric chart, Fig. 7.6.

Considering the design condition of the west zone of the example the room condition is shown at point A, the off-coil condition at point

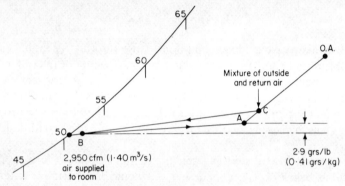

FIG. 7.6. Air process of west zone for maximum load.

B, and the mixture on coil at point C. If at a design day the sun did not shine the room sensible heat gains would be reduced to 37 850 BTU/h (*11 095 W*), whereas the latent gain would remain at 5400 BTU/h (*1580 W*) and the effect of fresh air would still be 12 800 BTU/h (*3750 W*) giving a grand total heat of 56 050 BTU/h (*16 425 W*).

Because the cooling coil is now required to perform less refrigeration effect, the coil dew point will be reduced. For a direct expansion system, the compressor would unload under the influence of lower suction pressure, the set point of the unloading mechanism being related to desired evaporating temperatures at partial load. When using chilled water entering at a constant temperature the water would not pick up so much heat and so leave at a lower temperature than design. This, as with the direct expansion system, would have the effect of lowering the coil dew point, although such a close control cannot be guaranteed.

Figure 7.7 shows the effect of lowering the coil surface temperature. Point C′ is the mixture condition of the fixed amount of fresh air with a reduced amount of return air passing over the cooling coil. This mixed air is then passed over the cooling coil and is cooled and dehumidified to point B'. This air mixed with room air at condition A results in condition D, which is passed into the conditioned space, the net result being that the required amount of dehumidification takes place with a reduced amount of sensible cooling to match the room partial load requirements. The advantage of this system is the

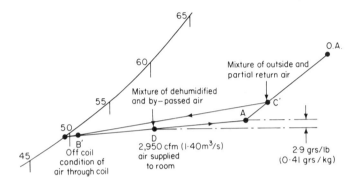

FIG. 7.7. Air process of west zone at partial load using face and by-pass damper system.

reduction of refrigeration machinery running costs and no requirements for reheat under normal circumstance.

A further way of controlling various zones is the blow-through or multizone system (see Fig. 13.14, Chapter 13). Mixed fresh and return air is first filtered and then passed through the cooling coil or heating coil at the dictate of zoning dampers at the discharge duct connection of the unit.

In summer, heat can be maintained in the hot deck so that the system acts as a reheat unit, although if there is a constant call for cooling the unit can be used without heat, as with the face and by-pass damper system. This system can operate at partial load conditions, one zone requires heating while another demands cooling.

The cost of a multizone system is dependent upon the layout of zones, many run-outs could prove costly, although these should be considered against the alternatives. Running costs are dependent upon the characteristics of the partial load requirements of the space, for at certain conditions the effect would be the same as the reheat system. Dehumidification can be a problem in areas of high occupancy since any air by-passing the cooling section would be a

TABLE 7.2

System	Characteristics
Reheat	Able to maintain low relative humidities. An excellent partial load performance. The only system permitting separate control of humidity.
Return air bypass (multistage packaged unit with coil face split)	Excellent control with DX coil, good with chilled water. The return air supplies heat, room relative humidity depends upon the relation between room latent load and coil dehumidifying capacity.
Volume control	Good control, but could impair room distribution, unless specific VAV terminals are used.
On-off control	Fairly good temperature control, but may result in high relative humidities.
Multizone system (dual-duct system)	Good results if used on reheat principles, but may result in poor RH control with cooling coil bypass.

mixture of room air and untreated outside air. An advantage of the multizone system is that one duct carries the conditioned air to each zone, the mixing taking place at the unit under the control of a thermostat located in the respective zone. By this method all equipment is kept in the plant room.

A variation of this system is the dual-duct system (see Chapter 14).

Chapter 8

EVAPORATORS

Having established a cooling load in terms of room sensible heat, room latent heat, fresh air load, and consequently the total cooling load, together with the amount of fresh air required for ventilation from a load estimate, the first consideration of an air conditioning system is the cooling coil. This will dictate the air volume to be handled by the plant and the operating level of the refrigeration plant.

There are two basic ways to examine what type of cooling coil should be used. Firstly, an off-coil condition can be assumed which satisfies the sensible and latent requirements of the room which will dictate the supply air volume; from this figure the mixture condition on coil can be determined when the fresh air volume is considered. This method then means that a cooling coil must be engineered to suit the assumed conditions which may result in a shallow coil to meet air conditions although a high temperature difference may be required between the coil dew point and the refrigerant medium to meet the total cooling duty requirement, resulting in expensive refrigeration plant. Alternatively, a deep coil may be called for although the large heat exchange surface may not be necessary for the cooling medium.

Secondly, a fixed coil can be considered which will dictate the air volume and the refrigerant medium temperature, giving a standard coil and economic temperatures.

Considering the first method, Fig. 8.1 shows how the arbitrary selection of an off-coil condition affects the ADP. The closer the off-coil condition to saturation the higher the ADP and higher the cooling medium temperature and the smaller the air volume handled by the apparatus. To achieve this, however, a deeper and more expensive cooling coil is required than with an off-coil condition further away from saturation. Figure 8.1 serves as an example to consider the economics of such a situation.

If the room sensible cooling (RSH) load was 216 000 BTU/h ($63.3\,kW$), including fan heat and duct gains, and the room latent heat gain (RLH) was 50 000 BTU/h ($14.6\,kW$), the room sensible heat factor (RSHF) would be

$$\frac{216\,000\text{ BTU}}{(216\,000 + 50\,000)\text{ BTU}} = 0.81 \quad or \quad \frac{63.3\,kW}{(63.3 + 14.6)\,kW} = 0.81$$

This would dictate the slope of line AB. Point B is the required room condition of 72°F ($22.2\,°C$), 50 per cent RH (60°F ($15.6\,°C$) WB) and point C is the outside design condition 82°F DB, 68°F WB ($27.8\,°C$ DB, $20\,°C$ WB). In the example, 5000 cfm ($2.36\,m^3/s$) of outside air is required to satisfy the room ventilation requirements, which would add 128 800 BTU/h ($37.59\,kW$) to the room cooling load requiring a total load of 394 800 BTU/h ($115.49\,kW$). The first step in this method would be to select an off-coil condition B, which arbitrarily has been fixed at 52°F ($11.1\,°C$) to give a 20°F ($11.1\,°C$) differential from room condition. The calculated air volume to satisfy the room load would be

$$\frac{216\,000\text{ BTU}}{1.08 \times 20\,°F} = 10\,000\text{ cfm} \quad or \quad \frac{63.3\,kW}{1.21 \times 11.1\,°C} = 4.72\,m^3/s$$

Knowing the air volume to be handled, the fresh air mixture (E), can be calculated along the line BC, which would be

$$72 + (82 - 72)\left(\frac{5000}{10\,000}\right) = 77\,°F$$

$$22.2 + (27.8 - 22.2)\left(\frac{2.36}{4.72}\right) = 25\,°C$$

Projecting these lines to the saturation line will give the ADP = 47°F ($8.3\,°C$). From this information the cooling coil can now be selected.

From these figures another factor is gained which is controlled from the face velocity and the geometry of the cooling coil, the by-pass factor (BF). If the coil was perfect and cooled all the air to saturation there would be no by-pass and 100 per cent contact. However, this is never the case and the BF can be analysed as that

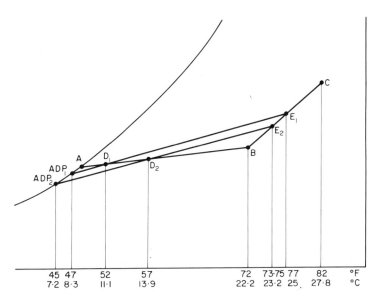

Fig. 8.1. Effect of ADP on off-coil control.

proportion of air which achieves no contact with the cooling surface to that which would have done had the coil been perfect, i.e.

$$BF = \frac{\text{off-coil condition—ADP}}{\text{on-coil condition—ADP}}$$

In the example considered this would be

$$\left(\frac{52-47}{77-47}\right) = \frac{5}{30} = 0.16 \quad or \quad \left(\frac{11 \cdot 1 - 8 \cdot 3}{25 \cdot 0 - 8 \cdot 3}\right) = \frac{2 \cdot 8}{16 \cdot 7} = 0.16$$

The cooling coil required for this example would probably result in say 500 fpm (2·54 m/s) coil face velocity, a four-row, 10 fins per inch 20 ft² (1·86 m²) coil with, in the case of a direct expansion coil, a refrigerant evaporating temperature of 35 °F (1·6 °C).

Consider now that the off-coil condition was put at 57 °F (13·9 °C) (D_2); the following results would be obtained:

supply air volume = 13 333 cfm (6·29 m³/s)
on-coil condition (E_2) = 73·75 °F (23·2 °C)
apparatus dew point (ADP) = 45 °F (7·2 °C)
by-pass factor (BF) = 0·47 (0·47)

As previously, the total cooling load would be 394 000 BTU/h (*115·49 kW*) and a coil selection would probably result in a two-row, six fins/in coil of 26·7 ft² (*2·48 m²*) face area having a refrigerant evaporating temperature of about 25 °F (*−3·9 °C*).

It can clearly be seen that arbitrary selection results in uneconomical equipment selection not only in terms of cooling coil performance, but in associated refrigeration equipment, the last example for instance could not use chilled water as a cooling medium because of the low temperatures involved.

The use of standard cooling coils can use the by-pass factor as the starting point or can be selected from standard tables giving various off-coil conditions corresponding to assumed on-coil conditions at given coil face velocities. This table method can be used quite easily but requires a certain amount of interpolation as it is not practical to show every possible increment of possible entering conditions. Nevertheless, it is a step forward from the hit and miss method of coil performance.

If one considers that a given coil has a fixed by-pass at a given face velocity, then economic coil selection can be made at the time of load estimation. The basic problem with all cooling coil sections using a mixture of fresh and return air is that the apparatus dew point cannot be selected until this mixture point is known, and similarly the on-coil mixture cannot be selected until the apparatus dew point and off-coil condition are known. However, if the coil by-pass factor is known or assumed these variables can be eliminated. The proof of this is shown in Fig. 8.2. Point B is the required room condition, C the outside air condition, A the ADP, E the off-coil condition; line BE is the slope of the RSHF and line EF is parallel to line BD. Line AE is a proportion of the BF to line AD. By constructing the two similar triangles ADB and AEF, the line AF is a proportion of the BF to line AB and line EF is a proportion of the BF to line BD.

The line BE can be constructed knowing the RSH and RLH, which are tangible figures from a cooling load estimate. Similarly, knowing the fresh air requirements needed to satisfy ventilation requirements the effect of the outside air in tangible figures can also be calculated. If the ratio of the BF of the outside air sensible and latent load was added to the RSH and RLH respectively a new RSHF called the effective room sensible heat factor (ERSHF) can be calculated. This by convenience would produce a line BA, with AF as a ratio of the BF to it.

8 EVAPORATORS

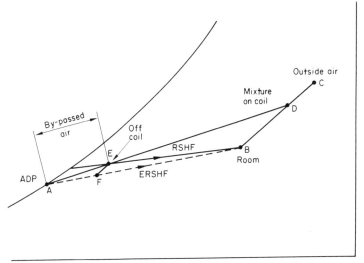

FIG. 8.2. Construction of effective sensible heat factor.

This ERSHF line which connects the room condition B to the ADP can be easily calculated. From this the air volume can be calculated and the mixture point D be found.

Considering the first example in Fig. 8.1 the following method would be adopted. From known data the BF for a four row 14 fins/in/cooling coil at 500 fpm (*2.54 m/s*) is 0·1.

$$
\begin{aligned}
\text{Room sensible heat (RSH)} &= 216\,000 \text{ BTU/h} \\
\text{BF} \times \text{outside air sensible heat} &= \\
0{\cdot}1 \times 5000 \times 1{\cdot}08 \times (82 - 72) &= 5\,400 \text{ BTU/h} \\
\text{Effective room sensible heat (ERSH)} &= 221\,400 \text{ BTU/h}
\end{aligned}
$$

$$
\begin{aligned}
\text{Room latent heat (RLH)} &= 50\,000 \text{ BTU/h} \\
\text{BF} \times \text{outside air latent heat} &= \\
0{\cdot}1 \times 5000 \times 0{\cdot}68 \,(80 - 58) &= 7\,480 \text{ BTU/h} \\
\text{Effective room latent heat (ERLH)} &= 57\,480 \text{ BTU/h} \\
\text{Effective room total heat (ERTH)} &= 278\,880 \text{ BTU/h}
\end{aligned}
$$

$$\text{ERSHF} = \frac{221\,400}{278\,880} = 0{\cdot}795$$

Room sensible heat $(RSH) = 63\,300\,W$
$BF \times$ outside air sensible heat
$0{\cdot}1 \times 2{\cdot}36 \times 1{\cdot}21 \times (27{\cdot}8 - 22{\cdot}2) = \underline{1600\,W}$
Effective room sensible heat $(ERSH) = 64\,900\,W$

Room latent heat $= 14\,600$
$BF \times$ outside air latent heat $=$
$0{\cdot}1 \times 2{\cdot}36 \times 2{\cdot}98 \times (11{\cdot}4 - 8{\cdot}3) = \underline{2160\,W}$

Effective room latent heat $(ERLH) = 16\,760\,W$

Effective room total heat $(ERTH) = 81\,660\,W$

$$ERSHF = \frac{64\,900}{81\,660} = 0{\cdot}795$$

From construction on the psychrometric chart the ADP is found to be $47{\cdot}5\,°F\,(8{\cdot}6\,°C)$. Using the ERSH as the datum the air volume can be calculated as follows.

$$\text{cfm} = \frac{ERSH\,(BTU/h)}{(1 - BF) \times 1{\cdot}08 \times (\text{room} - ADP)\,°F}$$

$$= \frac{221\,400}{0{\cdot}9 \times 1{\cdot}08 \times 24{\cdot}5} = 9300\,\text{cfm}$$

$$m^3/s = \frac{ERSH\,(kW)}{(1 - BF) \times 1{\cdot}21 \times (\text{room} - ADP)\,°C}$$

$$= \frac{64{\cdot}9}{0{\cdot}9 \times 1{\cdot}21 \times 13{\cdot}6} = 4{\cdot}39\,m^3/s$$

At this point one would examine the standard coils available and check the actual BF or the coil at 9300 cfm $(4{\cdot}39\,m^3/s)$. If the actual BF of the coil is within 15 per cent of that assumed no correction need be made. However, if the BF of the assumed coil was $0{\cdot}12$, as would be the case with a four-row, 14 fin/in/coil at 600 fpm $(3{\cdot}05\,m/s)$ face velocity, a correction should be made as follows:

$RSH = 216\,000\,BTU/h$
$BF \times OASH = 1{\cdot}2 \times 5400 = \underline{6480\,BTU/h}$
$ERSH = 222\,480\,BTU/h$

8 EVAPORATORS

$$RLH = 50\,000\,BTU/h$$
$$BF \times OALH = 1\cdot2 \times 7480 = \underline{8\,980\,BTU/h}$$
$$ERLH = \overline{58\,980\,BTU/h}$$
$$ERTH = \overline{281\,460\,BTU/h}$$

$$ERSHF = \frac{222\,480}{281\,460} = 0\cdot79$$

$$ADP = 47\cdot4\,°F$$

$$RSH = 63\,300\,W$$
$$BF \times OASH = 1\cdot2 \times 1600\,W = \underline{1\,920\,W}$$
$$ERSH = 65\,220\,W$$
$$RLH = 14\,600\,W$$
$$BF \times OALH = 1\cdot2 \times 2150\,W = \underline{2\,590\,W}$$
$$ERLH = 17\,190\,W$$
$$ERTH = \overline{82\,410\,W}$$

$$ERSHF = \frac{65\,220}{82\,410} = 0\cdot79$$

$$ADP = 8\cdot55\,°C$$

$$\text{Air volume} = \frac{222\,480}{(1 - 0\cdot12)(72 - 47\cdot4)1\cdot08} = 9540\,\text{cfm}$$

$$= \frac{65\cdot22}{(1 - 0\cdot12)(22\cdot2 - 8\cdot55)1\cdot21} = 4\cdot50\,m^3/s$$

No further calculation is necessary other than to add to the ERTH the balance of the outside air effect not yet considered in order to establish the total cooling load which in this case would be:

$$ERTH = 281\,460$$
$$(1 - 0\cdot12)(5000 \times 1\cdot08) \times 10 = 47\,500$$
$$(1 - 0\cdot12)(5000 \times 0\cdot68) \times 22 = \underline{65\,820}$$
$$\text{Total} \quad 394\,780$$

$$ERTH = 82\,410\,W$$
$$(1 - 0\cdot12)(2\cdot36 \times 1\cdot21)(27\cdot8 - 22\cdot2) = 14\,080\,W$$
$$(1 - 0\cdot12)(2\cdot36 \times 2\cdot98)(11\cdot4 - 8\cdot3) = \underline{19\,000\,W}$$
$$\text{Total} \quad \overline{115\,490\,W}$$

or the previous total.

The refrigerant evaporating temperature of the 15·5 ft² (1·44 m²) face area selected coil would be 35°F (1·7°C).

It should be remembered that to use this method the actual off- and on-coil temperatures have not been considered. The calculation of these would be as follows:

$$\text{actual off-coil DB temp.} = 72 - \frac{216\,000}{1\cdot08 \times 9540} = 51\,°F$$

$$= 22\cdot2 - \frac{63\cdot3}{1\cdot21 \times 4\cdot50} = 10\cdot5\,°C$$

$$\text{on-coil DB temp.} = 72 + (82 - 72)\frac{5000}{9540} = 77\cdot25\,°F$$

$$= 22\cdot2 + (27\cdot8 - 82\cdot2)\frac{2\cdot36}{4\cdot50} = 25\cdot1\,°C$$

The WB temperatures can be selected from the psychrometric chart.

The temperature difference from off-coil to room in this case is $72 - 51 = 21\,°F$ ($22\cdot2 - 10\cdot5 = 11\cdot7\,°C$). If the duct gains and fan gains account for 16 000 BTU/h (4·70 kW) then the net room gain would be 200 000 BTU/h (58·6 kW) and the actual temperature difference of supply air at the diffuser to room would be:

$$\frac{200\,000}{1\cdot08 \times 9540} = 19\cdot45\,°F$$

$$\frac{58\cdot6}{1\cdot21 \times 4\cdot50} = 10\cdot8\,°C$$

Should the condition arise where because of a low ceiling height the air cannot be distributed at more than 15°F (8·3°C) temperature difference, the air volume to be handled by the apparatus would be

$$\frac{200\,000}{1\cdot08 \times 15} = 12\,350\,\text{cfm}$$

$$\frac{58\cdot6}{1\cdot21 \times 8\cdot3} = 5\cdot84\,m^3/s$$

Rather than redesign the economical coil selection, $12\,350 - 9540 = 2810\,\text{cfm}$ ($5\cdot84 - 4\cdot50 = 1\cdot34\,m^3/s$) of room air could be by-passed

8 EVAPORATORS

FIG. 8.3. Direct expansion cooling coil. (Courtesy: Buffalo Forge Co.)

around the cooling coil which would provide the small temperature difference required for room distribution.

This method of cooling coil selection holds good for both direct expansion and chilled water coils. The direct expansion coil selection is achieved from tabular data showing, for the fixed coil selected, refrigerant evaporating temperatures for total cooling loads against apparatus dew points. The chilled water selection uses a similar method by considering the effect of the temperature difference between ADP and the entering chilled water. By dividing the total cooling load by the ADP minus EWT a Q-factor is established, which from charts will dictate the required water quantity for alternative coil water circuitry. Economic entering-water temperatures can be achieved by changing the Q-value to give water temperature splits of say 10°F (5·5°C), which provides an economical split in terms of water quantities handled.

Fig. 8.4. Chilled water cooling coil. (Courtesy: Buffalo Forge Co.)

At this point one should consider what type of coil should be used against different applications, since many standard coils are available to meet varying load conditions per cfm of air handled. Table 8.1 gives a guideline for the designer so that unnecessary trial calculations are averted. The most common form of construction of cooling coils is copper tubes with plate or spiral fins of aluminium or copper. Aluminium fins are more often used for air-conditioning applications because of economy, but, in areas of corrosive atmosphere or applications using sprayed or wetted coils, copper fins should be used. Fins are mechanically bonded to the tubes to give good heat exchange between tube and fin. In the case of all copper coils electro-tinning is often used to ensure good contact and further protect the coils against corrosion. It can be anticipated that future cooling coils will be

TABLE 8.1
TYPICAL COIL SELECTIONS

Example	Type of application	Cooling coil and BF
Retail shop, small offices	Comfort with small total load and low sensible heat factor (packaged equipment)	3 row, 11 fin/in, 0·22–0·27 3 row, 13 fin/in, 0·20–0·24 2 row, 15 fin/in, 0·18–0·22
Department store, factory, bank	Comfort with large total load	4 row, 8 fin/in, 0·20–0·25 4 row, 14 fin/in, 0·10–0·13
Computer room, factory	High sensible heat loads, little outside air	4 row, 14 fin/in, 0·10–0·13
Department store, restaurant, factory	High proportions of outside air	6 row, 8 fin/in, 0·10–0·15 6 row, 14 fin/in, 0·03–0·05
Hospital, factory, primary air plants	All outside air	8 row, 8 fin/in, 0·04–0·07 8 row, 14 fin/in, 0·01–0·02

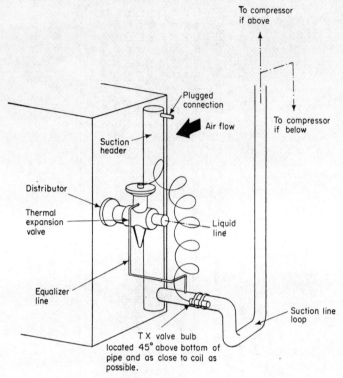

Fig. 8.5. Typical connections at direct expansion coil.

constructed from all aluminium, providing cheaper coils. However, while the technology for making such coils is now available in the factory, there are still problems on site, inasmuch as coil repairs cannot be met by brazing. At present such repairs would have to be made by epoxy resin adhesives.

Figure 8.5 shows typical connections at a direct expansion coil. Standard coils can be inverted to give opposite hand connection. However, to achieve proper counterflow (i.e. suction at air inlet, distributor at air outlet) the suction connection would be at the top of the header. Proper oil return is achieved by connecting the plugged connection now at the bottom of the header to the suction line, allowing oil to drain into the suction loop.

It is good practice with cooling coils to extend the condensate collector below the uninsulated return bends and connections, which include the thermal expansion valve and suction header.

Chapter 9

RECIPROCATING REFRIGERATION COMPRESSORS

Of the three basic types of refrigeration compressors, centrifugal, rotary and reciprocating, the latter is by far the most commonly used. Ranging in capacities from as little as a fraction of a ton R to 150 or even 200 tons R capacity they fulfil the requirements of all but the very large air-conditioning projects. The reciprocating compressor is economical in applications up to around 150 tons R, whereafter rotary or centrifugal compressors used for water chilling applications take over. They have the versatility of operation such that they can be used for direct expansion operation, for package systems or made up systems or for dry expansion or flooded evaporator chilled water applications; they can be used with air cooled, evaporative cooled or water cooled condensers whereas centrifugal or rotary systems are confined in common use to water cooled water chilling applications.

The reciprocating compressor relies on the action of pistons sliding back and forth within a cylinder driven by an eccentric shaft. Figure 9.1 shows a cross-section of a two cylinder compressor, one piston on the suction stroke, the other on the compression stroke. Clearly shown is the suction inlet to the compressor from whence the superheated refrigerant gas from the evaporator passes through the open suction valve disc (left-hand cylinder) into the cylinder chamber; on this stroke the discharge valve disc is kept closed by the action of valve springs. As the piston returns to the compression stroke so the suction valve springs close the suction valve disc and the discharge valve is forced open so that the compressed gas can pass into the discharge manifold and on to the condenser.

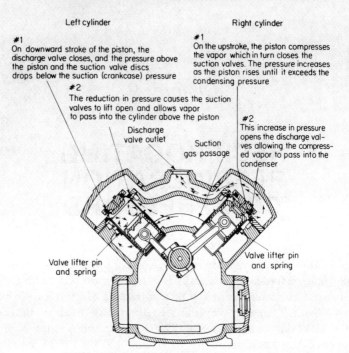

FIG. 9.1. Cross-section through reciprocating compressor. (Courtesy: Carlyle Air Conditioning Co. Ltd.)

The compressor valve system shown in Fig. 9.1 is usually confined to the larger compressor, say above $2\frac{1}{2}$ in (63 mm) bore, whereas on smaller compressors a reed valve is used. Relying on much the same action as the disc valve, above each cylinder is a valve plate which has two parts, one to the suction side of a split cylinder head which has the valve secured to its lower side and one on the discharge side of the cylinder head which has its valve on the upper side. A suction stroke will allow the valve on the bottom of the valve plate to open and the one on the upper side to close; on the compression stroke the reverse happens.

Compressors are usually multi-cylinder machines with as many as 12 or 16 cylinders, usually arranged in banks of two cylinders, such that one cylinder head covers two bores, and are arranged in a 'V' or 'W' configuration. For air conditioning purposes compressors often require to have capacity control, and to achieve economical operation this is done by isolating certain cylinders or banks of cylinders. Figure

9 RECIPROCATING REFRIGERATION COMPRESSORS

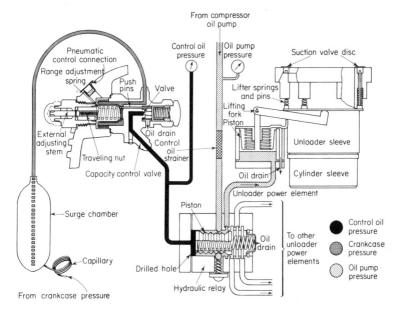

FIG. 9.2. Cylinder unloading mechanism for open compressor. (Courtesy: Carlyle Air Conditioning Co. Ltd.)

9.2 shows a method of cylinder unloading with the compressor shown in Fig. 9.1. Under light load conditions with a given evaporator and compressor the suction temperature and pressure would drop; this action can serve as a signal to unload cylinders. The pressure of the crankcase or suction pressure is sensed in a capillary tube which feeds into a surge chamber, which levels out the instantaneous pressure fluctuations caused by the action of the compressor. This pressure is fed to a capacity control valve which can be preset by the adjustment of an external spring to provide a signal from a given set point. Within the capacity control valve is a range adjustment spring, which is necessary to provide the range from fully loaded to fully unloaded conditions within the desired range of suction temperatures; normally this spring would be 7 lb (*31 N*) for R.12 which corresponds to 7°F (*3.9°C*) suction temperature range at normal air conditioning levels and 11 lb (*49 N*) for R.22 to produce the same results. The capacity control valve will change the control oil pressure to a hydraulic relay, which in response to this change will pass none or full oil-pump pressure to the unloader power element. This element keeps

FIG. 9.3. Open reciprocating compressor. (Courtesy: Carlyle Air Conditioning Co. Ltd.)

open the suction valve disc such that no compression will take place in the unloaded cylinders.

The added advantage of such an unloading system is that when the compressor is started there is no oil-pump pressure until full compressor speed is reached. This means that an automatic unloaded start is maintained because all controlled cylinders are in the unloaded position until the compressor is at speed. Because of this normal starting torque electric motors can be used.

The system can be adopted for external pneumatic or electric compensation control from a thermostat or humidistat. The pneumatic operation is accomplished by a control signal acting directly to the capacity control valve, this pressure directly opposing the spring pressure on the control valve bellows. Electric compensation is achieved by the use of a modulating motor and gear box which turns the control valve stem thus resetting the control point of the system.

Figure 9.4 shows the method of cylinder unloading using a reed valve system. In this position the cylinder is unloaded, the discharge being open to the suction manifold and so producing no compression.

9 RECIPROCATING REFRIGERATION COMPRESSORS

When the solenoid valve is de-energised it seats the by-pass valve and the compressor then carries out useful compression, the discharge pressure forcing open the check valve assembly. This system can be adapted for use with pressure actuated valves for control from suction pressure.

Reciprocating compressors can be divided into two categories each with two alternatives. There is the open compressor either belt or

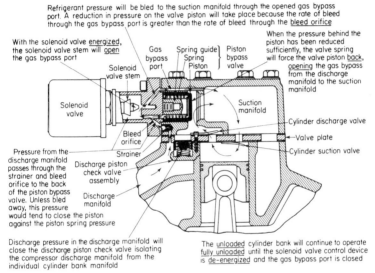

FIG. 9.4. Cylinder unloading mechanism for hermetic compressor. (Courtesy: Carlyle Air Conditioning Co. Ltd.)

direct driven and the hermetic compressor, either fully hermetic (welded) or serviceable hermetic (semi-hermetic).

The open type compressor is so called because one end of the crankshaft extends through the crankcase through a shaft seal to prevent leakage of refrigerant during operation and inward leakage of air on shut down. Most open compressors are driven by electric motors although they can be readily adapted to petrol, diesel or gas engine drive. Combustion engine drives are normally direct via a flexible coupling, and engine speed can be controlled to give close capacity control. Belt drives are common on smaller compressors with few cylinders which require a flywheel effect, the motor pulley providing this effect. Direct drive compressors are governed by the

FIG. 9.5. Hermetic compressor. (Courtesy: Carlyle Air Conditioning Co. Ltd.)

motor speed and are usually run at 1450 rev/min on 50 hertz operation or 1750 rev/min on 60 hertz operation. The belt driven application means that any speed within the limits of compressor operation can be used by changing pulley sizes to meet exact load design conditions.

The motor and compressor of the hermetic types are enclosed in a common housing with the crankshaft extended as the motor shaft. The advantage of such systems is the elimination of a shaft seal, the elimination of site alignment of couplings or belts, and compactness, which make them commonplace in package units. On serviceable, hermetic types the compressor section is basically the same as the open compressors whereby the cylinder heads' end-bell and crankcase-cover plates can be removed for internal servicing. However, the welded hermetic compressor and motor are completely sealed and field service cannot be carried out with such machines. Usually the compressor and motor are mounted in the vertical plane.

Motor cooling on hermetic compressors is achieved by passing the

9 RECIPROCATING REFRIGERATION COMPRESSORS

suction gas over the motor windings before compression, although there are some hermetic compressors which rely on external finning of the motor casing to carry heat away.

All reciprocating compressors have four factors which influence performance; they are:

1. Rotative speed.
2. Suction pressure.
3. Discharge pressure.
4. Type of refrigerant.

The rotative speed is basically a direct proportion on compressor performance such that at 700 rev/min a compressor will require half the power input and will provide half the refrigeration effect of a similar compressor operating under the same pressure conditions and with the same refrigerant as the compressor running at 1400 rev/min. Each compressor manufacturer lists the maximum speed at which his compressor will work. However, today's practice dictates that 1750 rev/min is the maximum economical compressor speed for the medium sized compressors, say up to 100 tons R capacity, which is the speed of a four-pole electric motor on 60 hertz operation. Small welded hermetic compressors are run at 3500 rev/min maximum, i.e. two-pole, 60 hertz motor speed, whereas very large open compressors may run at 700 or 950 rev/min. The minimum compressor speed may be governed by two levels, the lower speed being the minimum at which lubrication oil pressure is kept at a safe level and a somewhat higher speed is required to provide the control pressure for cylinder unloading.

For a given compressor its performance is most greatly affected by changes in suction pressure. Since the refrigerant effect which a compressor can produce is a factor of the weight of refrigerant handled, the weight handled at low suction pressures, when the refrigerant gas has a high specific volume, is low because the reciprocating compressor is a fixed displacement machine when running at constant speed. The lower the evaporating temperature the lower the evaporating pressure and so the lower the refrigeration effect the compressor can produce. The suction pressure at the compressor is also lower than the evaporator pressure since allowance must be made for the pressure drop in the suction line. It is normal practice to allow a pressure drop equivalent to $2°F$ ($1·1°C$), i.e. the saturated suction temperature would be $2°F$ ($1·1°C$) lower than the

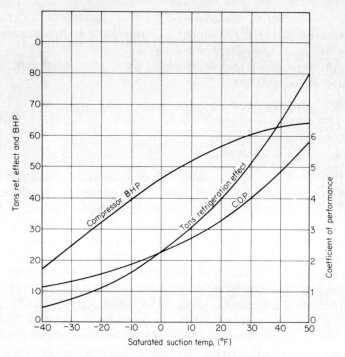

Fig. 9.6. Effect of suction temperature change at constant condensing temperature (100°F) (Courtesy: Carlyle Air Conditioning Co. Ltd.)

evaporating temperature. This terminology of temperature is used since it is this which governs heat transfer in the evaporator coil, and whilst the refrigerant is evaporating, pressure and temperature are in a fixed relationship. Compressors are selected at saturated suction temperatures for a given refrigerant. The amount of superheat contained in the suction gas will also affect the compressor performance since at a given pressure additional superheat will decrease the specific volume of the gas. Compressors are rated at certain superheat levels and manufacturers give corrections for other superheats, e.g. a compressor operating with refrigerant R.12 at a saturated suction temperature of 10°F ($-12 \cdot 2$°C) with an actual gas temperature of 30°F ($-1 \cdot 1$°C), i.e. 20°F ($11 \cdot 1$°C) superheat, would only produce 91 per cent of the refrigeration effect of the rated value if the basis of the ratings were based on 75°F ($23 \cdot 9$°C) actual gas temperature. Figure 9.6 shows the effect of compressor performance

9 RECIPROCATING REFRIGERATION COMPRESSORS 149

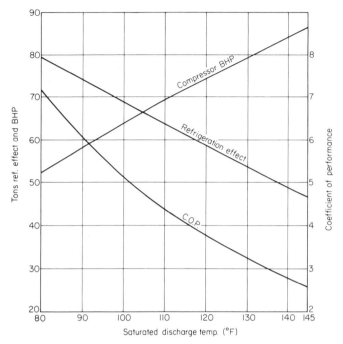

FIG. 9.7. Effect of discharge temperature change at constant saturated suction temperature (40°F).

at constant condensing pressure with varying saturated suction temperatures with 10°F (5·5°C) superheat.

The condensing temperature is governed by the temperature and nature of the condensing and the size of the condenser employed. Compressors are rated at given saturated discharge temperatures, which as with the saturated suction temperature should allow for a 2°F (1·1°C) loss in discharge line, therefore the saturated discharge temperature must be selected 2°F (1·1°C) higher than the condensing temperature. The condensing level would normally be about 25 to 30°F (14 to 17°C) above the wet bulb temperature of air in the case of water cooled systems with a cooling tower or evaporative condensers and 25 to 30°F (14 to 17°C) above the dry bulb temperature of the air with air cooled condensers. Since there is always a depression of wet bulb temperature below dry bulb temperature at design, water or evaporative cooled systems usually condense at lower levels than air cooled systems. Figure 9.7 shows a plot of compressor performance with a constant saturated suction temperature and superheat with varying saturated discharge temperatures or condensing temperature. This chart assumes 15 °F (*8·3 °C*) subcooling of the liquid refrigerant before it passes to the Expansion Device and is re-evaporated. The higher the condensing level the less the amount of latent heat in the refrigerant and the greater the total heat of the refrigerant, which affects the refrigeration cycle by producing more flash gas and so reduces the refrigeration effect of the system. The effect of subcooling is to decrease the total heat of the liquid refrigerant and produce more refrigeration effect. For example, R.22 condensing at 100 °F (*37·8 °C*) with no subcooling would have a total heat of 39·267 BTU/lb (*91·4 kJ/kg*), whereas at 115 °F (*45·6 °C*) it would have a total heat of 44·065 BTU/lb (*102·6 kJ/kg*); if 15 °F subcooling was available the total heat of the system condensing at 115 °F (*45·6 °C*) would be the same as at 100 °F (*37·8 °C*). At normal air conditioning levels a reduction of 1 °F (*1 °C*) in condensing temperature or increase of 1 °F (*1 °C*) subcooling would result in a $\frac{1}{2}$ per cent (*0·9 per cent*) increase in refrigeration effect.

The fourth major factor which affects compressor performance is the type of refrigerant used. Common refrigerants are R.12, R.22 and R.502, each of which produces different refrigeration effects at similar suction and discharge levels. R.22 is the most commonly used in air conditioning systems; however, the other refrigerants have certain advantages which must be considered. R.12 produces the least refrigeration effect but can be used at higher condensing levels than R.22 and R.502 and can be used as high as 145 °F (*63 °C*) saturated discharge temperature with a saturated suction of 0 °F (*−18 °C*). R.502 is most beneficial at low temperatures where it can be used as low as −60 °F (*−50 °C*) saturated suction temperature, although at low condensing levels. Table 9.1 shows the comparative capacities of the three basic refrigerants at three operating levels, which serves to show the advantages of these refrigerants.

As mentioned earlier, compressor speed has a great effect on the operating limits of a compressor. Also there are limits on the suction and discharge levels. Care must be taken not to exceed the maximum actual discharge temperature of 275 °F (*135 °C*) which effects the breakdown of lubrication oil. Chapter 2, The Refrigeration Cycle, discussed the calculation of actual discharge temperature, the main factors being compression ratio and the amount of superheat in the suction gas. In extreme conditions of high compression ratios, the

TABLE 9.1
COMPARISON OF COMPRESSOR PERFORMANCE WITH COMMON REFRIGERANTS

Refrigerant	40°F (4·4°C) sat. suction temp. 105°F (46·1°C) sat. discharge temp.			−40°F (−40°C) sat. suction temp. 80°F (26·7°C) sat. discharge temp.			10°F (−12·2°C) sat. suction temp. 120°F (48·9°C) sat. discharge temp.		
	Duty tons R	b.h.p.	b.h.p./ton	Duty tons R	b.h.p.	b.h.p./ton	Duty tons R	b.h.p.	b.h.p./ton
R.12	66·2	66·5	1·00	7·4	20·9	2·82	28·0	56·7	2·02
	100%	100%	100%	100%	100%	100%	100%	100%	100%
R.22	105·7	101·9	0·96	13·5	30·7	2·28	45·4	86·4	1·90
	160%	153%	96%	182%	147%	80%	162%	152%	94%
R.502	112·8	109·8	0·97	15·4	38·7	2·52	47·2	93·4	1·98
	170%	165%	97%	208%	185%	89%	168%	165%	98%

discharge temperature can be reduced by cooling the compressor cylinder heads with water cooled jackets, and oil coolers can be incorporated to keep the oil from breaking down. Manufacturers' data should always be consulted before applying high compression ratios. The application of suction/liquid heat exchangers should be closely examined as the high superheat suction gas could cause high discharge temperatures, particularly with R.22 systems.

Hermetic compressors are restricted more in performance because of the suction gas flow which cools the motor windings. Low suction temperature means low weight flow of refrigerant which may not be sufficient to cool the windings. High suction temperatures will not· provide sufficient cooling and normally the maximum saturated suction temperature is limited to 50°F ($10°C$) with 15°F ($8·3°C$) superheat; it should be noted that the compressor motor heat adds further superheat which can cause high discharge temperatures with fairly high compression ratios.

To safeguard compressor operation and to facilitate servicing the following devices should be considered:

Suction and Discharge Shut-off Valves, normally back seated so that pressure gauges can be fitted.

High and Low Pressurestats. Low pressurestat trip-out would indicate blocked filters, reduced air volume over the evaporator, blocked drier, lack of refrigerant, or low condensing temperature; this would normally be automatically reset in operation. High pressurestat trip-out is indicative of a blocked or fouled condenser, air cooled condenser fan failure, cooling-tower fan failure, or water pump failure, because of possible permanent damage to the compressor this safety device would always be manually reset.

Oil Level Sight Glass.

Safety Relief Valve in compressor crankcase.

Oil Safety Switch. This would be advisable on all compressors except the small welded hermetic type which do not provide such a facility. This should always be used on hermetic compressors on air cooled applications with long pipework runs or with air cooled water chilling applications.

Dual Pressure Switches are used with larger compressors to ensure proper differential between crankcase (suction pressure) and oil pressure.

Crankcase Heaters are recommended for air cooled application so

9 RECIPROCATING REFRIGERATION COMPRESSORS 153

that on compressor shut-down the crankcase is kept at a temperature above that of the condenser in order to prevent liquid refrigerant from migrating to the compressor sump. On start-up after a winter shut-down this should be put into operation some hours before starting the compressor so as to ensure that proper lubrication takes place.

Other accessories which should be considered are:

Discharge Line Mufflers.
Compressor Anti-vibration Mountings.
Oil Filter.
Suction Filter.
Oil Coolers.
Water Cooled Cylinder Heads.

Hermetic compressors in addition to the safety devices mentioned above should be provided with a motor-temperature sensing device which is in the form of an element fixed to the outside of the motor casing or, preferably, for faster response, an element embedded in the motor windings.

When two or more compressors are required to operate in parallel precautions must be taken to ensure that each compressor performs similarly and that at the shut-down of one compressor, oil is not drained from one compressor to another. This is achieved by having oil-equalising lines between each compressor sump, and crankcase pressure equalising lines, together with discharge line equalisation. Crankcase cover plates are available with pre-drilled connections to facilitate this. Care should be exercised when equalising hermetic compressors since failure due to compressor burn-out on one compressor will transmit damaging acid resulting from the burn-out to other equalised compressors and so cause damage there also.

With any refrigeration system using Halocarbon refrigerants a completely dehydrated installation is necessary, since the action of water and refrigerant produces damaging solutions which will attack the moving parts of the compressor and will cause damage to motor windings in hermetic machines.

Chapter 10

CENTRIFUGAL REFRIGERATION COMPRESSORS

Reciprocating compressor applications are limited to the maximum economical machine size of between 100 to 150 tons R capacity at air-conditioning levels. Above this, applications require two or more reciprocating compressors, and it is here that the role of the centrifugal compressor comes into its own. Above 100 tons R the majority of air-conditioning applications have multiple heat exchangers and so resolve themselves into water chilling applications. Because of this, centrifugal compressors are limited in the vast majority of cases to being a component part of a liquid chilling package. In addition, because of their size and relatively low head characteristics, they are also limited to water cooled applications. Therefore, practically all centrifugal compressors are assembled into water cooled liquid chilling packages with evaporator and condenser as an integral part. There are some notable exceptions, such as purpose-built compressors running at the very high speeds used for aircraft air conditioning, and in the chemical industry for such applications as chlorine condensing.

The basic refrigeration cycle used with reciprocating compressors applies to centrifugal machines, with the exception that because the evaporator is normally of the flooded type and is close coupled to the compressor the suction gas contains little or no superheat and is in the form of a saturated vapour. Because of the close coupled design and aerodynamically shaped suction elbows the suction pressure drop is so small as to be ignored, as is the discharge connection to the condenser. Instead of thermal expansion valves the refrigerant

10 CENTRIFUGAL REFRIGERATION COMPRESSORS

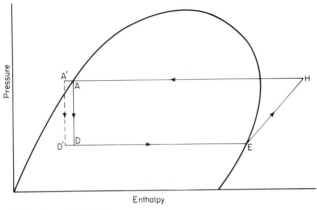

Single stage compression

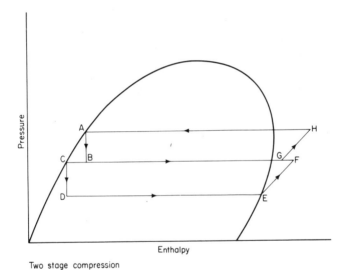

Two stage compression

FIG. 10.1 Pressure–enthalpy diagrams for centrifugal refrigeration systems.

expansion is normally carried out by a float valve which can handle the large volumes of liquid refrigerant associated with centrifugal machines.

With large applications where running costs are closely considered, the use of two-stage compression is an advantage. Figure 10.1 shows the basic refrigeration cycle of single and two-stage compression. The two-stage diagram shows how the condenser liquid passes through the high, side-float valve from points A to B. At point B the flash gas is removed and put into the interstage between the two compressor wheels. The liquid entering the low-stage float valve is at point C and is expanded to point D where it enters the evaporator. Comparison with the single-stage machine shows that the refrigeration effect available

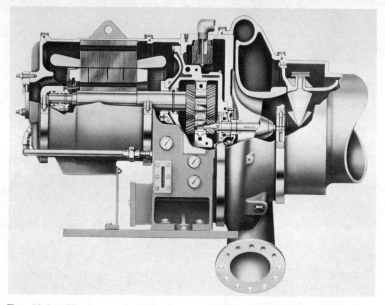

Fig. 10.2. Single-stage hermetic centrifugal compressor. (Courtesy: Carlyle Air Conditioning Co. Ltd.)

from D to E where the saturated vapour enters the compressor is greater. At point F the first stage of compression is complete and this gas is mixed with the flash gas from the high, side-float valve at the interstage point. The line GH represents the second stage of compression which is handling a greater amount of gas than the first stage. The line HA represents the action of condensing. The two-stage arrangement represents a running-cost saving of about 6 per cent. The evaluation of this cost saving must be considered against the higher capital cost of such equipment.

The single-stage cycle shows the added value of a sub-cooling circuit in the condenser which also increases the refrigeration effect, line A'D' representing the action of the single-float valve expansion.

10 CENTRIFUGAL REFRIGERATION COMPRESSORS

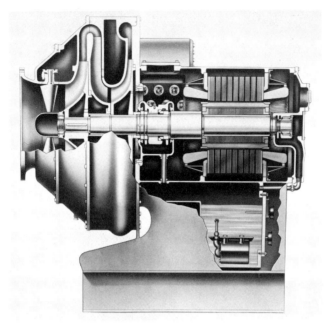

FIG. 10.3. Two-stage hermetic centrifugal compressor. (Courtesy: Carlyle Air Conditioning Co. Ltd.)

The addition of a sub-cooling circuit could add $2\frac{1}{2}$ per cent to the refrigeration effect.

Selection of refrigerants is governed by the characteristics of the centrifugal compressor, which is a high volume machine with the compression head dependent on the compressor speed. For many years common refrigerants were R.11, R.113 and R.114 which are low pressure refrigerants such that under certain conditions the refrigeration machine could be under a vacuum both on the evaporator and condenser side. This situation existed because the maximum compressor wheel speed available with direct driven electric motors was 3500 rev/min for 60 hertz machines and 2900 rev/min for 50 hertz machines. The development of reliable gears and even frequency converters, allows compressor speeds greatly in excess of normal synchronous speed up to 20 000 rev/min and over. Higher speeds mean higher compression heads which allows R.12 to be considered. Table 10.1 shows the pressures for a centrifugal machine evaporating at 35 °F (*1·7 °C*) and condensing at 100 °F (*37·8 °C*), for

TABLE 10.1
REFRIGERANTS AND THEIR USES IN CENTRIFUGAL REFRIGERATION MACHINES

Refrigerant	Evaporating pressure		Condensing pressure		Hermetic machines		Open machines	
	PSIA at 35°F	kN/m^2 at 1·7°C	PSIA at 100°F	kN/m^2 at 37·8°C	Speed range	Tonnage range	Speed range	Tonnage range
R.113	2·33	16·1	10·48	72·3	2900–3500	100–400	—	—
R.11	6·25	43·1	23·45	161·8	2900–12000	100–1500	2900–3800	100–1600
R.114	13·55	93·5	45·85	316·0	2900–3500	1600–2000	2900–3800	1600–2000
R.12	47·26	326·0	131·86	909·0	10000–24000	100–1100	4500–6400	1000–10000

10 CENTRIFUGAL REFRIGERATION COMPRESSORS

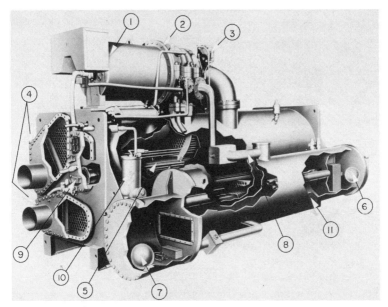

FIG. 10.4. Cut away of high-speed, two-stage, hermetic centrifugal liquid chilling package for operation on refrigerant R.12. The hermetic motor (1) is directly connected to the two-stage compressor (2) which has variable inlet guide vanes automatically controlled from motor (3). The condenser and evaporator are of uni-shell construction with external water boxes (4). The subcooling circuit (5) within the condenser improves refrigeration effect as do the two-stage expansion float valves (6 and 7) where flash gas from the first expansion (6) is fed to the compressor interstage by pipe (8). The lubrication system is shown (9) as is a dehydrator (10) which provides a manual purge and keeps the system dry and air free. The storage vessel (11) acts as a receiver for service and during normal operation serves as an economiser housing the two-stage expansion system. (Courtesy: Carlyle Air Conditioning Co. Ltd.)

the various common refrigerants. Also shown are the tonnage range and speed range associated with such refrigerants. There are no exact parameters for speed against refrigerant use since different manufacturers use various wheel geometries which are designed for operation at certain speeds with a given refrigerant.

Besides allowing higher heads, geared machines mean more compact units. High-speed machines, however, are restricted in size because of tip speed limitations of the impeller.

Centrifugal machines can be categorised into two types, open and

FIG. 10.5. Two-stage open centrifugal system with condenser and evaporator as separate vessels. This machine would operate at relatively low speeds with refrigerants R.11 and R.114 in the larger sizes. Clearly shown is the impeller shaft seal and shaft for connection to prime mover. (Courtesy: Carlyle Air Conditioning Co. Ltd.)

hermetic. The majority of machines currently manufactured are of the hermetic type, essentially because they fall into the most common size range of 100 to 2000 tons R. Hermetic machines have electric motor drives which are most commonly cooled with refrigerant, although water-cooled models are still made, the motor shaft is extended directly to the impeller or through a gear for high speed machines. Geared machines are normally of the single-stage type.

Open machines have an extended shaft and seal which can be coupled to many prime movers, such as electric motors, gas engines, diesel engines, steam or gas turbines, either direct coupled or through a gear drive. The open machine lends itself favourably to total energy systems, for example a boiler can produce high pressure steam where the kinetic energy of the steam is used to power a turbine-driven refrigeration plant and the heat energy of the steam can be used in an indirect absorption refrigerant system.

Centrifugal machines for water chilling applications use horizontal

10 CENTRIFUGAL REFRIGERATION COMPRESSORS

shell and tube condensers and evaporators, having finned copper tubes rolled into tube sheets at each end of the heat exchanger with intermediate support sheets. Evaporators being of the flooded type (i.e. water in tube) must have a means for preventing liquid refrigerant from entering the compressor since no superheating can take place; this is achieved by eliminator blades located above the tubes in the evaporator shell. Refrigerant is introduced from the float chamber into a perforated channel which distributes it evenly through the length of the heat exchanger. The condenser is of similar construction, hot gas from the compressor passing through a longitudinal distribution plate giving even coverage over the condenser tubes holding the condenser water. Liquid refrigerant gravitates to the bottom of the condenser shell and is sub-cooled by the lower tubes where the lower temperature condenser water enters. From the bottom of the condenser shell the liquid refrigerant feeds into the float chamber. Both condenser and evaporator are available with multiple pass arrangements such that proper selection of water boxes allows from one to four water passes thus providing maximum flexibility of water quantity and temperature rise with optimum heat transfer.

On the smaller sized units, although units are being developed up to 1100 tons R, the evaporator and condenser are housed in a common shell with a division plate between the upper condenser section and lower evaporator section. The heat interchange across the division plate provides some sub-cooling of the refrigerant liquid and at the same time provides superheat to the evaporator vapour.

Figure 10.6 is a performance curve for a centrifugal compressor showing compressor head against volume. Unlike the reciprocating compressor which concerns itself with compression ratios, it is the pressure difference between inlet and outlet of the compressor which has an influence on its volumetric efficiency or refrigeration capacity and coefficient of performance. The head is the pressure which would be exerted by a column of refrigerant gas of uniform density measured in feet. A centrifugal compressor running at a given speed and volume loading will theoretically produce the same head in feet no matter what gas or vapour is compressed. The equivalent pressure difference will be dependent on the density of that vapour or gas such that a dense vapour will require a low head for a given pressure difference. Since density is proportional to molecular weight a refrigerant with high molecular weight requires a lower head, and since horsepower is

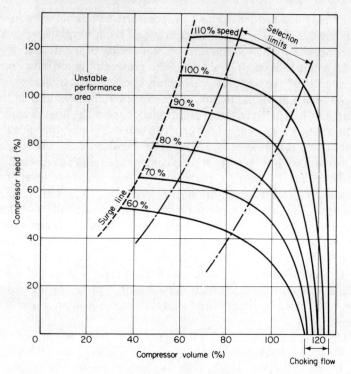

Fig. 10.6. Typical centrifugal compressor performance curve.

proportional to head a compressor using a dense refrigerant produces a higher coefficient of performance. From the performance curve it can be seen that the higher the impeller speed the higher the head. At each impeller speed a maximum volume or 'choking flow' is reached where the head drops off rapidly, this is referred to as the 'stonewall' and compressor selections are made within this limit.

As refrigerant flow is reduced through the impeller at partial load conditions a condition results where the compressor can no longer produce the required head and this causes a partial or complete reversal of flow through the impeller. As the refrigerant flows backwards so the increased weight causes the head to increase and allows positive flow through the impeller to be restored. This phenomenon is referred to as 'surge' and is found where the compressor is operating in an unstable condition; it is this condition which limits the minimum capacity of the compressor. No mechanical defects should be produced by short durations of surging although it is associated with a high increase in operating noise level and wide fluctuations of discharge pressure. However, continuous surging should be avoided since it may cause undue wear to the impeller thrust bearing. A simple method of minimising surge is to allow condenser water temperatures to fall at partial load such that the required head is kept lower than design. The condenser water temperature should not be allowed to fall too low as this could cause low evaporating conditions. A guide would be to set the condenser water temperature control to 20°F ($11 \cdot 1°C$) below the design entering water temperature or 65°F ($18 \cdot 3°C$) minimum. This would mean that at partial loads or at low ambient conditions the lower water temperatures would also produce lower running costs. In all cases, however, it is advisable to check with the manufacturer to assess the minimum condenser water temperature that his machine can withstand under the application operating conditions.

There are two basic methods of capacity control of the centrifugal refrigeration machine, these being speed control and refrigerant flow control. Speed control is most economically achieved with open machines using steam turbines where the steam flow to the turbine controls the speed and thus the refrigeration effect produced by the impeller. Variable speed electric motors prove very costly as can magnetic or slip couplings whose drive loss should be closely considered in assessing operating costs. At partial load the volume of refrigerant required to be handled is reduced and the compressor at constant speed can produce a higher head at reduced refrigerant flow. To achieve control of the leaving chilled water it is necessary to reduce the available head of the compressor. For the major part of load reduction the reduced volume against suitable operating head can be achieved by speed reduction. However, at a certain minimum head, that required to prevent the evaporating temperature from falling below danger point, the refrigerant flow could occur beyond the surge line in the unstable region. Beyond this point speed reduction is no longer of service and lower operating conditions must be met by artificially maintaining a load on the compressor by by-passing hot gas from the condenser to evaporator. At the point where hot gas by-pass occurs no further reduction in absorbed power can be achieved as is the case when the compressor can be controlled with speed reduction.

The alternative capacity control method is regulation of refrigerant

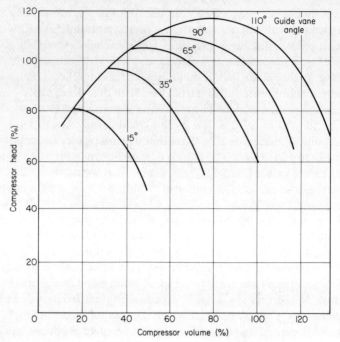

Fig. 10.7. Typical centrifugal compressor performance curve (fixed speed and variable inlet guide vanes).

flow. The simplest and cheapest method is the throttling of the suction pipe by means of a butterfly valve either manually or automatically controlled from leaving chilled water. The disadvantage of this control is that it is limited at fairly high capacities since it cannot maintain high head at low flow without the introduction of hot gas by-pass to ensure the compressor performs in the stable region. In open machine operation a combination of variable speed and suction damper throttling can produce relatively good results.

The most common form of centrifugal capacity control in current use is the use of variable inlet guide vanes at the compressor. These are located at the inlet to the impeller such that they produce a pre-whirl effect which will provide some increase in compressor head whilst the refrigerant flow is being throttled. Figure 10.7 shows a plot of the performance of a centrifugal compressor at constant speed with variable inlet guide vanes. Even at low flow rates relatively high head is maintained. The horsepower required at partial load is reduced but

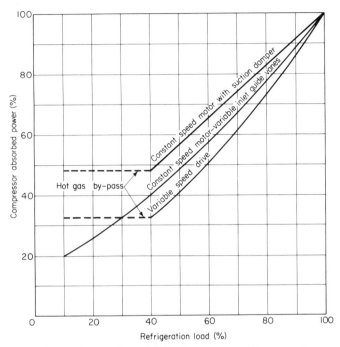

FIG. 10.8. Comparison of partial load power requirements (constant condensing temperature).

not so rapidly as variable speed control until such time as hot gas by-pass has to be introduced. A distinct advantage of this method of control is the availability to reach refrigeration capacities as low as 10 per cent whilst maintaining adequate head to keep the machine from entering the unstable region and causing surging.

Figure 10.8 shows a comparison of absorbed power against refrigeration load for the various methods of capacity control. It will be noticed that speed control provides the lowest power up to about 40 per cent capacity only requiring 32 per cent power at the impeller; care should be taken, however, to compare this with the efficiency of steam turbines at the lower speeds incurred.

Where refrigeration machines are using refrigerants R.11 or R.113 they could well be operating under a vacuum, and any leakage of the refrigeration machine would be inward, with the possibility of air and water being drawn into the machine. Air in the machine will cause high condensing temperatures, causing increased operating costs and possible plant shut-down on safety controls. Water mixing with the refrigerant will cause damaging acids which attack metal surfaces and motor winding in hermetic machines. A purge unit, normally automatic, is usually installed as a standard feature with such machines. This unit automatically removes air and water (non-condensables) from the system with little or no loss of refrigerant. Frequent operation of a purge system will indicate a leaky system which must be found and eliminated.

Machines using R.12 are positive pressure systems, where any leakage is outwards and so a purge system is unnecessary. However, such a high pressure refrigerant will evaporate rapidly at normal ambient temperatures and therefore for servicing purposes it is often considered necessary to provide an auxiliary storage vessel or receiver to where the refrigerant can be pumped out and stored to avoid wastage.

With all centrifugal machines, as with reciprocating compressors using Halocarbon refrigerants, a completely dehydrated system is essential.

Safety controls inherent with centrifugal machines are low chilled-water temperature thermostat and low refrigerant temperature or pressure cut-out, both these providing safety against the freezing of water in the evaporator tubes; an additional safety device is often considered for the same purpose, this being a chilled-water flow switch. High pressure cut-out should be provided to safeguard against blocked condenser tubes, restricted water flow, fouling of condenser tubes or failure of condenser water pump or cooling-tower fan. Safety relief valves should be provided as a last safety measure on both evaporator and condenser. Further protection is required for the lubrication system, which would consist of pressure cut-outs in the oil lines; in addition, with hermetic machines, motor temperature sensing elements and bearing temperature cut-outs are also used. All of these functions are normally provided by the machine manufacturer in a control box located on the machine or purge unit, each device having a warning or indicator light. Other services which should be interlocked into the safety circuit are chilled water pump, condenser water pump and lubrication pump, all of which should have overload protection. Units having variable inlet guide vanes can use an additional feature which is a current limiting device, such that at even as low as 40 per cent capacity the guide vanes will not open above the amperage setting that the device is set to. This device can prove

10 CENTRIFUGAL REFRIGERATION COMPRESSORS

useful during intermediate seasons, so that demand charges are kept down. On start-up after the initial run-up period, with the vanes closed for no load start, the compressor will tend to bring the chilled water temperature down to set point as quickly as possible which may require 100 per cent power, even for only a short period. Automatic control of the refrigeration machine is operated by a thermostat sensing the chilled water temperature leaving or entering the chiller. This will govern the capacity reducer of the machine whether motorised variable inlet guide vanes, a suction throttling valve or speed control is used. Controls are mostly electronic although pneumatic operation is not uncommon and solid state control has recently been introduced; this provides the speediest control response without overcorrection.

Gauges are to be provided for the centrifugal machine which should indicate the evaporating temperature and pressure; the condensing temperature and pressure; oil temperature and oil pump pressures, entering and leaving chiller; and condenser water temperatures and pressures.

The lubrication of the system must be kept at a reasonable constant temperature and an oil heater should be provided for operation during shut-down; during operation heat must be dissipated from the oil and this is usually carried out by using a very small portion of the recirculating chilled water or alternatively by bleeding off water from the condenser circuit, which is not wasteful since the bleed-off required for proper water treatment of the cooling tower water is normally well in excess of this requirement. The flow of water is controlled by a manual throttle valve and is isolated by a solenoid valve when the plant is not in operation.

With electric driven machines and all hermetic machines the choice of starter must be closely considered. There are four basic methods of motor starting available—firstly, Direct-On-Line, which is by far the most economical, but is generally restricted to high voltage machines operating above 2200 volts because of the high inrush currents of 500 to 600 per cent of normal full load current. However, this starting method provides the highest torque and speediest compressor start. Star-Delta, usually closed transition, is probably the most common method of starting at normal voltages. This method is effective in limiting starting currents which means that the motor will not be subject to much overheating and should provide longer motor life. Auto-transformer is another method of reduced voltage start, which limits the current inrush until the motor has attained speed, when on the action of an automatic timer the motor is disconnected from the timer and reconnected across the line. The fourth method of starting is the primary resistor which is normally cheaper than the auto-transformer starter and obtains a reduced voltage start by imposing a resistor in the circuit until the motor reaches speed, when the resistors are taken out of circuit by the action of a timer which puts the machine back on line. This latter method requires about 250 per cent to 300 per cent of the full load current for start up. All starter panels would normally include an oil pump starter which is brought on line before the main compressor motor.

Practically all centrifugal machines are started manually and are allowed to run continuously; however, systems having more than one machine can be arranged for automatic start with features allowing for lag–lead control.

When two or more machines are used on one application there are two basic methods of connection. It is not normally practical to connect more than two machines in series because to avoid high pressure drops on the heat exchanger single pass arrangements are required, which rarely gives as good heat transfer as multiple pass arrangements. Where machines are used in series it is good practice to arrange the condenser water flow in a counterflow direction such that the higher chilled water temperature machine, which would normally provide more refrigeration effect and require more power, would receive the higher temperature condenser water resulting in approximately the same refrigeration effect and power requirement as the low temperature machine which receives the lower temperature condenser water.

For multiple systems which do not require a high overall chilled water temperature drop parallel arrangements are used. Partial load operation should be considered because on the shut down of one machine a by-pass effect will be realised such that return chilled water at a high temperature will mix with chilled water from the operating machine(s). However, in multiple machine operation it is normal practice to run all machines to as low as 35 per cent duty where advantage is gained of lower per cent operating cost. Since centrifugal machines are fully modulating at this point of 35 per cent load the chilled-water leaving temperature from a given machine will have fallen to a temperature 65 per cent of the thermostat differential below the design leaving chilled-water temperature. The mixing of this water with a reduced return water temperature is not often detrimental in air-conditioning applications since the heat transfer requirement of chilled water coils has been greatly reduced at this condition and a higher chilled-water temperature could well suffice.

Whichever connection is made, series or parallel, the cycling of machines is achieved by a thermostat in the common-entering, chilled-water line controlling the starting and stopping of machines which are individually controlled from leaving-water temperature. The control thermostat should be set to correspond to a return chilled-water temperature of 35 per cent load to cut out the first machine and 35 per cent load of the remaining machine(s), etc., and cut in at temperatures corresponding to 45 per cent load. This operation will ensure economical running and avoid rapid cycling of machines.

Chapter 11

ABSORPTION CYCLE REFRIGERATION EQUIPMENT

As with centrifugal equipment, absorption machines for air conditioning use are essentially confined to liquid chilling applications. Equipment available is divided into two categories of widely varying capacity ranges from 3 to 20 tons R capacity, direct fired units with either air or water cooled condensers are available, and from 100 to 1000 tons R capacity it is the water cooled, indirect fired units which are used. The wide gap between the two types of system sizes is caused by economics of heat exchangers, and duties between 20 and 100 tons R are achieved by multiple, direct-fired units.

The basic absorption cycle relies upon the same principle as a mechanical refrigeration cycle in that a refrigerant—water—will boil at low temperatures when subjected to low pressures, in this case a high vacuum, and in addition that certain solutions, normally lithium bromide or ammonia, have the ability to absorb water vapour.

Before examining this cycle it should be stressed that it is the water that is the refrigerant and in the examples considered, lithium bromide which is the absorbent.

Water will evaporate more quickly if the surface of a given volume is extended. Rather than making a large vessel with a large surface area this is best achieved in much the same way as an air washer or cooling tower by spraying the water. Figure 11.1 shows how the heat from water to be chilled can be picked up by the action of water boiling under a vacuum, the chilled water coils being immersed in the sprayed water. Because of its affinity to lithium bromide the water vapour is carried away from the evaporator to the absorber where it is

11 ABSORPTION CYCLE REFRIGERATION EQUIPMENT

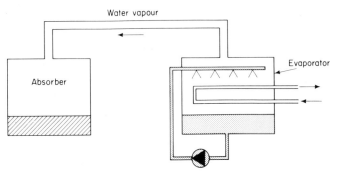

FIGURE 11.1.

mixed to provide a lithium bromide solution. If these were the only cycle components, once the lithium bromide had become diluted its capacity for absorbing water vapour would be reduced and an equilibrium position would be reached whereby no further evaporation of the water could take place and no further useful water cooling could take place.

Figure 11.2 shows how the weak solution from the absorber can be pumped to a generator where, with the addition of heat, the water vapour can be boiled out of the lithium bromide to produce a strong solution which can be returned to the absorber where it is sprayed to increase the surface area and as in the evaporator, increase the capacity to absorb the water vapour from the evaporator.

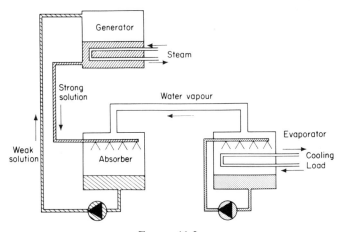

FIGURE 11.2.

This secondary cycle maintains the absorbent at an operating level; what is now required is a water supply to replace the evaporated water vapour from the evaporator which is rejected at the generator. If the rejected water vapour from the generator were passed to a fourth vessel, a condenser as shown in Fig. 11.3, the water vapour from the generator at a high temperature can be condensed and returned to the

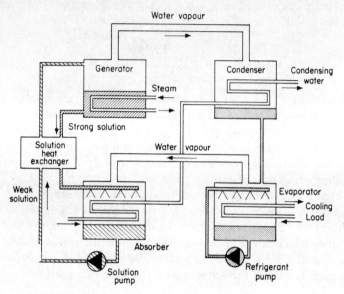

FIGURE 11.3.

evaporator to complete the cycle. In addition to eliminating the need for make-up water this fourth vessel provides a vacuum-tight system.

When the lithium bromide solution absorbs water, heat is generated which consists of the heat of condensation of the absorbed water plus a reaction heat, much the same as when water is added to an acid, between the lithium bromide and water vapour. To increase the capacity of the lithium bromide to accept the water vapour it should be kept cool; this is achieved by passing the condenser water first through the absorber and then onto the condenser.

Because the generator is hot and the absorber cool, the cycle efficiency can be increased by a heat exchanger which heats the weak solution pumped from the absorber to the generator and cools the strong solution returning.

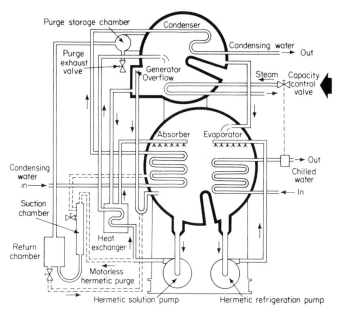

FIG. 11.4. Cross-section through absorption unit. (Courtesy: Carlyle Air Conditioning Co. Ltd.)

In the diagrams used to describe the absorption cycle the flow of water vapour is restricted between evaporator and absorber, and generator and condenser by the size of the pipes connecting the vessels. In practice this is overcome by housing the evaporator and absorber in one shell and generator and condenser in a second shell or, alternatively, all in one common shell with a division plate between the components. Figure 11.4 shows a two-shell design which has division baffles between the component heat exchangers. The baffles are of double thickness, the air gap between acting as an insulation. This division of two shells provides an economical split between the high temperature, high pressure elements of the generator and condenser and low temperature, low pressure elements of the evaporator and absorber. Figure 11.5 shows how all four heat exchange elements can be placed in a single shell. There are obvious capital cost advantages of such an arrangement but the close proximity of the high and low temperature elements complicates the problems of insulation and isolation.

Improvement in cycle performance in the development of

absorption machines has been brought about by the arrangements of components, the insulation between elements and the efficiency of the solution heat exchanger. Since 1945, when the first major absorption cycle units were introduced, there have been countless changes of design by manufacturers in an attempt to reduce the heat input per unit of refrigeration, and it can be anticipated that the vast amount of

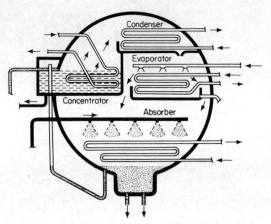

FIG. 11.5. Cross-section through single-shell design absorption unit. (Courtesy: The Trane Co.)

research into these products will bring new and better arrangements with lower energy consumption rates.

Regardless of the absorption machine's physical arrangement one major factor must be considered which is that the vessel or vessels must be held at a high vacuum. If the physical sizes of the components are realised this is no easy achievement. Any air leaking into the system will reduce the machine capacity, since the evaporation of the water relies on the high vacuum to produce low boiling temperatures. Secondly, there is an increased possibility that solution crystallisation will occur. The reduction in capacity reflects lighter loads on the generator and condenser such that the solution will be concentrating at higher temperatures and lower pressures with resulting overstrength solutions. If the solution overconcentrates it may crystallise when cooled in the solution heat exchanger. The third consideration of air leakage is one of corrosion. A vacuum-tight system is oxygen free, and oxygen with the lithium bromide solution produces a

11 ABSORPTION CYCLE REFRIGERATION EQUIPMENT

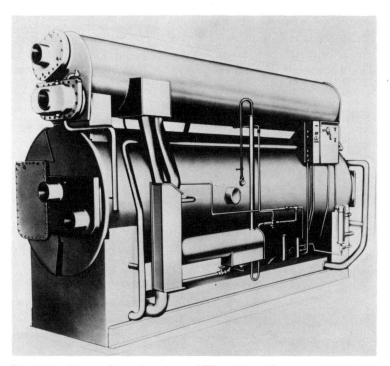

FIG. 11.6. Large absorption water chilling plant. (Courtesy: Carlyle Air Conditioning Co. Ltd.)

corrosive action which will attack the metal components of the system.

All lithium bromide absorption systems must have an efficient purge system to overcome the problems of air leakage. As with the absorption unit arrangement the method of purging has been developed from the early machines which had two-stage steam and water jet systems with development to a self-contained purge system, which had its own water pump; this was developed further to another self-contained system using a single jet of lithium bromide as the purging medium. The single-stage purge was possible because of the lower solution vapour pressure which increased the purging capacity of standard ejectors.

Currently, purge systems are confined to two types. The first to be developed was the mechanical vacuum pump. This is operated manually and consists of a single-stage reciprocating or vane type pump which is connected via an oil tap to a purge chamber where air or non-condensables are accumulated. This system operates quite satisfactorily although there is a possibility of oil or oil vapour being sucked back into the machine where it will contaminate the system.

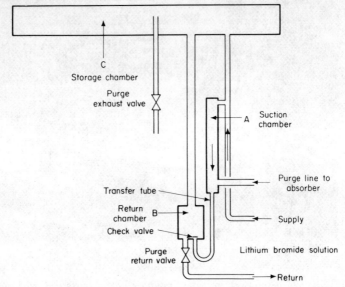

FIG. 11.7. Diagrammatic arrangement of automatic purge unit. (Courtesy: Carlyle Air Conditioning Co. Ltd.)

Frequent oil changes may be required because of the emulsifying effect of the water vapour.

A successful alternative to the mechanical purge pump has been the development of an automatic continuous purge system which operates whilst the machine is running (Fig. 11.7). Weak lithium bromide solution is taken from the discharge of the solution pump and metered into a suction chamber A. Since this chamber serves as a low pressure area, non-condensables are drawn from the absorber. Solution and non-condensables flow through a check valve into the return chamber B, where non-condensables bubble into the storage chamber C whilst the solution returns to the absorber through a return valve. As the non-condensables fill the storage chamber so the solution level is depressed; the displaced solution flows back to the absorber until a predetermined level is reached where a level switch

11 ABSORPTION CYCLE REFRIGERATION EQUIPMENT 177

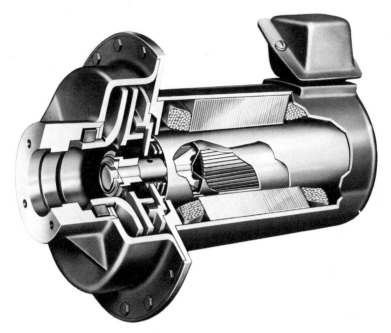

FIG. 11.8. Hermetic pump used on absorption machines. (Courtesy: Carlyle Air Conditioning Co. Ltd.)

energises an indicator light signalling the need to purge. With the purge return valve closed the non-condensables can be purged at the exhaust valve, the storage chamber being refilled with solution from the absorber via the return chamber. When purging is completed the exhaust valve is closed, the return valve opened and normal operation can be resumed.

Until the development of hermetic solution and refrigerant pumps a constant source of trouble was that of making an adequate pump seal for the motor drive. Seal breakage, copper plating and burn-outs brought the final development which was a double rotary ceramic seal which, although expensive, could withstand the rigours of this duty; even then the pumps had to be water cooled so that any inward leakage would be water not air, and seals had a recommended replacement of two years service. The hermetic pump eliminated the need for a seal and overcame leakage problems. The totally enclosed motor has its rotor encased in a stainless steel shell and the stator is wound around a further stainless steel cover. Solution circulating between the stainless steel enclosures provides adequate motor cooling and no external connections are required.

The lithium bromide absorption cycle can be shown on an equilibrium diagram for the solution (Fig. 11.9). The diagram shows four plots, the solution strength, measured as a percentage of lithium bromide present in the solution by weight, and the vapour pressure

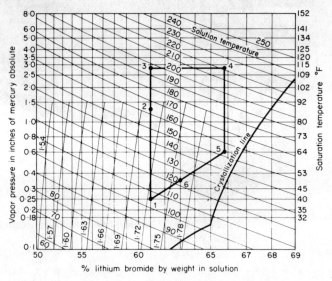

FIG. 11.9. Lithium bromide equilibrium chart. (Courtesy: Carlyle Air Conditioning Co. Ltd.)

measured in inches of mercury absolute, are the horizontal and vertical scales respectively. Depending upon the solution strength and vapour pressure the lithium bromide solution will be at equilibrium at a certain temperature and these constant temperature lines are shown as almost horizontal lines falling away as the solution increases. At certain conditions the solution will start to crystallise; this occurs as the solution starts to change from a liquid to a solid. On the bottom right-hand section of the chart is shown the crystallisation line beyond which operation should not take place. However, unlike a freezing process, crystallisation of salts means that the crystals formed are 100 per cent salt, which weakens the balance of the solution which has not crystallised, and therefore it is not until the solution becomes 100 per cent salt that complete solidification occurs.

11 ABSORPTION CYCLE REFRIGERATION EQUIPMENT

Crystallisation will cause no damage to heat exchange parts since, unlike water, it contracts on solidification; however crystallisation will make the solution pump inoperative and satisfactory design demands that cycle operation be maintained outside the crystallisation zone. The units plotted on the chart, shown as almost vertical lines, are the constant specific gravity lines. Whilst playing no part in analysing the absorption cycle, testing solutions samples is best carried out by a hygrometer, and knowing the temperature of the sample will enable the chart to be used to establish the lithium bromide concentration of the sample.

Shown in heavy lines on the equilibrium chart is the cycle diagram. The chart shown is for an absorption machine which is required to chill water to 44 °F (*6·7 °C*) when condenser water is available at 85 °F (*29·5 °C*) and with steam available at 12 psig (*184 k N/m²*). To achieve 44 °F (*6·7 °C*) leaving chilled-water temperature the evaporator temperature on the outside of the chilled water tubes must be lower, say 42 °F (*5·5 °C*), which from steam tables can be found to be a vapour pressure of 0·27 in. Hg (*915 N/m²*); the absorber must be at a lower vapour pressure, say 0·25 in. Hg (*845 N/m²*), for the water vapour to flow from evaporator to absorber; this corresponds to a saturation temperature of 38 °F (*3·3 °C*). If condenser water is available at 85 °F (*29·5 °C*) allowing for a 10 °F (*5·5 °C*) rise at the absorber with a 10 °F (*5·5 °C*) leaving difference to achieve heat transfer the solution can be maintained at 85 °F + 10 °F + 10 °F or 105 °F (*29·5 °C + 5·5 °C + 5·5 °C or 40·5 °C*). The position A on the equilibrium diagram is plotted at the vapour pressure of 0·2 in. Hg (*845 N/m²*) and 10 °F (*40·5 °C*) solution temperature, which is the point where the solution concentration is 61 per cent by weight.

The line 1–2 represents the flow of the weak solution through the solution heat exchanger. The point 2 depends upon the temperature of the strong solution coming from the generator. Before the balance of the system is considered the pressure in the condenser and thus the generator must be considered. Heat removed in the condenser is approximately 70 per cent of that removed in the absorber. Therefore if full flow from the absorber is directed to the condenser the temperature rise will be 7 °F (*3·9 °C*) (absorber 10 °F (*5·5 °C*)). The water leaving the absorber is 95 °F (*35 °C*), the rise through the condenser is 7 °F (*3·9 °C*), the leaving condenser water temperature therefore would be 102 °F (*38·9 °C*). Allowing for a 10 °F (*5·5 °C*) leaving temperature difference the condenser would be at 112 °F

($44.4\,^\circ C$). From steam tables the saturated vapour condition which corresponds to $112\,^\circ F$ ($44.4\,^\circ C$) can be found as 2·8 in. Hg ($9490\ N/m^2$). Point 3 can be fixed as a continuation of line 1–2 where the vapour pressure is 2·8 in Hg ($9490\ N/m^2$). The heat added by the generator is first sensible which is the line 2–3, representing the raising in temperature of the solution until it reaches the vapour pressure in the condenser. At point 3 the water vapour begins to boil out as further heat is added and solution concentration takes place along the line 3–4 at constant vapour pressure.

To establish the point 4 the temperature of the heating medium must be considered. The maximum steam pressure associated with absorption is 12 psig ($184\ kN/m^2$), which is considered for the example, this corresponds to a saturation temperature of $244\,^\circ F$ ($117.7\,^\circ C$). Economics dictate that the generator heat exchanger is small because advantage can be taken of the large temperature difference available. Allowing a temperature difference of $24\,^\circ F$ ($13.3\,^\circ C$), the solution in the generator can be heated to $220\,^\circ F$ ($104.4\,^\circ C$), which fixes point 4, where the solution reaches its strongest concentration at 65·8 per cent.

The line 4–5 is the effect of the solution heat exchanger corresponding to line 1–2. This solution temperature drop will be almost equal to the temperature rise along 1–2 with allowances for the different specific heats of the different solution concentrations.

The line 5–1 completes the cycle with solution at condition 5 being sprayed into the absorber. To assist the spraying of the solution, some weak solution is constantly recirculated by being added to the strong solution. This quantity is fixed by the manufacturer and point 6 represents a typical mixed solution spray condition entering the absorber.

The capacity control of absorption machines is best achieved by intentionally weakening the solution to the absorber by either reducing the steam (or hot water) flow rate or by diluting the solution to the absorber by a controlled by-pass. The decrease in steam flow rate reduces the capacity of the generator to concentrate the solution and so the absorber solution is weakened to provide less refrigeration effect. This method is shown in Fig. 11.5. Steam control valves must be capable of controlled operation down to practically no flow so that light, partial-load operation can be achieved.

On larger systems, where large steam control valves or even more expensive three-way hot water valves are required to control machine

11 ABSORPTION CYCLE REFRIGERATION EQUIPMENT

capacity, the solution by-pass control method proves more economical both from a running cost and capital cost standpoint. This latter system allows full-flow steam to the generator so that what solution is passed to the generator is concentrated to a greater extent. This is then diluted with the by-passed weak solution from the absorber. At zero load the absorbed solution is completely by-passed from the generator and a no-load condition exists. Steam supply to a machine using solution by-pass control is governed by a steam trap in the condensate piping. Whichever control system is used the steam trap should be oversized to three times the normal design flow rate to assure proper drainage during start-up when the steam rate is high. Both control systems are operated by a sensing element located in the leaving chilled-water line.

Whichever control system is used the condenser water temperature must remain reasonably constant. Whilst the initial temperature selection can be made over quite a wide range, once the machine has been commissioned and solution concentration set, rapid temperature drops can cause solution solidification which must be avoided. The best method to achieve condenser water temperature control is by a three-way by-pass control around the cooling tower.

Thus far only the indirectly heated absorption cycle using lithium bromide has been considered. Direct gas-fired absorption machines using either lithium bromide or ammonia solutions are in wide use for the smaller water-chilling applications. Instead of using steam or hot water as the heating medium to the generator, a gas burner provides the same effect, capacity control being achieved by regulating the gas flow to the burner in just the same manner as the steam or hot water valve in the indirectly heated systems.

Lithium bromide systems using direct gas-fired burners are available with either water cooled or air cooled condensing and absorber cooling and just as in the larger systems good condensing temperature control must be maintained to avoid crystallisation.

The ammonia system, whilst being very similar to the lithium bromide system, has two major differences. Firstly, it is the ammonia which is the refrigerant and the water the absorbent (see Fig. 11.10). Because ammonia evaporates at much lower temperatures than water the pressures in the evaporator and, consequently, other heat exchangers are above atmospheric pressure, which is the second major difference. An ammonia–water system chilling water from 55°F to 45°F (*12.8°C to 7.2°C*) would be operating at about 50 psig

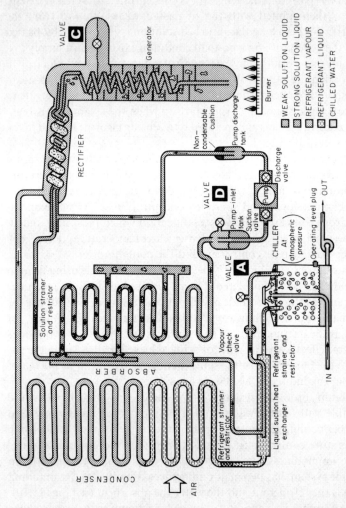

Fig. 11.10. Ammonia–water absorption cycle. (Courtesy: Carlyle Air Conditioning Co. Ltd.)

($447\,kN/m^2$) in the evaporator and absorber, and at about 300 psig ($2170\,kN/m^2$) in the generator and condenser. Since there is no danger of crystallisation of the ammonia solution, control of the condensing medium—air—is not so critical, and such units can operate at a relatively low ambient condition. Unlike the lithium bromide system the very rapid evaporation of the ammonia will

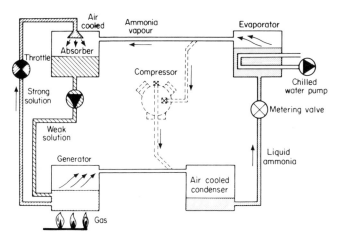

FIG. 11.11. Direct-fired absorption chiller compared to a mechanical refrigeration system.

provide the motivation to force the ammonia vapour from the evaporator to the absorber and relatively small-bore pipework can be used, thus separating the heat exchange elements. The weak solution in the absorber is pumped to the generator where the added heat from the gas flame increases the pressure. This pressure enables the high temperature, high pressure ammonia vapour to travel to the condenser, from where it continues to the evaporator through a metering valve in much the same way as in a mechanical refrigeration system. A comparison of the ammonia–water cycle and mechanical refrigeration cycle is shown in Fig. 11.11.

Comparison between an electrically driven mechanical refrigeration system and a large absorption system is shown in Table 11.1. It can clearly be seen that the electrical power requirements for the absorption machine are much lower, and depending upon the cost between steam raising and electrical power consumption the

TABLE 11.1
POWER REQUIREMENTS PER UNIT REFRIGERATION EFFECT

	Absorption Refrigeration	Centrifugal Compressor
Condenser water pump	0·012	0·007
Cooling tower fan	0·041	0·021
Sol. and ref. pumps	0·003	—
Compressor	—	0·222
Sub total Electrical power	0·056	0·250
Steam or hot water	1·430	—
Total power	1·486	0·250
Total heat rejected	2·486	1·250

economics of the absorption machine (traditionally higher in capital cost although cheaper in maintenance costs) and mechanical refrigeration owning and operating cost can be made. In comparing the systems the size of cooling towers and condenser-water pumping and piping costs must also be considered.

The absorption machine can provide advantages in certain applications over the mechanical system.

1. Where steam or hot water is available at an economical rate during the cooling season. A good example is a hospital which requires steam for heating during the winter and relatively light loads during the summer periods for laundry and sterilisation. Where light loads give inefficient boiler operation the spare capacity can be used for absorption refrigeration and so bring the boiler to a better operating efficiency. A similar example would be a factory requiring process steam throughout the year.
2. For small, multiple-room applications with each room having widely varying loads, a direct gas-fired absorption machine can provide full modulation down to very light loads and so offer a better controlled system than a reciprocating liquid chiller which can only reduce in capacity in stages.
3. Where electrical supply is limited or mechanical refrigeration increases peak load demand.
4. For total energy systems the absorption machine can be

11 ABSORPTION CYCLE REFRIGERATION EQUIPMENT 185

combined with a steam turbine driven centrifugal plant, where the kinetic energy of the steam is used to motivate the steam turbine and the thermal energy used for the absorption system. In such applications, the absorption unit tonnage would be about twice that of the steam driven centrifugal, although the available steam pressures would dictate the exact relationship.

Chapter 12

CONDENSERS AND COOLING TOWERS

Having first selected an evaporator to meet the requirements of the air conditioning load and second, a compressor, the heart of the refrigeration system, the final major component to be selected is the condenser. The requirement of the condenser is to remove the heat of compression and refrigeration effect from the superheated refrigerant gas leaving the compressor and desuperheat and liquefy this gas for reintroduction to the evaporator. The total heat rejection (THR) that the condenser must perform can be expressed as follows:

$$\text{THR} = \text{evaporator load} + \text{absorbed power}$$
$$= \text{tons R} + (\text{kW} \times 0\cdot 285)$$
$$= \text{tons R} + (\text{bhp} \times 0\cdot 212)$$

or

$$\text{BTU/h} = \text{BTU/h} + (\text{kW} \times 3414)$$
$$\text{BTU/h} = \text{BTU/h} + (\text{bhp} \times 2540)$$

or

$$kW = evaporator\ kW + compressor\ kW$$

Once the desired heat rejection capacity of the condenser is known it is the discharge level of the compressor, required to give the desired refrigeration effect, and the temperature of the condensing medium, either water, air or a combination of both, which dictate the size of the condenser. Whatever condensing medium is used, it is the ambient air conditions which dictate the available temperature difference from compressor saturated discharge temperature, to perform heat transfer. With air cooled systems it is the dry bulb temperature of the air and, with water or evaporative cooled systems, the wet bulb

12 CONDENSERS AND COOLING TOWERS

temperature which affect performance. At normal design conditions the dry bulb temperature is always higher than wet bulb, therefore, for a given saturated discharge temperature, water or evaporative cooled condensers would have a wider temperature difference available for heat transfer than air cooled systems, and so require less heat exchanger surface. In addition, the film heat transfer resistance of water to heat exchanger tube is much less than air to heat exchanger tube, which again has the effect of reducing the heat exchanger surface.

Rather than increase the surface of air cooled condensers it is normal practice to allow the saturated discharge temperature to rise to attain the required total heat rejection. The effect of this is to raise the power requirements per ton of refrigeration for the air cooled system over the water or evaporative cooled systems. Table 12.1 shows the effect of saturated discharge temperature to compressor capacity and power requirements at a given saturated suction temperature or evaporator performance level. If ambient conditions of 85 °F (*29.4 °C*) DB and 70 °F (*21.1 °C*) WB are considered, the probable saturated discharge temperature of a water or evaporative cooled system would amount to 100 °F (*37.8 °C*). This is made up allowing a 2 °F (*1.1 °C*) discharge-line pressure drop to provide a condensing temperature of 98 °F (*36.7 °C*), a 10 °F (*5.5 °C*) leaving-temperature difference between water and condensing temperature and a 10 °F (*5.5 °C*) rise of the water passing through the condenser, resulting in a water temperature of 78 °F (*25.7 °C*) entering the condenser. The 8 °F (*4.5 °C*) difference between water leaving the cooling tower and design wet bulb temperature, called the approach, is the temperature difference to perform heat transfer at the cooling tower.

The comparable operating level for an air cooled system would be about 118 °F (*47.7 °C*); allowing a 2 °F (*1.1 °C*) discharge-line pressure loss this gives a compressor saturated discharge temperature of 120 °F (*48.8 °C*) and allows 118 °F − 85 °F = 33 °F (*47.7 °C − 29.4 °C = 18.3 °C*) temperature difference between condensing temperature and entering air temperature. Examination of Table 12.1 shows that at 100 °F (*37.8 °C*) saturated discharge temperature the available refrigeration effect for a given compressor is 40 tons R, whereas at 120 °F (*48.8 °C*) the figure is only 34.9 tons R. For the same evaporator performance of, say, 40 tons R, the air cooled system would have to be 15 per cent larger than the comparable water cooled

TABLE 12.1
SATURATED DISCHARGE TEMPERATURE vs CAPACITY kW INPUT BASED ON 40°F (4·5°C) SATURATED SUCTION TEMPERATURE

Discharge temperature			Capacity		kW input	kW input per capacity		
°F	°C	Tons R	kW	Per cent	kW input	kW/ton R	kW/kW	Per cent
100	37·8	40·0	140·8	100	31·0	0·78	0·22	100
105	40·6	38·8	136·4	97	32·4	0·83	0·238	107
110	43·4	37·5	132·0	94	33·7	0·90	0·255	116
120	48·9	34·9	122·7	87	36·4	1·04	0·297	134
130	54·4	32·1	112·8	80	39·0	1·22	0·346	156
140	60·0	29·1	102·3	73	41·5	1·42	0·405	182
145	62·8	27·4	96·4	68	42·8	1·56	0·44	200

12 CONDENSERS AND COOLING TOWERS

system compressor. In addition, the power requirement would be 134 per cent of the comparable water cooled system.

Once the increased power requirements of the air cooled system are realised these operating costs should be compared with capital and maintenance costs for the alternative systems. The capital cost of the refrigeration plant for a given tonnage is about the same, the less expensive air cooled condenser compared to water cooled condenser, cooling tower and pumps or evaporative condenser, compensating for the larger compressor and motor. Therefore, it must be the maintenance costs which dictate the popularity of air cooled systems. The elimination of water, often prohibited by local authorities for use in air conditioning systems, with its inherent problems of frost protection or winter bleed down, and water treatment, make the air cooled system very attractive for the small system.

The last two decades have seen the acceptable size of single air cooled condensing systems raised some twenty-fold. This increase in size has been brought about by increasing user confidence; reduced maintenance costs, and installation labour costs in comparison to water cooled systems; and to a lesser degree, engineering development. The use of 300 ton (about 1000 kW) refrigeration effect systems is practical from a manufacturing standpoint, and only the physical size of the condensers themselves is the restrictive factor in growing size, since handling such large apparatus can prove difficult, particularly when associated with a rooftop location. It is therefore current practice to limit single air cooled condenser systems to about 85 tons (300 kW) refrigeration effect, unless special transport, lifting and rigging facilities are available.

For the most part, air cooled condensers, although available as a single-piece item, find themselves as either pre-matched to a compressor to form an air cooled condensing unit, or as part of a total refrigeration system in an air cooled single-piece air conditioning plant, or air cooled liquid chiller. An additional application is when used in heat pumps.

When rating an air cooled condenser system the amount of subcooling allowed for in the compressor ratings should be known. From the P–H Chart (Fig. 12.1) it can be seen that the refrigeration effect for a given compressor operating at air cooled condensing levels will increase $\frac{1}{2}$ per cent per 1 °F (*0·9 per cent per 1 °C*) of subcooling.

The condenser should be capable of as much subcooling as possible to increase compressor capacity and decrease the bhp/tons R running costs. However, if a given condenser cannot be operated at the subcooling called for, the compressor ratings should be reduced. To obtain the required evaporator capacity the compressor must operate at lower discharge pressures and thus require a larger condenser surface because of the lower temperature difference. It should be noted here, however, that condensers having large subcooling ratings

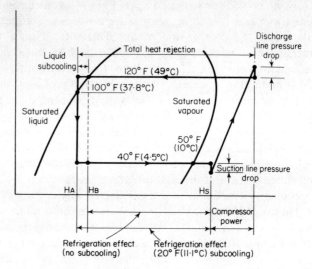

FIG. 12.1. P–H chart showing increase in refrigeration effect due to subcooling.

have extended surfaces or subcooling circuits. The net effect, however, is that the more subcooling, the more economical the selection.

When selecting compressor/condenser combinations, which could involve considerable interpolation in ratings, a graphic technique can be used which immediately reflects the economics of this balance, and can often result in a smaller condenser. Sections of compressor and condenser ratings are shown in Figs. 12.2 and 12.3 from which the graphic selection (Fig. 12.4) is based. Example:

Required refrigeration effect: 25 tons R.
Refrigerant evaporating temperature: 44°F ($6.6\,°C$).
Outside ambient condition: 85°F ($29.4\,°C$).

For selection purposes it should be assumed that the pressure drop

12 CONDENSERS AND COOLING TOWERS

REFRIG-ERANT	TD	TOTAL HEAT REJECTION (TONS)									
		9AB			09DC						
		6	8	12	016	024	028	034	044	054	064	084
12 and 500	20	4.2	6.3	8.3	10.5	14.5	18.3	23.5	30.9	36.5	46.9	61.8
	25	5.2	8.0	10.4	13.1	18.2	22.9	29.4	38.5	45.8	58.8	77.1
	30	6.2	9.5	12.5	15.7	21.8	27.5	35.1	46.3	55.0	70.3	92.6
	35	7.2	11.1	14.5	18.3	25.5	32.1	41.2	54.0	64.2	82.3	108.0
	40	8.2	12.7	16.7	20.9	29.2	36.7	47.0	61.7	73.4	94.0	123.4
22	20	4.7	7.3	9.5	11.3	15.7	19.8	25.3	33.2	39.6	50.6	66.4
	25	5.9	9.1	11.9	14.0	19.6	24.8	31.7	41.6	49.7	63.4	83.1
	30	7.1	10.9	14.3	16.8	23.6	29.8	38.1	49.8	59.6	76.1	99.7
	35	8.2	12.7	16.6	19.7	27.4	34.8	44.3	58.3	69.5	88.7	116.6
	40	9.3	14.7	19.0	22.4	31.4	39.7	50.7	66.5	79.5	101.4	133.0

TD (Temperature Difference) = Condensing Temp. Minus Entering Air Temp.(°F)

           A   B   D C

FIG. 12.2. Air cooled condenser ratings (15°F (8·3°C) subcooling). (Courtesy: Carlyle Air Conditioning Co. Ltd.)

in the suction line is equivalent to 2°F (1·1°C) saturated suction temperature (this is the normal method of pipe sizing). The resulting saturated suction temperature (SST) = 44°F − 2°F = 42°F (6·6°C − 1·1°C = 5·5°C). From the compressor selections (based on 15°F (8·3°C) subcooling) the points X and Y are plotted showing refrigeration effect at 42°F (5·5°C) SST. and selected saturated discharge temperatures (SDT).

For selection purposes the discharge temperatures are taken at 20°F (11·1°C) and 35°F (19·4°C) above the outside ambient condition, i.e. 105°F and 120°F (40·4°C and 48·8°C). The power requirements of the compressor at the selected conditions are

06EW033				
SST	SDT	Cap.	kW	THR
	80	35.4	19.6	41.0
Q Y	90	32.5	21.6	38.6
	100	29.8	23.4	36.4
	105	28.5	24.3	35.4
42°F (5·5°C)	110	27.2	25.1	34.3
	120	24.8	26.8	32.4
	130	22.5	28.4	30.6
X P	140	20.4	30.0	29.0
	145	19.4	30.9	28.2

FIG. 12.3. Compressor ratings (15°F (8·3°C) subcooling). (Courtesy: Carlyle Air Conditioning Co. Ltd.)

converted into tons of refrigeration, and added to the refrigeration effect to give the total heat rejection; at the selected discharge condition the respective total heat rejection figures are plotted, points P and Q. (It can be seen from the compressor ratings in Fig. 12.3 that the total heat rejection figures are presented.) Since the line PQ is a function of compressor saturated discharge temperature this must be

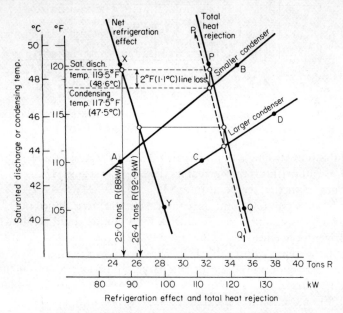

FIG. 12.4. Compressor/condenser balance.

adjusted to the condensing temperature. As with suction lines, discharge lines are sized to the equivalent to 2°F ($1 \cdot 1$°C) at the saturated condition. Therefore, line $P_1 Q_1$ is constructed allowing for the line loss.

From the total heat rejection tables of the condensers, lines AB and CD coincide with line $P_1 Q_1$, and the balance points are found. From these intersections the condensing temperature and total heat rejection of the balance can be read. By constructing a vertical from the balance points to the line PQ, the compressor saturated discharge temperature can be read. At this level from the line XY, the balance point showing the refrigeration effect can be found. From this information the question of economics asks which is the best selection

12 CONDENSERS AND COOLING TOWERS

in terms of capital cost against evaporator load and compressor power. At 119·5 °F (*48·6 °C*) SDT the compressor provides 25·0 tons R (*88 kW*) absorbing 26·7 kW [1·07 kW/tons R (*0·304 kW/kW*)] whereas at 113·5 °F (*45·2 °C*) SDT, the same compressor gives 26·4 tons R (*92·9 kW*) absorbing 25·7 kW [0·975 kW/tons R (*0·277 kW/kW*)]. These running costs should be compared with the difference in capital cost between the two condensers.

It should be noted here that both compressor and condenser have been selected for 15 °F (*18·3 °C*) subcooling. However, if the same compressor with the same condensers having no subcooling were selected the balance points would occur at 116·5 °F (*46·9 °C*) SDT, giving 23·7 tons R (*83·4 kW*), absorbing 26·2 kW [1·105 kW/tons R (*0·314 kW/kW*)] and at 111 °F (*43·9 °C*) SDT, giving 25·0 tons R (*88 kW*), absorbing 25·3 kW [1·01 kW/tons R (*0·288 kW/kW*)]. The net result is that with the same equipment there is a 5 per cent increase in capacity coincident with a 3 per cent reduction in kW/tons R absorbed power.

This example shows that to achieve the desired refrigeration effect, with 15 °F (*8·3 °C*) subcooling, a smaller condenser will suffice. However, there is a penalty to pay for the higher discharge temperature, for at 25 tons R (*88 kW*) with the smaller condenser, the compressor absorbs 26·7 kW, as against 25·3 kW for 25 tons R (*88 kW*) for the larger condenser. Assuming that the compressor runs for 12 hours a day for 100 days a year, and the cost of electricity is 3·0p per kW hour, the annual running costs would be £36·00 per kW. In this example, the smaller condenser has 3 × $\frac{3}{4}$ hp fan motors against 3 × 1 hp fan motors for the larger, thus giving a saving of $\frac{3}{4}$ hp or 0·56 kW. Therefore, the net increase in running costs would be 26·7 − 25·3 − 0·56 kW or 0·84 kW × £36·00 i.e. £30·24 per year. In this instance the probable difference in capital cost to the buyer would be £400 which without interest on the capital saved would take 13 years to become economical.

A further saving with air cooled condensers having subcooling coils is the elimination of a liquid receiver. The subcooling coil is designed to hold liquid refrigerant and acts as a reservoir, thus avoiding the necessity of a critical refrigerant charge. In these circumstances, the condenser should be supplied with shut-off valves so that the refrigerant can be pumped down during service. The addition of a liquid receiver means that liquid refrigerant cannot be held in the condenser to enable it to perform the amount of subcooling required.

If the system is such that because of large refrigerant charges and the question of losing refrigerant charge in the event of condenser damage is raised, then a receiver should be installed having a by-pass for normal running (Fig. 12.5).

In the example discussed, the elimination of a liquid receiver would add to the economics of the subcooled system and would provide an approximate saving of £150 to the buyer. Therefore, eliminating the

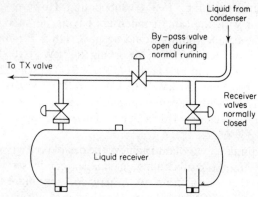

FIG. 12.5. Receiver by-pass for optimum subcooling.

receiver together with the utilisation of a smaller condenser, it would take 18 years for the non-subcooled system to become economical.

The selection method above holds good for an evaporative condenser where the duty is based on entering wet bulb temperature against condensing temperature in lieu of temperature difference. It can also be used for water chilling applications for either air or evaporating condensers.

Under normal circumstances one would not consider this method of selection below 15 tons R. Below this figure there are numerous selections of air cooled condensing units where the manufacturer has made his own balance of compressor/condenser. Because of the simplicity of such systems which eliminate discharge lines and interconnecting electrical wiring, compressor foundations, etc., they are invariably cheaper in capital cost. Also, below 15 tons R there are probably more air cooled condensing units available than air cooled condensers, giving a wider number of selections. Above 15 tons R condenser/compressor selections as discussed above should also be

12 CONDENSERS AND COOLING TOWERS

compared with the combination ratings of air cooled condensing units, together with capital and running costs. As with air cooled condensers, condensing units with high subcooling coil ratings should not be installed with liquid refrigerant receivers.

For the refrigeration system to function properly, the condensing pressure and temperature must be maintained within certain limits.

FIG. 12.6. Air cooled condenser. (Courtesy: Carlyle Air Conditioning Co. Ltd.)

Abnormally high condensing temperatures cause lack of capacity, extra power consumption, and overloading of the compressor motor with the possibility of permanent damage to the compressor and motor. Safety limit controls normally protect against such conditions. Low condensing pressures will cause insufficient pressure for liquid feed devices, thus starving the evaporator, with a loss of capacity resulting. This will also cause low suction pressures which in turn will cause the compressor to trip on the low pressurestat.

As the condensing pressure drops, the refrigeration capacity will increase within the limits of the expansion valve. Since the expansion

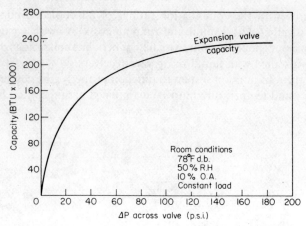

Fig. 12.7. Thermal expansion valve capacity vs pressure drop.

valve capacity is dependent upon the pressure drop across the valve, the lower the pressure drop, the lower the capacity. From Fig. 12.7 the expansion valve capacity is plotted against pressure drop, and it can be seen that as the pressure drops off, so the load falls off at an ever-increasing rate.

Superimposing suction and condensing temperatures in Fig. 12.8, together with system capacity, it will be noted that the system capacity

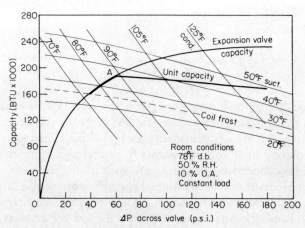

Fig. 12.8. Thermal expansion valve capacity vs drop in condensing temperature.

12 CONDENSERS AND COOLING TOWERS

increases to point A. Up to this point the expansion valve operates in a partially throttled position. Beyond this point the expansion valve is fully opened and any further decrease in condensing pressure results in a greater loss of duty. It should also be noted that the evaporating temperature is also dropping. This would initially cause a greater dehumidification on the evaporator coil (in the case of DX systems) and eventually will cause a frost or freeze-up condition. Therefore, to maintain proper evaporating temperatures and system capacities the condensing pressure must be kept within certain limits.

The method of head pressure control varies widely, depending on the type of condenser used; the temperature difference of condensing temperature to design outside air; and the load variation of the system (i.e. if compressor unloaders are used). The types of head pressure control for air cooled condensers can be categorised into two systems, (i) refrigerant side control, and (ii) air side control.

REFRIGERANT SIDE CONTROL

This is accomplished by reducing the active amount of condenser surface by flooding the coil with liquid refrigerant. This type of control requires the use of a liquid receiver and an excessive charge of refrigerant. This method should not be used where subcooling coils are used on condensers.

METHOD 1A

This type of control is used for:

(1) Refrigerant systems where the air temperature at the condenser remains above the evaporator temperature during off cycles.
(2) Refrigeration systems with thermostatic control of the compressor. The 'W' valve shown in Fig. 12.9 is actuated by the receiver pressure and modulates to maintain the condensing pressure at the control valve setting. The hot refrigerant gas which by-passes the condenser through the 'W' valve is prevented from flowing into the liquid outlet of the condenser by the 'O' non-return valve.

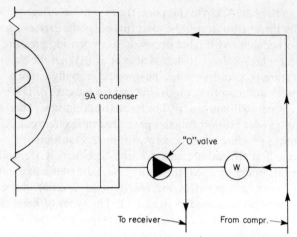

Fig. 12.9. Refrigerant side control (Type 1A). (Courtesy: Carlyle Air Conditioning Co. Ltd.)

METHOD 1B

This type of control is used for:

(1) Refrigeration systems where the air temperature at the condenser is below the evaporator temperature during the off cycle.
(2) Water chiller systems requiring freeze-up protection and instant start-up to prevent cut-out on the manual reset low pressure switch.

A condenser temperature lower than the evaporator temperature during off cycles causes refrigerant to migrate to the condenser, consequently suction pressure cannot build up to close the low pressure switch and start the compressor on automatic pump-down compressor control. This system completely isolates the condenser during off cycle, preventing refrigerant migration to it. This enables the suction pressure to build up and allows the compressor to start up under all out-door temperature conditions.

In Fig. 12.10 the 'C' non-return valve in the liquid line between the condenser outlet and receiver prevents refrigerant flowing to the condenser through the liquid outlet. On start up, the 'W' valve is open and allows the compressor to impose its full discharge pressure on the liquid in the receiver. When this pressure is up to normal the 'R' valve in the discharge line opens allowing discharge gas to flow to the

12 CONDENSERS AND COOLING TOWERS

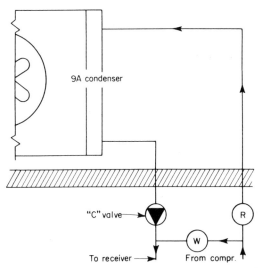

FIG. 12.10. Refrigerant side control (Type 1B). (Courtesy: Carlyle Air Conditioning Co. Ltd.)

condenser. When the discharge pressure falls below a present level, the 'R' valve closes, halting hot gas flow to the condenser.

Once the system is in operation the 'W' and 'R' valves modulate to maintain the proper refrigerant pressure at the expansion valve.

METHOD 1C

The basic principle of this pressure stabiliser is a heat transfer device which transfers heat from the discharge line to the liquid leaving the condenser. The heat exchange is controlled from a regulating valve installed between the condenser and receiver. This valve is set at the desired operating pressure, and throttles from the open to closed position as the head pressure drops. The throttling action of the valve forces liquid to flow through the heat exchanger section of the liquid stabiliser. The heat picked up by the liquid raises the receiver pressure and prevents further flow of refrigerant from the condenser. This results in a flooding of the condenser and a reduction of effective condensing surface, which will maintain satisfactory discharge and receiver pressures. The system is provided with a check valve in the heat-transfer-liquid section which remains closed during warm weather operation to ensure against liquid refrigerant reheating (see Fig. 12.11).

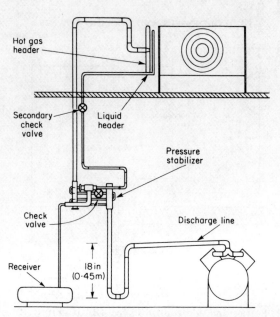

Fig. 12.11. Refrigerant side control (Type 1C). (Courtesy: Searle Manufacturing Co. Ltd.)

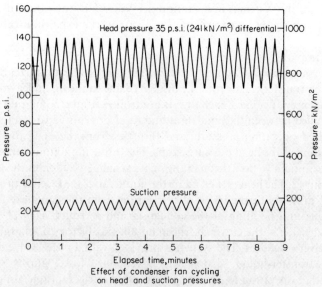

Fig. 12.12. Fan cycling head pressure control (one fan).

AIR SIDE CONTROL

This is obtained by reducing the air flow across the condenser coil, thus reducing its cooling effect.

METHOD 2A
Fan cycling can accomplish head pressure control from the response of head pressure or outdoor ambient temperatures. Fan cycling in response to head pressure controlled from a pressure switch results in the rapid cycling of the fan motor and is not recommended, for it results in expansion-valve hunt and may cause a burn-out of the condenser-fan motor (see Fig. 12.12).

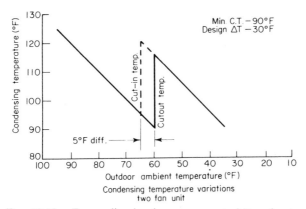

FIG. 12.13. Fan cycling head pressure control (two fans).

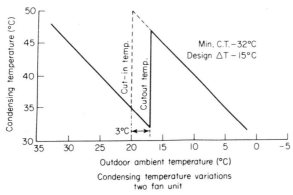

FIG. 12.13a. Fan cycling head pressure control (two fans).

202   APPLIED AIR CONDITIONING AND REFRIGERATION

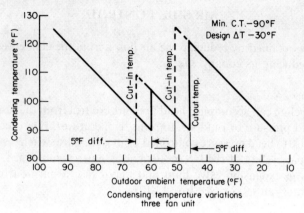

FIG. 12.14.   Fan cycling head pressure control (three fans).

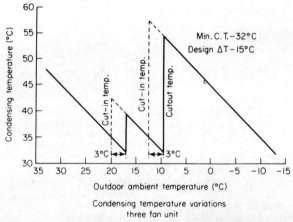

FIG. 12.14a.   Fan cycling head pressure control (three fans).

Fan cycling in response to outside air temperature eliminates rapid cycling of the fan motor but is limited to use with multiple fan units. One fan must be running at all times since a completely static set of fans would raise the head pressure above the high pressure setting and cause shut-down of the plant. Figures 12.13 and 12.13a show condensing temperature variations with a two-fan unit, and Figs. 12.14 and 12.14a show this operating with three fans working.

METHOD 2B

Modulating damper control may be used on single fan units or on

12 CONDENSERS AND COOLING TOWERS 203

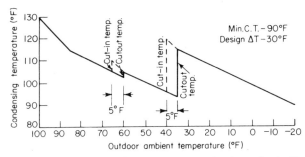

FIG. 12.15. Fan cycling head pressure control (three fans). Two fans cycling, third fan with damper or solid state fan speed control.

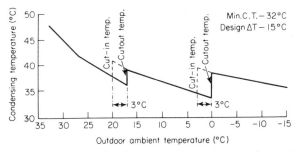

FIG. 12.15a. Fan cycling head pressure control (three fans). Two fans cycling, third fan with damper or solid state fan speed control.

multiple fan units in combination with fan cycling. The damper is mounted on the active fan section to reduce the air flow and so achieve infinite steps of control (Figs. 12.15 and 12.15a). This system can be operated electrically from a pressure control device controlling an electric motor linked to the dampers or directly by use of a head pressure controller operated by condensing pressure (Fig. 12.16). Care should be taken when using propeller fans as the fan power increases with added resistance. The fan motor must therefore have adequate horsepower for operation with the dampers throttled.

METHOD 2C

Fan speed control can be accomplished by multi-speed motors, or more economically with solid state controllers. This type of control can be used with single fan condensers or with multiple fan condensers. The resulting performance will give infinite steps of control as will the damper operated system (Figs. 12.15 and 12.15a).

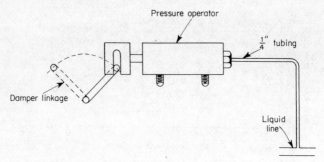

Fig. 12.16. Modulating damper pressure control device.

This method of control is extremely economical and does not have the maintenance problems of the damper system.

When multiple circuit condensers are used, the air slide control method should be operated from outside air conditions, since from refrigerant pressure only one circuit can be detected. When multiple circuits are used with compressors having capacity control, problems could arise with air side controls. However, if the temperature difference at design is high this can be overcome. Single circuit condensers with unloading compressors should be controlled from condensing temperature.

After a shut-down period, with outdoor conditions below that of the evaporator, refrigerant migrates to the condenser. To avoid the low pressurestat stopping the compressor on start-up because the suction temperature cannot build up, a timer can be inserted in the control circuit, wired in parallel to the low pressurestat. In start-up this timer will be closed, and will allow the system to operate even though the low pressurestat is open. After the system has been operating long enough to maintain normal conditions (say 4 minutes) the timer circuit will open, allowing the control circuit to return to its normal operation, for by this time the low pressurestat will have closed. Table 12.2 summarises the various methods of control. In this exercise it is assumed that the minimum allowable condensing temperature is 90°F at 100 per cent capacity, 80°F at 73 per cent capacity and 70°F at 50 and 25 per cent capacity. *Table 12.2a summarises the various methods of control. In this exercise it is assumed that the minimum allowable condensing temperature is 32°C at 100 per cent capacity, 27°C at 75 per cent capacity and 22°C at 50 and 25 per cent capacity.*

TABLE 12.2

Type of control	Condenser type	Design ΔT (°F)	Minimum outdoor temperature (°F) at compressor capacity (per cent)			
			100	75	50	25
None	All	30	60	57	55	62
		25	65	61	57	64
		20	70	65	60	65
Fan cycling	Single fan or Single shaft centrifugal fans[a]	30	60	Not recommended		
		25	65			
		20	70			
	Two fans (cycling one fan on outside air temperature)	30	33	38	42	56
		25	43	45	47	58
		20	52	52	51	61
	Three fans (cycling two fans on outside air temperature)	30	3	15	26	48
		25	18	30	34	52
		20	22	36	41	56
Variable speed control or modulating damper with fan cycling	All	30	−20	−14	−8	−2
		25	−20	−9	2	13
		20	−20	−5	10	25
Refrigerant side control (Type 1A)	All	All	40	40	40	40
Refrigerant side control (Type 1B)	All	All	−30	−30	−30	−30
Refrigerant side control (Type 1C)[b]	All	All	50	0	−20	−20

[a] Where centrifugal fans are used having large motors not capable of solid state control it is necessary to use damper controls. However, care should be taken as most centrifugal fan condensers have the fans on a common shaft and, therefore, to avoid unbalance all fan outlets must be controlled, simultaneously. This would be categorised as single fan operation.

[b] The effective minimum outdoor temperature is governed by the size of the pressure stabiliser. Should lower temperatures be required at 100 per cent compressor duty a larger control must be installed.

TABLE 12.2a

Type of control	Condenser type	Design ΔT (°C)	Minimum outdoor temperature (°C) at compressor capacity (per cent)			
			100	75	50	25
None	All	20	12	12	12	17
		15	17	15·75	14·5	18·5
		10	2	19·5	17	20
Fan cycling	Single fan or Single shaft centrifugal fans[a]	20	12	Not recommended		
		15	17			
		10	22			
	Two fans (cycling one fan on outside air temperature)	20	−3	0·75	4·5	13·25
		15	2	4·5	7	14·5
		10	7	8·25	9·5	15·75
	Three fans (cycling two fans on outside air temperature)	20	−18	−9·5	−3	
		15	−13	−6·25	−0·5	10·75
		10	−8	−3	2	12
Variable speed control or modulating damper with fan cycling	All	20	−30	−28	−26	−24
		15	−30	−25	−20	−20
		10	−30	−22	−14	−6
Refrigerant side control (Type 1A)	All	All	4·5	4·5	4·5	4·5
Refrigerant side control (Type 1B)	All	All	−35	−35	−35	−35
Refrigerant side control (Type 1C)[b]	All	All	10	−18	−29	−29

[a] *Where centrifugal fans are used having large motors not capable of solid state control it is necessary to use damper controls. However, care should be taken as most centrifugal fan condensers have the fans on a common shaft and, therefore, to avoid unbalance all fan outlets must be controlled, simultaneously. This would be categorised as single fan operation.*

[b] *The effective minimum outdoor temperature is governed by the size of the pressure stabiliser. Should lower temperatures be required at 100 per cent compressor duty a larger control must be installed.*

CONDENSER DESIGN LIMITATIONS

The requirements to operate condensers in ambient temperatures as high as 125 °F ($52°C$) have brought with them many design problems for the product engineer designing air cooled liquid chillers and condensing units. An economical design temperature difference at 95 °F ($35°C$) ambient, at which condensers are nominally rated, with say, a 41 °F ($5°C$) saturated suction temperature for a condensing unit, or 41 °F ($5°C$) leaving chilled water temperature for an air cooled chiller, would be at 27 °F ($15°C$), or a saturated condensing temperature of 122 °F ($50°C$). When such standard designed machinery is imposed in ambient conditions of 125 °F ($52°C$), both the required heat rejection and condensing to entering air temperature difference fall by 15 per cent to give a rise of some 23 °F ($13°C$). The resulting saturated condensing temperature would rise therefore to about 150 °F ($65°C$).

Whilst in itself such a saturated condensing temperature is not a problem, the associated high discharge temperature at the compressor can create many adverse effects. For refrigerants R.12 and R.500 the maximum temperature recommended for continuous operation in the presence of oil, copper and steel is 250 °F ($120°C$) and for refrigerants R.22 and R.502 about 300 °F ($150°C$). At about 265 °F ($130°C$) continuous operation damage to compressor valves can occur and at about 275 °F ($135°C$) lubricating oils can suffer thermal breakdown.

The problem can be demonstrated most easily by the use of the P-H diagram (Fig. 12.17). It can clearly be seen that for the same saturated condensing temperature, an air cooled chiller operating at a saturated evaporating temperature, some 10 °F ($6°C$) lower than the equivalent air cooled condensing unit to achieve similar dehumidifier coil dew points, would have a compressor discharge temperature of 267 °F ($130·6°C$) against 258 °F ($125·6°C$) for the air cooled condensing unit.

These rules have been established for hermetic compressors using suction gas to cool motor windings. The actual superheat imparted by cooling the windings would amount to about 20 or 30 °F (10 or $15°C$) at the operating levels considered which, when added to the superheat imposed by the expansion device, means that a practical allowance of 40 °F ($20°C$) superheat must be made when calculating the actual refrigerant discharge temperature.

What can be clearly seen is that using high condensing temperatures in association with high compression ratios results in an acute danger of permanent compressor damage. Having made a compressor/condenser balance it behoves the designer to check the actual discharge temperature before making a final plant selection.

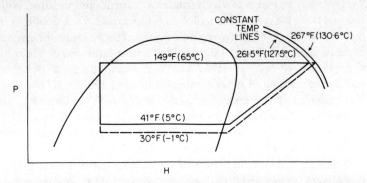

FIG. 12.17. P–H chart showing effect of suction temperature on compressor discharge temperature.

Even after the designer has assured himself that operational limits will not be exceeded, he has had to assume clean condensers without blockage, proper condenser air flow, and that the design entering air temperatures will not be exceeded. When using refrigeration plant at high ambient conditions, it is good practice to add to the system safety controls a discharge line thermostat, such that in adverse conditions the plant can be stopped before damage occurs, or such that the action of the thermostat will cause the refrigeration plant to unload this, providing partial refrigeration rather than stopping the plant completely at times when it is obviously required to operate at maximum load.

As stated previously, the major factors governing condenser performance and selection are (a) the available saturated discharge temperature from the compressor to provide the necessary refrigeration effect, (b) the temperature of the condensing medium and (c) the amount of subcooling available or required. Since water or evaporative cooled condensers depend upon the wet bulb temperature of the ambient air the resulting condensing temperatures are lower than corresponding air cooled systems operating in the same

12 CONDENSERS AND COOLING TOWERS

climate, with resulting increase in coefficient of performance. Subcooling is achieved by special circuits located in the larger shell and tube water cooled condensers, and evaporative condensers have them available as accessories. A subcooling circuit is only feasible if there is a high enough temperature difference between condensing medium and the condensing refrigerant. For example, it is impossible to achieve 15°F *(8·3°C)* subcooling if the design condensing level is 85°F *(29·4°C)* and the entering wet bulb to an evaporative condenser is 70°F *(21·1°C)*, i.e. only 15°F *(8·3°C)* temperature difference. It is normal practice to have a leaving temperature difference of 10°F *(5·5°C)* to achieve economical heat exchange, although this level must differ from application to application. Where compressor ratings are provided with subcooling allowance, corrections should be made to make the condenser and compressor performances compatible by correcting the compressor ratings $\frac{1}{2}$ per cent per 1°F *(0·9 per cent per 1°C)* subcooling in accordance with the amount of subcooling provided by the condenser.

Water cooled systems are essentially always of the recirculated water system type using a cooling tower or spray pond to recool the water. Local regulations almost always forbid the use of once through or waste water systems, although recirculated well-water is often used for condensing purposes. The amount of water to be recirculated by the two systems is as shown on p. 210, assuming that the heat rejection is equivalent to 15 000 BTU/h tons R *(4·4 kW)* effect.

However, despite the amount of water circulated it is the actual water usage which is more important in considering economies. The once through system requires 1·25 imp. gal/min *(0·095 litres/s)* ton of water per ton R refrigeration effect, whereas the recirculated system only requires the amount of water evaporated at the cooling tower with allowance for windage loss and bleed off associated with water treatment. At the operation level the latent heat of evaporation of water approximates to 1000 BTU/h *(646 W/kg)*. Therefore, the amount of water evaporated would be 15 000 ÷ 1000 = 15 lb/h/ton R *(4400 ÷ 646 = 6·8 kg/h/ton R)*. Assuming the bleed-off rate and windage loss to be the same as the amount evaporated the actual water usage would amount to 30 lb/h or 0·05 imp. gal/min/ton R *(13·6 kg/h or 0·003 79 litres/s/ton R)* refrigeration effect. Comparison with the once-through system of 1·25 imp. gal/min/ton R *(0·095 litres/s/ton R)* shows a usage ratio of 30 to 1, which is very much in favour of the recirculated system.

	Recirculated system		Once-through system	
	°F	°C	°F	°C
Condensing temperature	95	34·9	95	34·9
Leaving temperature difference	10	5·5	10	5·5
	—	—	—	—
Leaving water temperature	85	29·4	85	29·4
Entering water temperature	75	23·9	65	18·3
	—	—	—	—
Water rise	10	5·5	20	11·1
	—	—	—	—

$$\text{gal/min tons R refrigeration effect } \frac{15\,000}{10 \times 600} = 2\cdot5 \text{ imp. gal/min } (0\cdot19\ litres/s) \qquad \frac{15\,000}{20 \times 600} = 1\cdot25 \text{ imp. gal/min } (0\cdot095\ litres/s)$$

12 CONDENSERS AND COOLING TOWERS

Water cooled condensers can be divided into three categories: (a) tube-in-tube; (b) shell and coil; and (c) shell and tube.

The tube-in-tube condenser (Fig. 12.18) is usually limited to systems of 4 or 5 tons R, and is available in two configurations. Most common is the coiled shape where the concentric tubes, the centre one of copper, often finned, and the outer of steel, have water passing on a

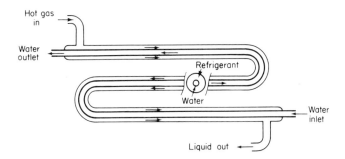

FIG. 12.18. Tube-in-tube condenser.

single-pass principle through the centre tube with the refrigerant on the outside. The coiled arrangement allows the compressor to be housed inside the bundle to make a compact unit. The alternative tube-in-tube arrangement has vertical headers whereby the concentric tubes pass horizontally between them. This arrangement can be adapted whereby on each header at the centre line of each tube there is a plug which can be removed for mechanical cleaning of the condenser, otherwise cleaning of tube-in-tube condensers is accomplished by chemical means. The size of such condensers is restricted because of the number of tubes required for larger systems and the pressure drops associated with continuous single-pass tube systems.

The shell and coil condenser (Fig. 12.19) is common on systems up to 20 tons R. It consists of a steel shell enveloping one or more finned copper tube bundles. The refrigerant gas is fed into the shell where it is condensed by water within the copper tube(s)—liquid refrigerant dropping to the bottom of the shell. These condensers can be horizontal or vertical in arrangement and because of their compactness are ideally suited to packaged systems. As with tube-in-tube systems cleaning must be accomplished by chemical means.

The chemical cleaning of condensers is achieved by using an

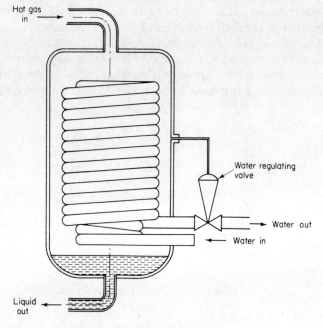

Fig. 12.19. Shell and coil condenser.

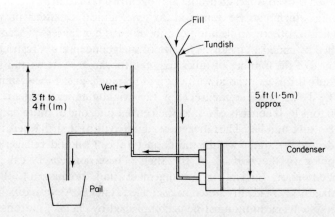

Fig. 12.20. Chemical cleaning of water cooled condenser.

inhibited hydrochloric acid solution. The inhibitor is added so that the steel is not attacked. While cleaning with acid it is good practice to protect all adjacent surfaces from spillage. A warm solution will speed up the cleaning process, which can be achieved by a gravity flow method as shown in Fig. 12.20, or a forced circulation method. It should be noted that the vent pipe is used to exhaust gases formed by chemical reaction. The condenser should be filled by this method and

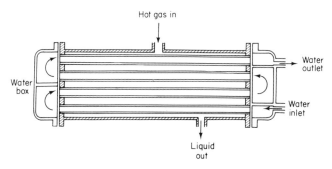

FIG. 12.21. Shell and tube condenser.

allowed to stand for a number of hours, normally overnight. After soaking the condenser should be drained and flushed with clean water.

For the larger refrigeration system, shell and tube condensers (see Fig. 12.21) are used. They consist of a cylindrical shell enveloping straight copper tubes (or cupro-nickel in marine applications) supported at their ends by tube sheets, and water boxes are bolted to the tube sheets. Water enters the water box at one end of the condenser and baffling controls the number of water passes before the water leaves at the other water box for single, three, or five-pass condensers, on the same end for double, four or six-pass. This type of condenser can has subcooling coils fitted, these being located at the bottom of the condenser where the refrigerant liquid is stored. To achieve proper and adequate subcooling a correct refrigerant charge must be used. An advantage of this type is that it can be mechanically cleaned.

An important factor in the selection of water cooled condensers is the amount of scaling or fouling which is caused by the build-up of precipitated mineral solids from the condenser water. This fouling restricts the heat flow and to a lesser degree the water flow, both of which reduce system capacity. Condensers are usually rated with fouling factors of 0·0005, 0·001 and 0·002, the units being ft²/h °F/BTU (*0·000 088, 0·000 176 and 0·000 352 m² °C/W*). Comparison of these figures with water film resistance, metal resistance and refrigerant film resistance, could mean that as much as 60 per cent more water flow is required through a given condenser having a 0·002 (*0·000 352*) fouling factor in comparison to 0·001 (*0·000 176*) fouling factor, to achieve the same heat rejection at constant condensing temperature and entering water temperature. The water velocity through the condenser tubes will affect both the water film resistance and also the rate of scale build up. Because of this, condensers are selected at a minimum velocity of 3 ft/s (*0·9 m/s*) below which scaling occurs rapidly since there is little or no cleansing by water velocity and high water film resistance also occurs. The maximum velocity through a condenser is usually fixed by the manufacturer and is normally a function of water pressure drop and erosion, and the maximum velocities vary between 10 and 12 ft/s (*3·0–3·7 m/s*).

Naturally, the quality of the condenser water plays a major part in the scaling of tubes, and adequate consideration should be paid to water treatment of recirculating cooling tower water. High water temperatures will also affect the rate of fouling. The following is presented as a guide to fouling factor selection:

Fouling factor	Usage
0·000 5 (*0·000 088*)	Once-through systems using town water or well water.
0·001 (*0·000 176*)	Systems using cooling towers with adequate water treatment or spray pond applications.
0·002 (*0·000 352*)	Systems where water treatment cannot be guaranteed or areas with particularly hard water, or where concentrations of airborne wastes are high.

The selection of the above factors does not mean that 0·0005 (*0·000 088*) fouling factor cannot be selected for a hard-water area using a cooling tower without water treatment. However, such a system would require very frequent descaling to avoid loss of refrigeration effect, high running costs and possible tripping of the high-pressure cut-out. The majority of cases fall into a category of 0·001 (*0·000 176*) fouling factor. Under no circumstances should

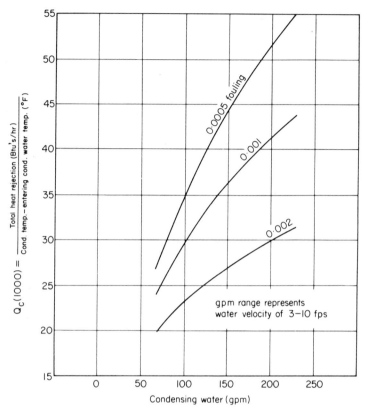

Fig. 12.22. Typical condenser ratings.

condensers be selected for clean tubes, the minimum factor being 0·0005 (0·000 088) no matter what the application.

Selection of water cooled condensers varies from manufacturer to manufacturer, but four figures must be known no matter what the selection method, these being total heat rejection, the condensing temperature, entering water temperature and fouling factor. Many ratings are of the tabular form where for each condenser size values of water quantities against total heat rejection are shown for various fouling factors and temperature difference between condensing temperature and entering water temperature. Such ratings require interpolation for accurate calculations. An alternative selection method uses values of condenser load factor (CLF) which equals the total heat rejection divided by the temperature difference between condensing and entering temperatures. Such ratings can be in graphical form having plots of CLF against imperial gallons per minute for a given condenser size, the plots being for various fouling factors. Alternatively, the same information can be shown in tabular form. A further factor affecting performance is the number of passes in the condenser. To achieve the same refrigeration effect a greater number of water passes would require less water quantity, but would have a higher pressure drop and probably high water temperatures enhancing the need for more frequent cleaning. Maximum pass selections are made when the water quantities are to be kept to a minimum because of well water, waste water, or applications with very remote cooling towers where it is economical to install smaller bore pipework.

The world-wide water shortage problem has brought water cooled systems to be almost always associated with cooling towers. Condenser water leaving the condenser is passed to the top of the tower where it is evenly distributed and broken into droplets by one of several methods such that it may be brought into contact with outside air where it may be cooled by the evaporation of part of the water. The evaporation of the water is enhanced by extending its surface into droplets and with most air conditioning systems air is either forced or induced through the droplets to promote speedier evaporation for a given tower cross-sectional area.

Figure 12.23 shows the psychrometric process through the cooling tower where entering air picks up essentially latent heat from the hot water entering the tower. The temperature drop in the water passing through the tower is called the cooling range and the temperature difference between leaving water temperature and entering wet bulb temperature is known as the approach. The rate of heat transfer through the cooling tower is dependent primarily on the enthalpy or total heat of the air entering the tower and for all practical purposes it is the wet bulb temperature which is the important factor.

The cooling tower performance is controlled by those factors which influence the evaporation of water into the air such as the velocity of the air through the tower, the surface area of water exposed to each unit volume of air, the vapour pressure difference between air and water, and the relative directions of water and air flow.

All the various types of cooling tower are classified by the method of air flow circulating through the tower. When air circulates through

12 CONDENSERS AND COOLING TOWERS

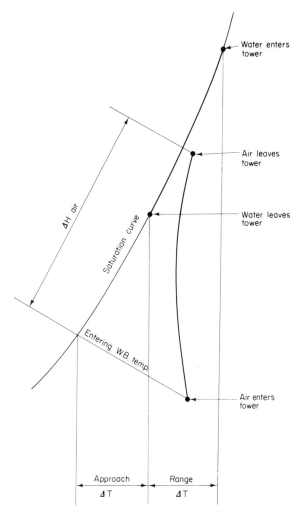

FIG. 12.23. Psychrometric process through cooling tower.

a tower by natural convection or with a chimney effect without mechanical assistance, the tower is classified as a natural draught or atmospheric tower. This type of tower is associated with water cooling problems at power stations, etc., and would normally be out of the range of air conditioning systems.

When air is circulated by means of a fan the tower is classified as a mechanical draught tower, and is further classified as induced

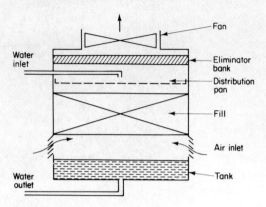

Fig. 12.24. Diagrammatic cross-section of induced draught cooling tower.

draught (Figs. 12.24 and 12.25) or forced draught (Figs. 12.26 and 12.27), depending on whether the fan draws air or blows air through the tower. Distribution of water through a tower is varied and can be sprayed through nozzles, distributed through a perforated pan, through notched channels, or by one of many other methods, all of which are intended to break up the water into droplets and spread it evenly across the tower. Most towers include a 'fill' which is often of redwood timber, protected steel or plastic. As the water passes over the fill it assists the breaking up of the water into droplets and provides a large area of wetted surface which enhances evaporation. Also the water, by taking an arduous path, takes a long time to reach the bottom of the tower and so has a greater time to be cooled.

Mechanical draught cooling towers are relatively compact since large volumes of air at high velocities are circulated by the fan. Because of the velocity some manufacturers, particularly with the blow-through type, install eliminators at the tower discharge to minimise windage loss and carry over.

The siting of cooling towers must be carefully considered. All towers are essentially for outdoor usage although some designs can be installed internally and have fans capable of handling the external resistance of the sheet metal trunking necessary to discharge the humid air leaving the tower to atmosphere. Care should be taken not to place towers too near walls or overhangs which could reduce the air flow through the tower and so impair performance. Similarly, the discharge air should not come into contact with surfaces where condensation could take place and cause damage. The tower should

219

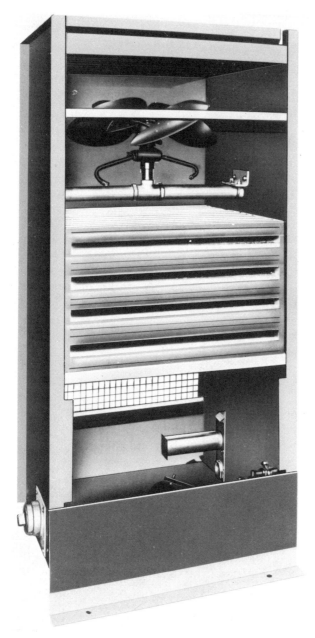

Fig. 12.25. Induced draught cooling tower (back panel removed to show components). (Courtesy: AORI Division of Marelli Ltd.)

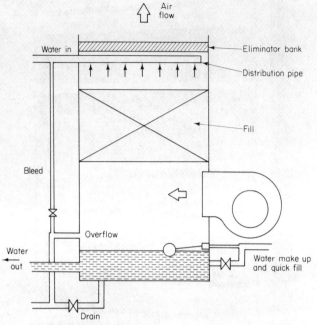

FIG. 12.26. Forced draught cooling tower.

FIG. 12.27. Forced draught cooling tower. (Courtesy: AORI Division of Marelli Ltd.)

12 CONDENSERS AND COOLING TOWERS

be located away from exhaust heat systems or contaminated air exhausts, and should be sited so that noise from the air discharge does not cause objection.

Cooling tower performance ratings are presented in many forms but the common requirements are the total amount of heat to be rejected (sometimes this can be expressed in refrigeration effect with corrections for actual heat rejection factors) the desired entering and leaving water temperatures (and thus water quantity) and design ambient wet bulb. From these figures the cooling range, approach and tower duty can be used to find the required tower size.

It is common practice to assess the condenser water requirements for various entering water temperatures and thus temperature range and water quantity. From these various figures the optimum cooling tower size can be selected. For example, a given condenser could provide the necessary total heat rejection 70 tons R at 100 °F *(248 kW at 37·8 °C)* condensing temperature provided with 140 imp. gal/min of water entering at 80 °F *(26·7 °C)* and leaving at 90 °F *(32·2 °C)*, or similarly, 98 imp. gal/min from 75 to 89·3 °F *(23·9 to 31·8 °C)*. From cooling tower selections it is found that the cooling tower size at 68 °F *(20 °C)* entering wet bulb ambient temperature for the 140 imp. gal/min selection is 60 per cent of the cooling tower size necessary for the 98 imp. gal/min selection. Depending upon plant layout, the pumping and pipework costs can be examined in terms of cooling tower costs and so an economical selection can be made. Whichever tower is selected it should never be sized small; often specifications state that cooling towers should be selected for maximum expected wet bulb temperatures rather than normal design temperatures. The logic is very sound since there are many factors which reduce tower capacity during normal operation, and an undersized system would cause high head pressure, rapid scaling and probable high maintenance costs.

As with air cooled condensing systems the head pressure or condensing temperature of an air conditioning system should remain reasonably stable to prevent low evaporating conditions and possible freeze up. With once-through systems an automatic water regulating valve (Fig. 12.28) located in the condenser discharge will provide a simple and adequate control, which also gives economical water requirements at partial load operation. However, with a cooling tower system there are two main factors to consider, firstly, low ambient temperatures which lower the leaving water temperature from the tower, and/or partial refrigeration plant operation. There are several ways of maintaining head pressure, all of which should, wherever possible, maintain constant water flow through the condenser.

The most common method of head pressure control is a cooling tower by-pass method which can best be achieved by use of a three-way valve. Alternatively, a butterfly valve can be located in the by-pass for systems where the ambient temperature does not fall too low,

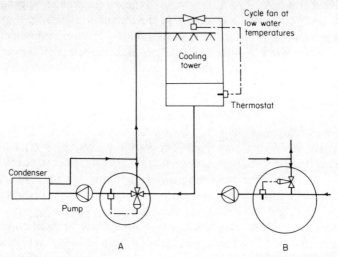

FIG. 12.28. Cooling tower by-pass head pressure control: (A) three-way valve; (B) butterfly valve.

for example all fresh air cooling systems, or where the refrigeration load is relatively constant. Both systems are best controlled from the water temperature leaving the condenser. A simple refinement can be added to either system whereby a thermostat located in the cooling tower simply stops the fan at low temperatures. This means that at light load and/or low ambient temperatures the control valve is not working in an almost closed position, with resulting better system control.

Other forms of head pressure control restrict the air flow through the cooling tower by means of dampers located in the fan inlet or discharge. The air volume handled is controlled from either the sump temperature or leaving condenser water temperature.

During winter operation it is common practice to use an immersion heater in the tower sump, controlled from the sump water temperature, to prevent freeze up during off cycles. If the system is not used during winter periods, it would normally be bled down to prevent damage due to icing conditions.

The evaporative condenser (Fig. 12.29) is basically a combination of an aircooled condenser and cooling tower. This type of condenser is physically very much like a cooling tower but instead of a 'fill' it has a tube bundle in which the refrigerant condenses.

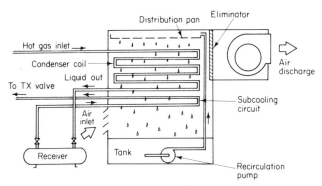

FIG. 12.29. Evaporative condenser.

Water is recirculated around the condenser from the sump to distribution pans at the top where the water is allowed to fall onto the condenser tubes. Air is mechanically forced or induced through the condenser so that the heat from the condenser is dissipated into the air stream by the evaporation of the recirculating water.

To minimise scaling and facilitate cleaning the condenser tubes are usually constructed from copper or galvanised steel prime surface rather than finned tubing. The problems of scaling are just the same as those associated with a cooling tower and water cooled condenser, and particular care should be taken to ensure adequate and proper water treatment in association with bleed-off rates that minimise the concentration of residual solids.

The amount of water required for make up, bleed-off and windage loss would approximate to that of a cooling tower [i.e. 0·05 imp. gal/min/ton R (*0·003 79 litres/s/ton R*)]. Windage loss is minimised by the use of eliminator blades located in the air discharge from the condenser. The circulation pump would normally be supplied with the evaporative condenser and be located outside the casing below the normal water level. The amount of water circulated would be about the same as a comparable cooling tower size, about 2·5 imp. gal/min/ton R (*0·19 litres/s/ton R*).

As with a cooling tower it is the entering air wet bulb temperature which plays the major part in condenser performance, and ratings are made for temperature difference between condensing temperature and entering wet bulb temperature against refrigeration effect or total heat rejection. Ratings are much simpler than cooling towers since the variable of water temperature range is eliminated. The selection of an evaporative condenser would be most economically achieved with a compressor balance exactly as used for air cooled condenser applications.

Head pressure control is achieved by varying the air volume across the condenser with dampers located in the fan discharge, and operated directly from a condensing pressure signal. During low temperature operation the circulating pump can be stopped and water bled down to provide an air cooled system. The pump should not, however, be cycled to provide head pressure control, since the constant wetting and drying of the condenser coils will cause rapid scale build up.

Most evaporative condensers are suitable for outdoor installation, although many are sited internally with ducting to the atmosphere.

For applications with many small water cooled condensers, an evaporative condenser can be converted to a closed circuit water cooler whereby the condenser water is passed through the condenser coil, with water being distributed over the coil in the normal manner. While much more expensive than a cooling tower, the advantages of keeping many condensers scale-free with obvious savings in maintenance costs could warrant the application of closed circuit coolers.

Chapter 13

AIR MOVING DEVICES

The motivation of any air system is produced by a fan, which is exclusively driven by an electric motor, either directly or via a belt drive. Fans are essentially low pressure devices capable of handling relatively large volumes.

The pressures which a fan must overcome are measured in inches water gauge, and the total pressure of a unit is produced by two quotients, static head and velocity head. Figure 13.1 shows how these pressures are established by the use of a manometer.

Fans can be categorised into three basic groups:

(a) Propeller.
(b) Axial flow.
(c) Centrifugal.

The propeller fan has not the capacity of producing high pressures and is limited to about 1·00 inches water gauge (in. w.g.) ($250\ N/m^2$), although is rarely used above 0·5 in. w.g. ($125\ N/m^2$). It has, however, a capacity for very large air volumes and coupled with its economy finds itself in applications such as air cooled condensers, unit heaters, and room extract devices. It is rarely found in air distribution systems.

The axial flow fan can be operated at pressures compatible to air distribution systems, but it creates relatively high noise levels. Certain industrial applications can take advantage of its excellent performance of high air volume produced from a convenient and small casing; however, commercial applications invariably call for sound attenuation to be added to the system.

Two basic axial flow types are available, the tube-axial fan being the basic type, and the vane-axial having guide vanes before or after the

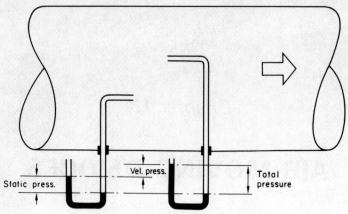

Fig. 13.1. Static and velocity pressures.

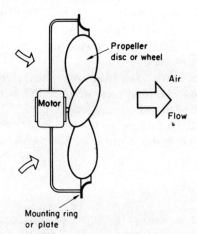

Fig. 13.2. Propeller fan.

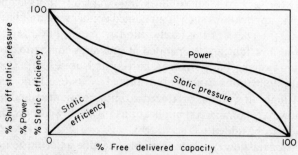

Fig. 13.3. Propeller fan characteristics.

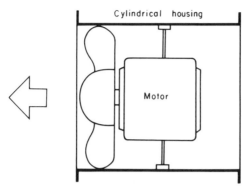

Fig. 13.4. Tube-axial fan.

fan wheel. The stationary aerofoil vanes recover a portion of the energy of the tangentially accelerated air and so make this type of fan more efficient in operation.

The most common fan used in air conditioning systems is the centrifugal fan, which can be categorised by the shape of its fan blades. Five basic arrangements are shown in Fig. 13.7. Of these types those in most frequent use are the forward curved, and backward curved, the latter using aerofoil blades when operating costs in large systems warrant their added capital cost.

The advantages of the forward curved fans are the ability to run at

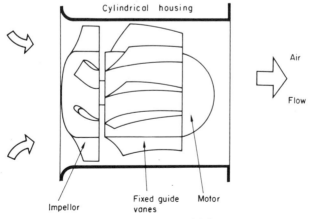

Fig. 13.5. Vane-axial fan.

228   APPLIED AIR CONDITIONING AND REFRIGERATION

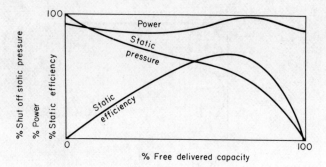

Fig. 13.6.   Axial-flow fan characteristics.

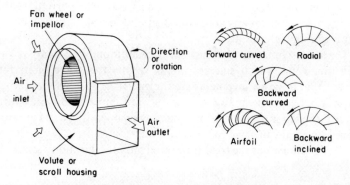

Fig. 13.7.   Centrifugal fan and fan blades.

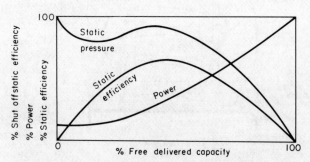

Fig. 13.8.   Forward curved centrifugal fan characteristics.

relatively low speeds compared with other types in order to achieve the same air volume, and the need for a smaller fan diameter for a given duty. Figure 13.8 shows how the power increases continually with increasing air quantity, whilst the static pressure falls from 100 per cent to zero with a characteristic dip. It can clearly be seen that pressures of about 85 per cent of maximum for a given fan speed can produce about 10, 30 and 50 per cent air volume, whilst the peak static

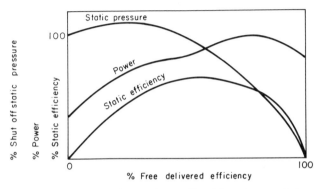

FIG. 13.9. Backward curved centrifugal fan characteristics.

efficiency occurs at about 40 per cent air volume. Fans operating below the 50 per cent free delivered capacity can be unstable and fans badly selected are known to change their characteristics; it is essential therefore that fans are selected in their stable range. Motors must be carefully selected when using this type of fan since a reduction in static pressure caused by the removal of some system resistance, sày filters, could cause overloading. It is normal practice to select fans with clean filters so that any build up of resistance will decrease the motor horsepower.

Figure 13.9 showing the backward curved fan characteristics indicates that the static pressure rises continuously without a dip in the curve, also the power curve peaks at a high capacity resulting in a maximum power demand for a given fan speed. This type of fan is said to have non-overloading characteristics and a motor can be selected for the maximum power requirements providing the fan speed is maintained.

Although the centrifugal fan has the desirable characteristics for

Fig. 13.10. Typical D.I.D.W. backward curved centrifugal fan. (Courtesy: Buffalo Forge Co.)

use in air distribution systems its arrangement is not as desirable as the axial flow fan whose air flow is in a continuous direction. The centrifugal fan requires the air that enters at the side to be propelled in a forward direction. This leads to complicated duct connections, circular at the side inlet and rectangular at the outlet. Further complications exist with double inlet fans. It is, therefore, common to locate fans in a chamber from where they can receive free air and be readily serviced. Other arrangements, where cabinets are used, are discussed later.

13 AIR MOVING DEVICES

The power output of a fan is termed the air horsepower (A_{hp}) or air power (A_p), and represents the actual work of the fan

$$A_{hp} = \frac{M_a P_t}{33\,000}$$

where: M_a = air mass (lb/h), P_t = total air pressure (ft hd).
Using common units this can be rearranged:

$$A_{hp} = 0{\cdot}000\,157 \times \text{cfm} \times p_t$$

where: p_t = total air pressure (in. w.g.),
or

$$A_p = M_a P_t$$

where: M_a = air mass (m^3/s), p_t = total air pressure (N/m^2), and A_p = power (W).

The actual power required at the fan shaft (S_{hp} or S_p) in relation to air power is the mechanical or total efficiency:

$$\text{total efficiency} = \frac{A_{hp}}{S_{hp}} \times 100$$

or

$$= \frac{A_p}{S_p} \times 100$$

The air power is a function of the fan total pressure, i.e. static pressure plus velocity pressure, since the latter plays a less important role in fan engineering, it is the static efficiency on which more emphasis is made

$$\text{static efficiency} = \frac{A_{hp}}{S_{hp}} \times \frac{p_s}{p_t} \times 100$$

where: p_s = fan static pressure
or

$$\text{static efficiency} = \frac{A_p}{S_p} \times \frac{p_s}{p_t} \times 100$$

Combining the above equations, the more commonly used forms are

$$\text{total efficiency} = \frac{\text{cfm} \times p_t}{6350 \times S_{hp}}$$

TABLE 13.1
FAN LAWS

Variable	Law	Formula
Fan speed	Capacity varies as the speed in same distribution system and fan size	$\dfrac{\mathrm{cfm}_1}{\mathrm{cfm}_2}$, or $\dfrac{m^3/s_1}{m^3/s_2} = \dfrac{\mathrm{rev/min}_1}{\mathrm{rev/min}_2}$
	Pressure varies as the square of the speed in same distribution system and fan size	$\dfrac{P_1}{P_2} = \left(\dfrac{\mathrm{rev/min}_1}{\mathrm{rev/min}_2}\right)^2$
	Power varies as the cube of the speed in same distribution system and fan size	$\dfrac{\mathrm{hp}_1}{\mathrm{hp}_2}$, or $\dfrac{W_1}{W_2} = \left(\dfrac{\mathrm{rev/min}_1}{\mathrm{rev/min}_2}\right)^3$
Fan size	Capacity varies as the cube of the size at constant speed	$\dfrac{\mathrm{cfm}_1}{\mathrm{cfm}_2}$, or $\dfrac{m^3/s_1}{m^3/s_2} = \left(\dfrac{D_1}{D_2}\right)^3$
	Pressure varies as the square of the size at constant speed	$\dfrac{P_1}{P_2} = \left(\dfrac{D_1}{D_2}\right)^2$
	Power varies as the fifth power of size at constant speed	$\dfrac{\mathrm{hp}_1}{\mathrm{hp}_2}$, or $\dfrac{W_1}{W_2} = \left(\dfrac{D_1}{D_2}\right)^5$
Fan size	Capacity varies as the square of the size at constant top speed	$\dfrac{\mathrm{cfm}_1}{\mathrm{cfm}_2}$, or $\dfrac{m^3/s_1}{m^3/s_2} = \left(\dfrac{D_1}{D_2}\right)^2$
	Power varies as the square of the size at constant top speed	$\dfrac{\mathrm{hp}_1}{\mathrm{hp}_2}$, or $\dfrac{W_1}{W_2} = \left(\dfrac{D_1}{D_2}\right)^2$
	Fan speed varies inversely as the fan size at constant top speed	$\dfrac{\mathrm{rev/min}_1}{\mathrm{rev/min}_2} = \dfrac{D_2}{D_1}$
	Pressure remains constant at constant top speed	$P_1 = P_2$

Note: For corrections in air density refer to Chapter 1.

13 AIR MOVING DEVICES

and

$$\text{static efficiency} = \frac{\text{cfm} \times p_s}{6350 \times S_{hp}}$$

or

$$\text{total efficiency} = \frac{m^3/s \times p_t}{S_p}$$

and

$$\text{static efficiency} = \frac{m^3/s \times p_s}{S_p}$$

Fan laws are used to predict the performance of all fans under changing operating conditions or fan size, where fans are geometrically similar. When changing fan sizes, however, it is recommendable to use fan data for that particular fan since the fan laws assume a constant efficiency. Table 13.1 lists these laws.

The most noticeable change in modern air conditioning design has been the advent of pre-engineered air handling equipment. The days are gone when the contractor or consultant designed his own system to meet the needs of his air conditioning plant. Not too many years ago it was commonplace to see a specially made filter bank, a transformation piece to a cooling coil, suitably raised to allow for condensate pans and drains, a transformation piece to a reheater coil and a further square to round transformation piece to a single inlet centrifugal fan, which by necessity meant a change in air direction, and a possibility of 'breaking the fan's back' to accommodate the plant within a limited space. All these sections meant that a great deal of time was being spent by a design engineer in the detail of each section, the matching of flanges on transformation pieces to match bought-out heat exchange equipment and fans, not to mention the special foundations for the fan motor and drive.

To select one piece of apparatus to meet all heat exchange requirements, and have prepared drawings for plant requiring minimal foundations together with a saving of plant room space makes this apparatus commonplace in today's plant rooms.

The main element of any air handling device must be the fans, which are cabinet mounted double inlet, double width centrifugal types, often with two or more wheels coupled on a common shaft. This configuration lends itself to a long, horizontal cross-section, which is also an economical shape for heat exchangers having horizontal tubes with few return bends.

Most commonly used is the forward curved fan, which has smaller and a greater number of blades per wheel than backward curved or airfoil fans. This results in slower rotative speeds and often smaller wheels giving quieter operation and smaller sizes, making it more adaptable for air handling units.

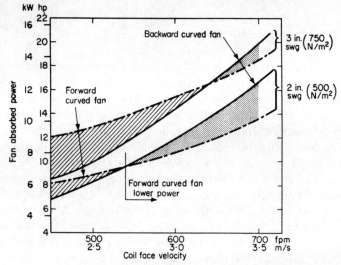

Fig. 13.11. Comparison of forward and backward curved types of fan showing power against air volume.

The backward curved fan has the advantage of non-overloading characteristics, that is at a given rotative speed despite the air volume, controlled from system resistance, the horsepower absorbed will reach a maximum. However, long before this maximum horsepower is reached the maximum efficiency is attained and for standard units enveloped in a cabinet this maximum is never reached because of the velocity limitations of heat exchangers. In addition, the efficiency of the backward curved fan peaks more sharply than the forward curved type, so that over the selection band the forward curved fan has probably a better characteristic. Figure 13.11 shows a comparison of the two types of fan showing power against air volume. As the static pressure of the fan increases so the forward curved fan tends to lose advantage. Where only one fan is available for selection in any given unit size it is common to design the fan for optimum efficiency at face velocities of 550 to 650 fpm (*2·8 to 3·4 m/s*) and static pressures of $2\frac{1}{2}$ to 3 in. w.g. (*625 to 750 N/m²*).

Today's practice dictates that above 5·5 in. w.g. (*1375 N/m²*) the backward curved fans are more efficient. Therefore, when using forward curved fans the higher the external resistance the higher the

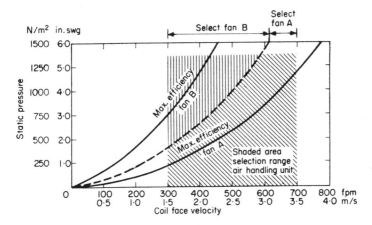

FIG. 13.12. Fan curves for plant with optional fan wheels.

face velocity of the heat exchange coil should be, up to a maximum of 5·5 in. w.g. (*1375 N/m²*) above which backward curved blades should be used, and units are available to overcome 8 to 9 in. w.g. (*2000 to 2250 N/m²*) using the latter type.

Some manufacturers, in addition to making standard cross-sections to meet the economics of coil and fan cabinet design, make available alternative fan wheels within the standard cabinet so that the engineer can take advantage of optimum fan performance for his particular requirements. Figure 13.12 shows a typical fan curve for an air handling plant having two optional fan wheels.

The cabinet fan design lends itself to flange mounted bearings which are of the self-aligning ball or roller type, which are simple to lubricate and easily interchanged in the event of failure. One bearing is capped, the other allowing the fan shaft to pass through for the fan drive. It is common to use tubular shafts which are lighter and stronger than their solid counterparts, the ends being swaged and plugged to suit standard bearings.

Fan motors up to 50 hp ($35\,kW$) can be mounted on the outside of the cabinet on adjustable bases for belt alignment. It is common practice to use variable pitch vee belt drives which allow accurate setting of the fan speed to meet air side requirements. This feature eliminates the need for a volume regulating damper in the main duct

FIG. 13.13. Central station air handling plant. (Courtesy: Carlyle Air Conditioning Co. Ltd.)

branch, which not only is subject to alteration, but adds resistance to the system, increasing absorbed power, and because the fan needs to run at a higher speed, and the damper could create air turbulence, a greater system noise is created.

The air volume handled is basically controlled by the face velocity of the air passing over heat exchange surfaces. The main factor is that a cooling coil, which by definition also dehumidifies, will not be able to hold any condensation on its fins by surface tension above a certain air velocity, usually about 500 fpm ($2 \cdot 5\,m/s$) depending on the coil geometry, and a carry-over of moisture can exist. In packaged air conditioning equipment these velocities are never exceeded, and even if the fan speed is increased to give greater air volume, the balanced refrigeration plant will mean less cooling per unit air volume, resulting in sensible cooling in most cases.

When made-up systems and coils in ducts are used, carry-over could not be tolerated because of leakages in sheet-metal ducts, etc.,

13 AIR MOVING DEVICES

and even today, because of past bitter experience, most consultants stipulate a maximum velocity of 500 fpm (2·5 m/s) unless high pressure loss eliminators are used. However, with standard air handling plants this can be overcome by the use of an extended condensate collector which allows the moisture to be carried over and collected. The judicious installation of a baffle ensures that any moisture from the top of the cooling coil is kept within the confines of the condensate pan.

It is, therefore, not only economical in first cost but also in efficiency to run an air handling plant at its maximum velocity.

Filter sections as standard are normally of vee-bank design so that an extended area is presented to the air flow to give low velocity and high efficiency. This type of bank, using standard size filter cells, means that a wide variety of media of the throwaway, cleanable and viscous types can be used. Roll type filters, and/or electronic filters can also be used, although these often require a transformation piece for connection to the air handling plant if the unit manufacturer does not also make these types of filters.

The vee-bank filter section can also be accommodated in a combination mixing box which serves as an economiser for mixing fresh and return air at mid-seasonal conditions.

Because a unit is of standard size the face area of a cooling coil is fixed, which means fixed performance. However, this gives one great advantage: standardised and performance tested equipment. By using standard coil geometry, normally units offer as many as six basic configurations, and the designer can select which coil suits his application, with the satisfaction that the ratings are based on practice and not tailor made with any in-built service factors to guarantee performance. Adding to the six basic coils (normally four, six or eight rows with two fin-spacing options) two methods of water circuiting or direct expansion, copper or aluminium fins on copper tube, the engineer can meet practically all air side requirements.

In practice, the fixed geometry coil means that design off and on coil conditions with a given air volume cannot always be met. However, the control of water quantities and temperature or refrigerant evaporating temperature will mean control of the apparatus dew point and the cooling coil performance can be met by adjusting the air volume. This adjustment of air volume often means better economics since an original concept of off-coil temperatures could result in a deep coil, which is not necessarily required to do a great deal of heat transfer, and by increasing the air volume by 5 or 10 per cent to cater for a high off coil condition, the number of rows could even be halved. Only tabulated selection data can show this easily.

Because the air handling plant itself acts as a support for all heat exchange elements, transformation of shape can be achieved by baffles. It is not necessary to install heating coils with such large surface areas as cooling coils since the effective temperature between heating medium and air is many times larger, nor is it necessary to use low face velocities to avoid carry-over. Consequently, heating coils are installed of the same length but shallower in depth and of less rows. Maintaining the length of coil keeps the coil economical with few return bends. Velocities through heating coils are designed for use between 500 and 1100 fpm (*2·5 to 5·5 m/s*). The velocity head change from say 700 fpm (*3·5 m/s*) at the cooling coil to 1100 fpm (*5·5 m/s*) would only constitute a pressure loss of approximately 0·01 in. w.g. (*2·5 N/m²*).

Two basic heating coil types are used, either the 'U'-bend type for steam or hot water, and the non-freeze or floating steam distribution tube-in-tube type. They are normally one or two rows deep with fin spacings from 3 to 14 per inch (*120 to 550 per metre*) in either copper or aluminium.

The availability of such a wide range of heat transfer coils does not stop here since unit arrangements make it possible to accomplish the heating requirement in four ways, using a preheat coil, a reheat coil, a combination of preheat and reheat coils and the use of the cooling coil instead of chilled water with hot water.

All heat exchangers are normally slotted into channel sections within the air handling plant which act as supports for the coils and baffles, so that the air cannot by-pass the coil via the return bends. This construction gives two basic advantages in that the coils can be easily withdrawn for repair or cleaning and all return bends are within the casing so that any condensate drops to the collection tray.

It is normal practice to insulate all the casing down-stream from the first heat exchanger, which also includes the fan section. Most important is the condensate collector where the base must be insulated from the cold condensate. Any external condensation under the unit would not be seen unless the unit was suspended, and could cause unseen damage.

Standard unit construction lends itself to the adaptation of face and by-pass dampers for cooling coil control or even a fixed by-pass when the design temperature difference between off-coil and room temperature cannot be tolerated.

Humidification is a simple matter with spray coil, atomising spray, or steam grid type humidifiers. The extended drip tray allows sprays to be directed to the heating coil for optimum effect without danger of condensation or water leakage.

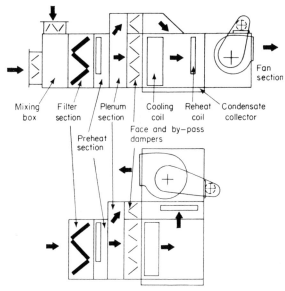

FIG. 13.14. Arrangements of horizontal and vertical draw-through air handling plants.

Figure 13.14 shows the assembly of all possible combinations of draw-through air handling plant in its two basic forms, horizontal and vertical arrangements. If the engineer considers that he can select coil connections and fan drives of either hand, as many as four fan discharge positions and either a vertical or horizontal arrangement, most physical requirements can be met.

Figure 13.15 shows the blow-through or multizone air handling plant, which has the same physical features as the draw-through unit. The addition of zoning dampers at the outlet side of the unit permits the selection of either cooling or heating to any given zone, such that one zone could have full cooling, one full heating and another a

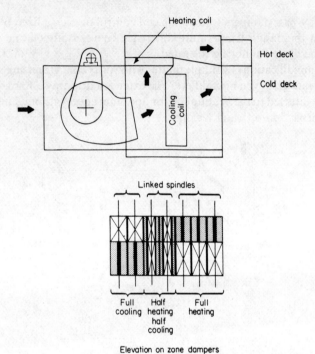

Fig. 13.15. Blow-through or multizone air handling plant.

mixture condition. The dampers would be controlled from a zone thermostat to give a desired mix to suit exact room requirements. The number of damper sections is generally limited to 8 in pitch, although zones can be further broken down by splitting these dampers into two halves.

For example, a small multizone unit would probably be provided with seven standard sections. If the design requirements were for three zones, one of 45 per cent capacity at full load, one at 35 per cent and one at 20 per cent, this could be achieved by splitting one damper section so that three sections satisfied the 45 per cent zone, $2\frac{1}{2}$ sections the 35 per cent load and $1\frac{1}{2}$ sections the 20 per cent load. Further themes can be worked by isolating a section. It is often good practice when balancing such a system to provide a volume damper in any leg very much out of balance, although the action of the dampers themselves will achieve this, but at the expense of tempering cooled air. Where zones call for more than one damper section to provide the

13 AIR MOVING DEVICES

necessary air volume these sections are linked so that they can be controlled from one motor.

When designing with multizone systems it should be remembered that dampers cannot be fully closed and there will always be about 5 per cent leakage through a closed damper. Therefore, the engineer should calculate his air side requirements across the cooling coil and allow the fan and sheet-metal ducting to handle 5 per cent in excess of his cooling requirements.

An advantage of multizone systems is that fresh air can be used for heating in summer and cooling in winter.

The elimination of zone dampers makes the blow-through unit suitable for a double duct installation. As with the draw-through units, a blow-through unit uses forward curved fans up to 5·5 in. w.g. ($1375 \ N/m^2$) and backward curved fans up to 9 in. w.g. ($2250 \ N/m^2$).

Chapter 14

TERMINAL DEVICES AND AIR CONDITIONING SYSTEMS

The performance of any large air conditioning system is dependent upon how its terminal devices, together with a primary plant, are able to meet the needs of the occupied zone and its behaviour at partial load. As previously discussed, zoning plays an important part in influencing the designer in selecting the correct system for the application.

Direct Expansion Systems are discussed in Chapter 16 and All-Air Systems in Chapter 7. The other major categories into which equipment and systems fall are Air and Water Systems and All-Water Systems. Both these arrangements require central water chilling and boiler plant with or without primary air stations serving various zones.

INDUCTION UNIT SYSTEMS

The need for individual room, or module, temperature control with widely fluctuating room gains is met by the induction unit system. Normally located below windows at the perimeter of a building the induction unit is fed from a primary air plant which provides adequate air for ventilation purposes at a dew point capable of absorbing the room latent heat gains and capable of providing the motivation for induced room air which passes over a secondary cooling coil controlled from the room, to cater for additional sensible gains not met by the primary air.

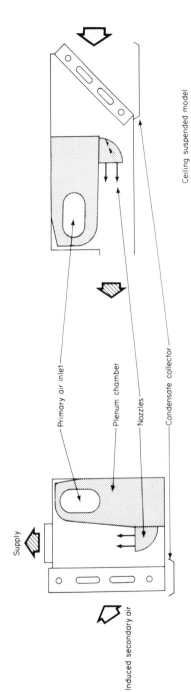

Fig. 14.1. Cross-section of typical induction units (water-control). (Courtesy: Carlyle Air Conditioning Co. Ltd.)

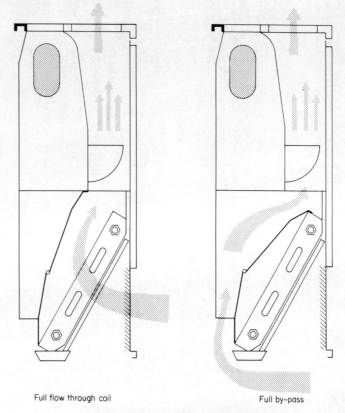

Full flow through coil   Full by-pass

FIG. 14.2. Cross-section through by-pass air controlled induction units. (Courtesy: Carlyle Air Conditioning Co. Ltd.)

The centralised primary air plant provides filtered and dehumidified air to each unit through high velocity sheet-metal ductwork which requires little space compared to conventional all-air low velocity systems. The quantity of air to each unit is governed as a minimum to serve the ventilation requirements of the area being served by one unit. However, this quantity is often exceeded since this primary air must also (in non-changeover systems) provide the heating to offset transmission losses in winter. On larger projects the induction system would normally be broken down into various zones, usually a zone for each building vertical face. When each module is considered the transmission gains or losses due to the outside ambient

14 TERMINAL DEVICES AND AIR CONDITIONING SYSTEMS 245

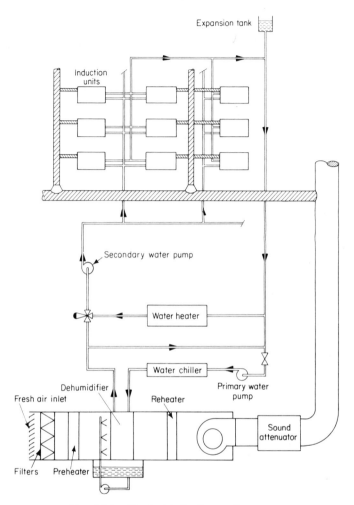

FIG. 14.3. Typical induction unit system.

temperature can be expressed as a ratio of the primary air volume to transmission loss. By considering the module with the lowest transmission gain or loss per degree temperature difference room to outside, the air volumes to other modules are adjusted so that they have the same ratio. This means that the primary air can be introduced into the various modules at a scheduled temperature related to outside temperature such that at all times, particularly during the heating season, the transmission losses or gains are catered for, and should there be no other factor influencing the room load, with the secondary coils off, constant room temperatures can be maintained from a single control in the plant room. Thus with a stabilised condition other room gains such as lighting, occupancy, solar gains, etc., are catered for individually by the secondary coil which always has chilled water circulating through it.

This method is found to be adequate and economical for climates where the design temperature does not fall below $20\,°F$ ($-6\cdot7\,°C$). (This temperature is, however, affected by the construction of the conditioned building.) For temperatures below this a changeover system can be considered whereby the secondary coils are used for heating and primary air is used for the additional cooling for lights, solar gain and occupants. The changeover point is quite critical and is evaluated at an ambient temperature where the transmission loss is greater than the summation of any solar gain, lighting gain and occupancy gain. The additional capital cost of changeover thermostats and plant room changeover requirements of extra pipework and valves must be carefully weighed against the saving in running costs for the duration of operation below the changeover point where the cooling is provided free by the primary air.

The primary air plant is sized to meet the summation of the air volumes of all induction units it serves plus an allowance for duct leakage (about 10 per cent). This plant would normally consist of a filter, preheater, sprayed cooling coil section to provide constant summer and winter dew points and a reheater section. The preheater would be controlled in association with the sprayed coil plant to provide a constant dew point. The reheater would be scheduled to the outside air temperature so that transmission losses are catered for at all times. The air moving device would have to operate at high static pressures to overcome the resistance of the primary air plant components, the high velocity ductwork and the pressure drop through the induction units, and would normally be of the backward curved aerofoil blade type.

Induction units are available in many cross-sectional shapes to meet architectural requirements, and for horizontal or vertical application. They all rely on the same principle, whereby air at high velocity will induce low velocity or still air into its path, the higher the air velocity the greater the inducement or induction ratio. The primary air to the induction unit is controlled at the plant room, room

14 TERMINAL DEVICES AND AIR CONDITIONING SYSTEMS 247

fluctuations are controlled by the secondary coil. This is achieved in two ways, either by adjusting the water flow through the coil or series of coils, or by-passing the air around the coil by use of a damper system actuated by the primary air pressure itself. This by-pass system, whilst more costly that the water control system can prove more economical when individual control to each induction unit is desired where the cost of many water control valves, thermostats and associated electrical wiring is considered. Since the secondary coil is providing only sensible cooling, both the water control and by-pass control provide similar results.

A condensate pan is normally provided with each induction unit which serves to catch any moisture which may collect during start-up when the room conditions could have risen. However, these are not normally connected to a drain since this small amount of moisture is allowed to re-evaporate during the normal cycle of operation. The provision of dehumidification from the primary air plant saves the cost and maintenance of many condensate drains.

The water chilling plant would be sized to meet the cooling and dehumidifying requirements of the primary air plants plus the diversified summation of the secondary cooling coil loads. Because the secondary coils are only required to achieve sensible cooling it is necessary to provide water at a temperature above the room dew point. This means that the chilled water would normally pass through the primary air plant at low temperature and could then be used for the secondary coil circuit. The large temperature difference on the chilled water system makes running costs slightly lower than if the secondary water was constantly by-passed.

The sizing of induction units is governed by the temperature difference between room air and entering chilled water temperature, the flow of chilled water (this is normally a given quantity evaluated to give most economical operation, and a fairly high velocity which keeps the water system free from blockage), the amount of primary air handled which will affect the induction ratio and the credit of the sensible cooling effect of the primary air itself.

Induction systems are often provided with a changeover system whereby hot water may be passed around the secondary circuit during the night, with the primary air system off, so as to provide convection heating at mid-seasonal or winter conditions.

The sizing of air conditioning apparatus can best be illustrated by considering a cross-section through a multistorey building (see Fig.

14.4) which for the example is considered as south exposure. The example assumes ambient design conditions of 85°F DB, 70°F WB in summer and 28°F DB in winter, and the respective room conditions to be held are 73°F DB, 50 per cent RH and 70°F. Conventional load estimates are made for nett room sensible heat gains and nett room transmission losses. The summer load estimate has been taken at 2 p.m., August, and windows are single glazed with light-coloured venetian blinds. Transmission coefficients for all building surfaces are 0·2 and 1·13 for glazing.

Modules type A

	Summer		Winter
Solar gain through glass			
$24\,\text{ft}^2 \times 71$	1704		—
Transmission through glass			
$24\,\text{ft}^2 \times 12 \times 1\cdot13$	326	$24\,\text{ft}^2 \times 42 \times 1\cdot13$	1139
Solar and transmission through walls			
$26\,\text{ft}^2 \times 20 \times 0\cdot20$	104	$26\,\text{ft}^2 \times 42 \times 0\cdot2$	218
Lights: $4\,\text{W}/\text{ft}^2$ fluorescent			
$75\,\text{ft}^2 \times 4 \times 4\cdot25$	1275		—
Occupants: 1 per $75\,\text{ft}^2$			
1×260	260		—
Nett room loads (BTU/h)	3669		1457

Modules type B

Nett room load module A	3669		1457
Solar and transmission gain through roof			
$75\,\text{ft}^2 \times 29 \times 0\cdot2$	435	$75\,\text{ft}^2 \times 42 \times 0\cdot2$	840
Nett room loads (BTU/h)	4094		2297

Modules type C

Nett room load module A	3669		1457
Transmission gain through floor			
$75\,\text{ft}^2 \times 12 \times 0\cdot2$	180	$75\,\text{ft}^2 \times 20 \times 0\cdot2$	300
Nett room loads (BTU/h)	3849		1757

It is assumed that the ventilation requirements are 30 cfm/person, i.e. 30 cfm/module. The latent gain to the room is that from one

14 TERMINAL DEVICES AND AIR CONDITIONING SYSTEMS

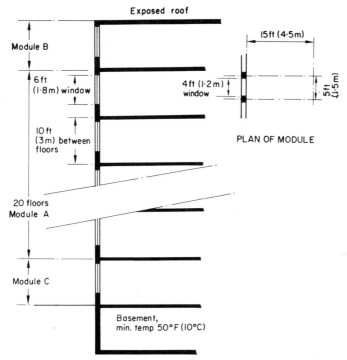

FIG. 14.4. Details of modules considered in induction system sizing example.

occupant, i.e. 190 BTU/h. At this stage the dew point of the primary air can be established, where:

$$W_{sa} = W_{rm} - \left(\frac{\text{room latent gain BTU/h}}{\text{vent. cfm} \times 0.68}\right)$$

W_{sa} = moisture content of supply air (grs/lb)
W_{rm} = moisture content of room (grs/lb)

From the psychrometric chart W_{rm} at 73 °F DB, 50 per cent RH can be found as 60·5 grs/lb, i.e.

$$W_{sa} = 60\cdot5 - \left(\frac{190}{30 \times 0\cdot68}\right)$$
$$= 51\cdot2 \text{ grs/lb}$$

From the psychrometric chart the dew point can be read as 49 °F. As stated previously, during winter operation the primary air must be capable of removing the nett room loss, and be controlled from the plant room as a function of ambient dry bulb temperature. It can be seen from the various modules that the top and bottom floors will require more air at a given temperature than the intermediate floors. These air quantities are a direct proportion to the ventilation air required at the intermediate zones to the ratio of heat losses.

$$\text{Module B, primary air} = 30 \times \frac{2297}{1457} = 47 \text{ cfm}$$

$$\text{Module C, primary air} = 30 \times \frac{1757}{1457} = 36 \text{ cfm}$$

The maximum primary air temperature during winter heating would be:

$$T_{pa} = T_{rm} + \frac{\text{nett heat losses (BTU/h)}}{\text{primary air (cfm)} \times 1 \cdot 08}$$

$$= 70 + \frac{1457}{30 \times 1 \cdot 08} = 115\,°F$$

The example assumes ambient design conditions of 29·4°C DB, 21·1°C WB in summer and −2·2°C DB in winter, and the respective room conditions to be held are 23°C DB, 50 per cent RH and 21°C. Conventional Load Estimates are made for nett room sensible heat gains and nett room transmission losses. The summer load estimate has been taken at 2 p.m., August, and windows are single glazed with light-coloured venetian blinds. Transmission coefficients for all building surfaces are 1·14 and 6·42 for glazing.

Modules type A

	Summer		Winter
Solar gain through glass			
$2 \cdot 16\,m^2 \times 223$	482		—
Transmission through glass			
$2 \cdot 16\,m^2 \times 6 \cdot 4 \times 6 \cdot 42$	89	$2 \cdot 16\,m^2 \times 23 \cdot 2 \times 6 \cdot 42$	322
Solar and transmission through walls			
$2 \cdot 34\,m^2 \times 10 \cdot 6 \times 1 \cdot 14$	28	$2 \cdot 34\,m^2 \times 23 \cdot 2 \times 1 \cdot 14$	62
Lights: 43 W/m² fluorescent			
$6 \cdot 75\,m^2 \times 43 \times 1 \cdot 25$	363		—
Occupants: 1 per module			
1×76	76		—
Nett room loads (W)	1038		384

Modules type B
Nett room load module A   1038   384
Solar and transmission gain through roof
     $6.75 m^2 \times 15.6 \times 1.14$ 120 $6.75 m^2 \times 23.2 \times 1.14$ 179

     Nett room loads (W)   1158   563

Modules type C
Nett room load module A   1038   384
Transmission gain through floor
     $6.75 m^2 \times 6.4 \times 1.14$ 49 $6.75 m^2 \times 11 \times 1.14$ 85

     Nett room loads (W)   1083   469

It is assumed that the ventilation requirements are 14 litres/s/person, i.e. 14 litres/s/module. The latent gain to the room is that from one occupant, i.e. 56 W. At this stage the dew point of the primary air can be established, where:

$$W_{sa} = W_{rm} - \left(\frac{\text{room latent heat gain W}}{\text{vent. air litres/s} \times 2.98}\right)$$

W_{sa} = moisture content of supply air (grs/kg)
W_{rm} = moisture content of room (grs/kg)

From the psychrometric chart W_{rm} at 23°C DB, 50 per cent RH can be found as 8·8 grs/kg, i.e.

$$W_{sa} = 8.8 - \left(\frac{56}{14 \times 2.98}\right)$$
$$= 7.5 \, grs/kg$$

From the psychrometric chart the dew point can be read as 9·7°C.

As stated previously, during winter operation the primary air must be capable of removing the nett room loss, and be controlled from the plant room as a function of ambient dry bulb temperature. It can be seen from the various modules that the top and bottom floors will require more air at a given temperature than the intermediate floors. These air quantities are a direct proportion to the ventilation air required at the intermediate zones to the ratio of heat losses.

$$\text{Module B, primary air} = 14 \times \frac{563}{384} = 20 \, litres/s$$

$$\text{Module C, primary air} = 14 \times \frac{469}{384} = 17 \, litres/s$$

The maximum primary air temperature during winter heating would be:

$$T_{pa} = T_{rm} + \frac{nett\ heat\ losses\ (W)}{primary\ air\ (litres/s) \times 1{\cdot}21}$$

$$= 21 + \frac{384}{14 \times 1{\cdot}21} = 43{\cdot}6\,°C$$

At this stage this temperature must be considered against the maximum temperature available from the reheater coil, and the maximum temperature at which the fan bearings can operate. Should there be a limitation on the supply air temperature then the air quantities must be increased and proportioned accordingly. It is fair to say that temperatures below 120°F (49°C) should not cause problems.

The air volume of the primary air plant can now be considered as:

20 × Module A at 30 cfm = 600   at 14 litres/s = 280
1 × Module B at 47 cfm = 47   at 20 litres/s = 20
1 × Module C at 36 cfm = 36   at 17 litres/s = 17
                         ———   ———
                         683 cfm   317 litres/s

It should be stressed that only one section of the building has been considered and that this value would be multiplied by the number of modules across the width of the south exposure.

Having established the primary air volume the apparatus can now be selected having an on-coil condition of the design ambient condition, 85°F DB, 70°F WB (29·4°C DB, 21·1°C WB), and a leaving air dew point of 49°F (9·7°C for the alternative example). Using an eight-row 14 fins per inch cooling coil the actual off-coil condition is determined as 49·6°F DB, 49·2°F WB (10·0°C DB, 9·8°C WB, for the alternative example).

Associated with induction unit primary air plants is the pressure loss of filters, heat exchangers, humidification plant, high velocity ducting and the induction units themselves; altogether this total pressure would mean between 5 in. w.g. and 12 in. w.g. at the fan. The resultant fan heat gain can be calculated from the fan horsepower and related as a sensible heat gain in the form of a temperature addition. The example chosen assumes a fan gain equivalent to 4·4°F (2·4°C) and a duct gain of 1°F (0·6°C). The effective primary air temperature

14 TERMINAL DEVICES AND AIR CONDITIONING SYSTEMS 253

to the induction units would result as $49 \cdot 6°F + 4 \cdot 4°F + 1°F$, i.e. $55°F$ (or $10 \cdot 0°C + 2 \cdot 4°C + 0 \cdot 6°C$, i.e. $13°C$).

In addition to the fan motor and duct gain, high velocity systems must associate themselves with ductwork leakage losses and these are taken to be 10 per cent. Therefore, the primary air plant must handle 683 cfm + 10 per cent, or some 750 cfm (*317 litres/s + 10 per cent, or some 349 litres/s*).

The total heat from the primary air can now be established as:

Q_{pa} = Primary air (cfm) × 4·45 × ($H_{ea} - H_{la}$)
H_{ea} = Enthalpy at 85°F DB, 70°F WB, 33·96 BTU/h
H_{la} = Enthalpy at 49·6°F DB, 49·2°F WB, 19·34 BTU/h

i.e.

Q_{pa} = 750 × 4·45 × (33·96 − 19·34)
= 48 900 BTU/h

The total cooling requirements for the secondary coils can be reduced by the cooling effect of the primary air entering at 55°F, i.e.

Module A = 3669 − [1·08 (73 − 55) 30] cfm
= 3175 BTU/h
Module B = 4094 − [1·08 (73 − 55) 47] cfm
= 3179 BTU/h
Module C = 3849 − [1·08 (73 − 55) 36] cfm
= 3149 BTU/h

The refrigeration plant nett capacity can now be established as:

Primary air plant = 48 900
20 Module A at 3175 BTU/h = 63 500
1 Module B = 3 179
1 Module C = 3 149
_____
118 728 BTU/h

The total heat from the primary air can now be established as:

Q_{pa} = *Primary air (litres/s)* × *1·19* × ($H_{ea} - H_{la}$)
H_{ea} = *Enthalpy at 29·4°C DB, 21·2°C WB, 61·06 kJ/kg*
H_{la} = *Enthalpy at 10·0°C DB, 9·8°C WB, 28·88 kJ/kg*

i.e.

Q_{pa} = *349* × *1·19* × *(61·06 − 28·88)*
= *13 365 W*

The total cooling requirements for the secondary coils can be reduced by the cooling effect of the primary air entering at 13°C, i.e.

$$\text{Module } A = 1038 - [1 \cdot 21\,(23 - 13)\,14] \text{ litres/s}$$
$$= 869\,W$$
$$\text{Module } B = 1158 - [1 \cdot 21\,(23 - 13)\,20] \text{ litres/s}$$
$$= 916\,W$$
$$\text{Module } C = 1083 - [1 \cdot 21\,(23 - 13)\,17] \text{ litres/s}$$
$$= 877\,W$$

The refrigeration plant nett capacity can now be established as:

Primary air plant =	13 365
20 Module A at 869 W =	17 380
1 Module B =	916
1 Module C =	877
	32 538 W

It should be noted here that the refrigeration load was taken for a south exposure at 2 p.m. Other exposures must also be considered and once the time of day and magnitude of their maximum gain are established, the summation of all the secondary coil duties at each selected time of day, together with the primary air load for each selected time of day (note the reduction in ambient temperatures for times other than 3 p.m.) must be considered and the maximum taken as the installed nett refrigeration load. The maximum load should not be computed as the summation of the maximum exposure loads for different times of day.

Once the nett refrigeration effect is established allowance must be made for pump gains and pipe losses in order to be able to select the most suitable refrigeration machine.

Figure 14.3 shows the piping arrangement to the primary air plant and the secondary coils. The secondary water must be circulated at a temperature above room dew point to avoid any dehumidification at the secondary coils. It is normal practice to allow all the water leaving the chiller to enter the primary air cooling coil, this water quantity being in excess or equal to the water quantity to be handled by the secondary pump. Using a cooling coil of known capacity will establish for the chosen water quantity the leaving chiller temperature, or entering coil temperature, and the leaving coil temperature, the leaving coil temperature being the minimum which can be

14 TERMINAL DEVICES AND AIR CONDITIONING SYSTEMS

accepted at the secondary coils. As the refrigeration requirements of the primary coil are reduced so the three way control valve to the secondary water circuit allows primary chilled water to by-pass into the primary water pump, and leaving secondary water to mix with the primary water to maintain a constant temperature to the secondary coils.

INDUCTION REHEAT SYSTEMS

The system provides the same features as a normal induction system except that cooling is provided by the primary air system and the secondary coils are used for heating throughout the year. The primary air system can be either high or low velocity since the induction effect is not normally required to be high because of the larger air volumes

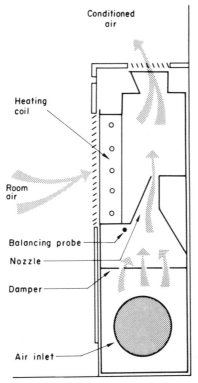

FIG. 14.5. Cross-section through induction reheat terminal. (Courtesy: Carlyle Air Conditioning Co. Ltd.)

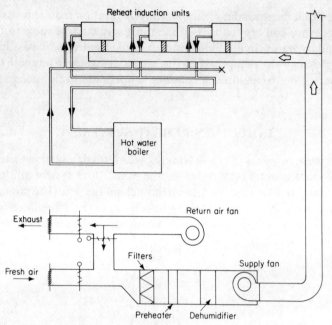

FIG. 14.6. Induction reheat system.

required for cooling purposes. Because of the larger air volumes a return air system would normally be incorporated to save the running costs of cooling all the primary air and only the air needed for ventilation purposes need be introduced as fresh air.

As with the induction system this is quiet in operation since all moving parts are located in the plant room and it is very flexible as it can provide for the needs of areas or zones having very different load conditions.

Because of the large amount of primary air this system is suited to applications having high latent gains which can be met at all times, fluctuations in room sensible heats being catered for by the secondary coils.

VARIABLE AIR VOLUME SYSTEMS

The air used in a variable air volume (VAV) system is cooled and dehumidified in a central station air handling apparatus, and is

14 TERMINAL DEVICES AND AIR CONDITIONING SYSTEMS

introduced to the conditioned area via terminals having the capacity to vary the quantity of supply air at the dictate of a room thermostat sensing changes in room load. Depending on the central station plant type, in winter and intermediate seasons, operation of the refrigeration system may be discontinued, providing the cold air supply temperature is maintained by control of the mixture of outdoor and return air.

Each VAV terminal meters the supply of cold air to the space to match or balance the cooling load of the space. It is therefore ideal for use with interior zones of a building or areas requiring cooling at all times of the year. Since the supply air quantity is directly proportional to the actual cooling load, operating costs of fans and refrigeration equipment are kept to a minimum, and the addition of heat to overcome excess cooling capacity is not required.

The successful application of a VAV system depends upon the use of suitable air distribution terminals that are capable of maintaining adequate air distribution over a wide range of discharge air quantities. This demands the use of a unit functionally designed for this service—one that has a high aspect ratio, that maintains high induction rates, and ensures rapid air mixing and good air circulation over the full range of modulating air quantities. Ordinary diffusers and side wall grilles are normally incapable of operating satisfactorily over the wide range in volume and with varying pressures required with most systems of this type.

In the interior of a building the air conditioning load consists of heat from occupants, lights and miscellaneous equipment. This is a consistent year-round requirement, independent of ambient temperature, always assuming that perimeter areas are maintained at the correct room temperature. As shown in Fig. 14.7, the interior cooling load varies only in magnitude depending upon occupancy and use. Therefore the variation of supply air quantity of cooled and dehumidified air will satisfy changes in load without the need to heat such air at partial loads. There could be a demand for some heating, however, on interior areas having an exposed roof. This transmission heat loss is rarely equivalent to the heat gain from lights and people; however some heating may be required to maintain correct room temperatures during prolonged unoccupied periods of plant shutdown at low ambient temperatures.

The effect of perimeter transmission gains or losses, and solar gain, gives a load pattern for exterior areas which differs from interior

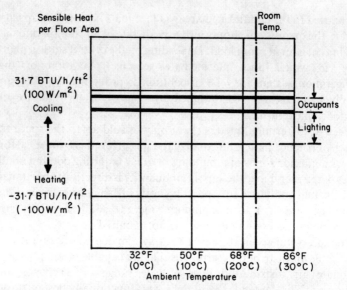

FIG. 14.7. Typical load profile—Interior zone air conditioning system.

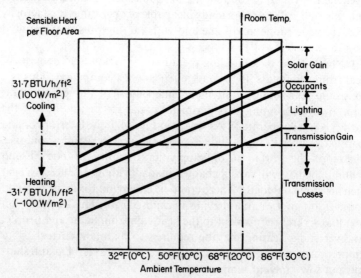

FIG. 14.8. Typical load profile—Perimeter zone office air conditioning system.

14 TERMINAL DEVICES AND AIR CONDITIONING SYSTEMS

areas. The transmission load is independent from the gain from lights, people and sun, and varies directly with the difference between outdoor and indoor temperature for any given area. Figure 14.8 shows how the ambient to room temperature difference has an effect on the transmission gain or loss in terms of room sensible heat. The addition of solar gain, lights and occupants could cause a net room cooling requirement even at ambient temperatures below normal design considerations. However, any air conditioning system should have the capacity to meet such a changing load which at any given ambient condition can vary between a heating and cooling requirement depending upon the variables of solar gain, lights and occupants.

Since variable air volume systems by concept are cooling only systems, which are served from the ceiling, they can adequately cope with the variables of solar gain, lights and occupants. However they do not provide the facility for simultaneously providing cooling and heating. In addition, a transmission loss at a perimeter wall would cause a down-draught which, even if there was a net cooling load on the space, would cause cold areas at floor level, the down-draught effect of the transmission loss being encouraged by the ceiling distribution system.

It is therefore necessary to provide an alternative source of heating at the perimeter, such that proper room temperature exists through the occupied space. Such heating can take various forms; it can be a wet-heat perimeter system, or a constant volume, variable temperature air system served to the room from below the window or in the ceiling adjacent to the perimeter. In any event, the heating system should be capable of offsetting the transmission losses for each perimeter area, and would normally be controlled on a schedule such that the heating system operates proportionally to the difference between room and ambient temperature. Figure 14.9 illustrates the heating requirement at the perimeter in conjunction with the VAV system. Providing the central plant is adequately designed, the cooling effect of the VAV system to satisfy the variable room gains will need no mechanical cooling since at low ambient conditions, free cooling can be achieved with the correct mix of outdoor air.

To maximise on economy of plant operation, the all-air VAV system should consist of a central station air handling apparatus with economiser dampers, a return air/exhaust fan, a medium pressure supply fan having some constant pressure device, a filtration and a

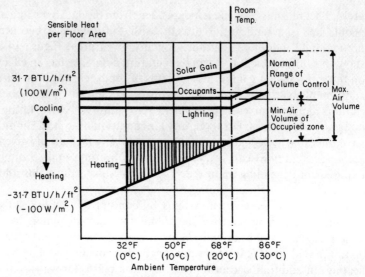

FIG. 14.9. Typical load profile—VAV system with wet perimeter heating.

cooling system. In addition, a medium or low pressure air distribution ductwork system is necessary to connect to the air terminal devices.

The central station apparatus would normally be located in a separate plant room, or, as is becoming more common, located on the roof within its own weatherproofed enclosure.

Since the VAV system caters for the actual room gain and no more, it takes maximum advantage of building diversity. Therefore, the larger the area a central station plant can cover, the greater the advantages of diversity. Where large buildings are concerned, calling for more than one plant because of the size of distribution ductwork, it is more favourable to distribute the air on a horizontal rather than a vertical basis. With other large air conditioning systems it has been common practice to treat the building by zoning, based on building orientation. This logic does not, however, take into consideration the shifting solar gain in terms of fan power used. Since over a year's operation it is the supply air fan which demands more energy than any other component, and a VAV system is capable of reducing this load, it becomes apparent that treating all orientations from one plant will cause the greatest diversity and thus the maximum energy saving. Figure 14.10 shows a typical central station system suitable for variable air volume application. The return and supply air fans are

14 TERMINAL DEVICES AND AIR CONDITIONING SYSTEMS 261

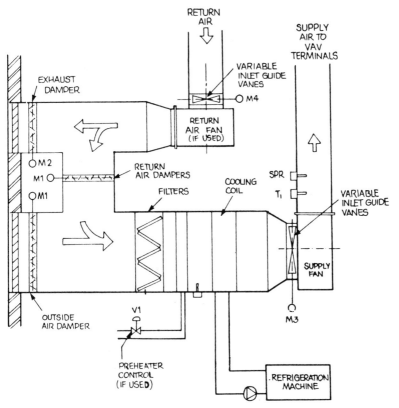

FIG. 14.10. Components of central station apparatus for VAV plant.

shown controlled by variable inlet guide vanes, which operate in sequence at the dictate of a static pressure controller in the discharge duct. Return air, exhaust and outside air dampers are controlled in sequence to provide maximum free cooling whenever the ambient temperature is lower than return air conditions, and to maintain constant dewpoint temperature by mixing return and outdoor air without the use of mechanical refrigeration when the outdoor air is below the desired dewpoint setting. Mechanical refrigeration will be called for to control the constant dewpoint temperature whenever free cooling cannot satisfy demand. An analysis of the system would dictate whether a preheater is required for the minimum fresh air setting to provide adequate room ventilation. To save energy, the VAV plant would not normally be switched on until a building is occupied, the separate heating system catering for the warm-up period, without the added load of heating ventilation air and the energy requirement of running the supply and exhaust fans. Under these circumstances the designer should evaluate whether the minimum room gains of lighting would compensate for the cooling effect of minimum fresh air.

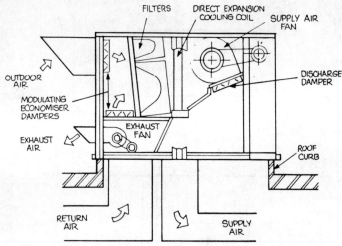

FIG. 14.11. Cross-section through a typical modular roof-top plant arranged for VAV application.

The use of weatherproofed roof-top units for VAV systems is becoming widespread and Fig. 14.11 shows such a system. The obvious advantage of such a system to the building owner is the minimisation of plant room space, thus maximising lettable floor area. To the mechanical contractor, however, such equipment represents a minimisation of his contractual risk in that he can purchase what can be termed a ready-made plant room, avoid many working drawings with associated builder's work details and electrical wiring diagrams, and minimise his site labour commitment. Such roof-top systems are limited only by their physical size and currently are capable of providing as much as 20 000 cfm ($9.5\ m^3/h$) of air at external static pressures compatible with VAV systems, and 45 tons R (160 kW) of cooling effect, which equates to a building having a floor area of between 15 000 and 20 000 ft² ($1400\ to\ 1850\ m^2$).

In order to realise the maximum potential benefits of the all-air

14 TERMINAL DEVICES AND AIR CONDITIONING SYSTEMS 263

variable volume system, the design engineer should be familiar with and provide for the maximum utilisation of the following major system characteristics:

1. Automatic room temperature control should be provided for individual rooms or modules at independently selected temperatures.
2. Zoning and extensive zoned ductwork are not required since the individually controlled variable volume terminals act as separate zones.
3. Because of the high induction rates and uniform draught-free air distribution, lower supply air temperature can be used without causing uncomfortable cold spots even at reduced volume. This keeps duct sizes and the number of terminals to a minimum.
4. Equipment sizes, initial costs, and operating costs can be kept at a minimum because the cooling capacity is automatically shifted around the building to follow the load changes, thus taking full advantage of load diversity.
5. Only the fan and refrigeration horsepower that is actually required to meet the load is consumed, resulting in minimum system operating costs. This is also a factor where operation on minimum emergency power may be required as in hospitals or windowless buildings.
6. System operating cost and performance are not materially affected by excess capacity of the system. The terminals will automatically adjust to the actual load, thus allowing wide design latitude without penalising operating costs when operating at reduced loads.
7. Depending upon the choice of variable air volume terminal, there should be virtually no air balancing for satisfactory performance. Many terminals are equipped with a volume regulator, which can be manually adjusted to a predetermined required air quantity. System balancing can be restricted to the adjustment of outdoor, return and exhaust air dampers.
8. No mechanical equipment requiring maintenance, other than controls, is located within the conditioned space.
9. Centralised apparatus location minimises operating and maintenance costs.

10. Centralised apparatus location minimises the cost of building service connections such as power, water and drain lines. Providing that the variable air volume terminals have controls that are powered by the supply air itself, then the cost of wiring to electric, or piping to pneumatic room controls is eliminated.
11. Centrally located variable volume apparatus permits an economical use of high efficiency filtration equipment to provide cleaner air and reduce cleaning and decorating costs.
12. Centrally located equipment permits the use of sprayed-coil dehumidifiers to provide better quality air supply and winter humidification.
13. Remotely located mechanical equipment provides for low system operating noise levels and minimum annoyance to occupants.
14. Refrigeration equipment may be selected using gas, steam, or electricity to provide the most economical operating costs.
15. Cooling can be made available from outdoor air, without refrigeration, during certain times of the year.
16. System changeover problems are eliminated because the heating system is controlled independently of the cooling system.
17. The system may be used in conjunction with a heat conservation cycle.

Consistent with good air conditioning design, a proper and detailed survey of any project is essential to establish the true heating and cooling requirements, and to establish where maximum load reduction can be achieved by addition of insulation, double glazing, shading devices, etc.

One of the advantages of a VAV system is that the central plant need only be sized to meet the building block load rather than a summation of the maxima of the loads each terminal has to cater for. It is normal to prepare a block load first to establish the parameters of room temperature differentials which will ultimately affect the sizing, although not necessarily the layout of sheet-metal ductwork and terminals.

The designer should be aware that internal areas will always require lighting during occupancy hours; however, consideration should be given to the perimeter areas, to establish whether lights will be on at

14 TERMINAL DEVICES AND AIR CONDITIONING SYSTEMS

times of solar gain. A very substantial saving in plant size and system operating costs can be realised if building occupants are encouraged to switch off lights when direct solar radiation exists in a given area. The saving is two-fold in terms of energy in that electrical energy is not used for the lights themselves and the added energy in terms of refrigeration plant and fan power, is not required to cool them. The reduction in plant size can also be significant in that catering for only one of either lights or solar gain, rather than both, could reduce equipment by as much as 30 per cent. It is therefore judicious to establish for each building face, the larger of the direct solar gain versus the lights plus the diffused solar gain when the area is shaded, such that the larger is used for load estimating purposes.

To establish guide lines with respect to the optimum equipment selection, the easiest way is to consider a typical application. It has been assumed that for the three plant selections chosen, the ducting sizing and arrangement remains constant, as does the layout of the terminals. This leads to a direct comparison of the central plant requirements. Three approaches have been taken. (A) considers a typical load estimate build-up using conventional logic with a room relative humidity of 50 per cent RH and a coil by-pass factor of 0·1, i.e. a 4-row, 14-fin cooling coil: (B) considers the same load estimate, but using a 6-row, 14-fin cooling coil in the same central station air handler; and (C) considers the effect on (B) if the room relative humidity were dropped to 45 per cent at design.

Condition (A)
AMBIENT DESIGN CONDITION

        82 °F DB   68 °F WB   81 grs/lb

        (27·8 °C DB   20 °C WB   11·6 g/kg)

ROOM DESIGN CONDITION

        72 °F DB   50% RH   58 grs/lb

        (22·2 °C DB   50% RH   8·3 g/kg)

Number of occupants—300
Ventilation requirements—5000 cfm   *(2·36 m³/s)*
Assumed coil by-pass factor—0·1

AIR CONDITIONING LOAD ESTIMATE

	BTU/h	kW
Solar and transmission gain	100 000	29·4
300 people (sensible heat)	81 000	23·8
Lights	255 000	75·0
Net room sensible heat	436 000	128·2
Fan heat (assumed $12\frac{1}{2}$ per cent)	54 400	16·0
By-passed outdoor air (sensible) $5000(82-72)\,0{\cdot}1\times 1{\cdot}08$	5 400	
$2{\cdot}36\,(27{\cdot}8-22{\cdot}2)\,0{\cdot}1\times 1{\cdot}21$		1·6
Effective room sensible heat	495 800	145·8
300 people (latent heat)	54 000	15·9
By-passed outdoor air (latent) $5000(81-58)\,0{\cdot}1\times 0{\cdot}68$	7 800	
$2{\cdot}36\,(11{\cdot}6-8{\cdot}3)\,0{\cdot}1\times 2{\cdot}98$		2·3
Effective room latent heat	61 800	18·2
Effective room total heat	557 600	164·0
Fresh air sensible heat $5000(82-72)\times(1-0{\cdot}1)\times 1{\cdot}08$	48 600	
$2{\cdot}36\,(27{\cdot}8-22{\cdot}2)\times(1-0{\cdot}1)\times 1{\cdot}21$		14·3
Fresh air latent heat $5000(81-58)\times(1-0{\cdot}1)\,0{\cdot}68$	70 200	
$2{\cdot}36\,(11{\cdot}6-8{\cdot}3)\times(1-0{\cdot}1)\,2{\cdot}98$		20·6
Grand Total Heat (GTH)	676 400	198·9
ERSHF	$\dfrac{495\,800}{557\,600}=0{\cdot}89$	$\dfrac{145{\cdot}8}{164{\cdot}0}=0{\cdot}89$

14 TERMINAL DEVICES AND AIR CONDITIONING SYSTEMS

	BTU/h	kW
Apparatus dew point	50 °F	*10 °C*
Effective coil temp. difference	$(72-50)(1-0\cdot1)$ $= 19\cdot8\,°F$	$(22\cdot2-10)(1-0\cdot1)$ $= 11\,°C$
Required air volume	$\dfrac{495\,800}{19\cdot8 \times 1\cdot08}$ $= 23\,200\,\text{cfm}$	$\dfrac{145\cdot8}{11 \times 1\cdot21}$ $= 10\cdot95\,m^3/s$
Room to supply air temperature	$\dfrac{436\,000}{23\,200 \times 1\cdot08}$ $= 17\cdot4\,°F$	$\dfrac{128\cdot2}{10\cdot95 \times 1\cdot21}$ $= 9\cdot67\,°C$

This would be the normal approach to establish the required total air volume and, using the resulting temperature difference, one would then establish the maximum air volume to each terminal based on its own maximum load. From this the ducting would be sized to establish the ducting pressure loss. Assuming this done, one could then establish the exact operating level of the plant to calculate fan power and heat exchanger requirements. Assuming the calculated duct pressure to be 1·52 in. w.g. (*378 N/m²*) then a typical build up of the required fan resistance would be, based on coil velocity of 500 fpm (*2·54 m/s*):

	in. w.g.	*N/m²*
Roll filter	0·35	*87*
Preheater	0·15	*37*
Coil	0·48	*120*
Ducting	1·52	*378*
Terminal pressure	1·50	*374*
	4·00	*996*

A central station fan selection for the above duty would result in an absorbed fan power of 21·4 bhp (*16·0 kW*), which is identical to that assumed in the load estimate.

Condition (B)
Using a deeper 6-row 14-fin coil, the by-pass factor would improve to 0·04 which would have the effect of reducing the air volume

requirements. Using the same air handling unit the coil resistance would increase because of the added 2 rows, however the other components of heater and ducting would reduce. Using only the room heat components a new load estimate can be prepared as follows:

	BTU/h	kW
Net room sensible heat	436 000	*128·2*
Fan heat (assumed)	47 600	*14·0*
By-passed outdoor air $5000(82-72)$ $\times 0.04 \times 1.08$	2 100	
$2.36(27.8-22.2)$ $\times 0.04 \times 1.21$		*0·64*
Effective room sensible heat	485 700	*142·84*
Room latent heat	54 000	*15·9*
By-passed outdoor air $5000(81-58)$ $\times 0.04 \times 0.068$	3 100	
$2.36(11.6-8.3)$ $\times 0.04 \times 2.98$		*0·92*
Effective room latent heat	57 100	*16·82*
Effective room total heat	542 800	*158·66*

ERSHF	$\dfrac{485\,700}{542\,800} = 0.90$	$\dfrac{142\cdot84}{158\cdot66} = 0\cdot90$
Apparatus dew point	50°F	*10°C*
Effective coil temp. difference	$(72-50)(1-0.04)$ $= 21.12°F$	$(22\cdot2-10)(1-0\cdot04)$ $= 11\cdot71°C$
Required air volume	$\dfrac{485\,700}{21.12 \times 1.08}$ $= 21\,300 \text{ cfm}$	$\dfrac{142\cdot84}{11\cdot71 \times 1\cdot21}$ $= 10\cdot05\,m^3/s$
Room to supply air temperature	$\dfrac{436\,000}{21\,300 \times 1.08}$ $= 19.0°F$	$\dfrac{128\cdot20}{10\cdot05 \times 1\cdot21}$ $= 10\cdot54°C$

14 TERMINAL DEVICES AND AIR CONDITIONING SYSTEMS 269

Using the above values the coil face velocity reduces to 460 fpm ($2\cdot33\,m/s$) and with the same ducting previously designed the duct resistance would reduce to $1\cdot28$ in. w.g. ($319\,N/m^2$), giving the following fan selection:

	in. w.g.	N/m^2
Roll filter	0·35	*87*
Reheater	0·12	*31*
Coil (6/14)	0·63	*157*
Ducting	1·28	*319*
Terminals	1·50	*374*
	3·88	*968*

The same fan as previously selected would now require an absorbed power of only $14\cdot0$ kW—a saving of $2\cdot0$ kW absorbed at maximum load.

Condition (C)
Both the foregoing examples considered the room condition at 50 per cent Relative Humidity. A further reduction in air volume can be achieved with plants running at lower apparatus dew points, which will achieve a lower room relative humidity. This logic is quite a sound one, since at most partial loads it is the sensible heat which is reduced and not the latent heat, therefore a lower coil dewpoint, with reduced air volume, would be more capable of satisfying room comfort conditions. In addition, at lower design relative humidities, the sensible heat gain from people is somewhat reduced, since metabolic heat is dissipated more by evaporation in the drier atmosphere. Using the previous example with a room relative humidity of 45 per cent in lieu of 50 per cent, the following load estimate can be made. It should be noted that the room moisture level of 52 grs/lb ($7\cdot5\,g/kg$) will require greater latent cooling of the outside air, whilst the sensible cooling will remain as previously calculated.

	BTU/h	kW
Solar and transmission gain	100 000	*29·4*
300 people (sensible heat)	75 000	*22·1*
Lights	255 000	*75·0*

	BTU/h	kW
Net room sensible heat	430 000	*126·5*
Fan gain	40 100	*11·8*
By-passed outdoor air	2 100	*0·64*
Effective room sensible heat	472 000	*138·94*
300 people (latent heat)	60 000	*17·64*
By-passed outdoor air 5000(81 − 52) × 0·04 × 0·68	3 900	
2·36 (11·6 − 7·5) × 0·04 × 2·98		*1·15*
Effective room latent heat	63 900	*18·79*
Effective room total heat	536 100	*157·73*
Fresh air sensible heat	48 600	*14·30*
Fresh air latent heat	94 100	*27·67*
Grand total heat	678 800	*199·70*

ERSHF $\quad \dfrac{470\,300}{536\,100} = 0.88 \qquad \dfrac{138 \cdot 34}{157 \cdot 73} = 0.88$

Apparatus dew point $\quad 47 \cdot 1\,°F \qquad 8 \cdot 4\,°C$

Effective coil temp. difference $\quad (72-47 \cdot 1)(1-0 \cdot 04) \quad (22 \cdot 2-8 \cdot 4)(1-0 \cdot 04)$
$= 23 \cdot 9\,°F \qquad = 13 \cdot 28\,°C$

Required air volume $\quad \dfrac{472\,200}{23 \cdot 9 \times 1 \cdot 08} \qquad \dfrac{138 \cdot 94}{13 \cdot 28 \times 1 \cdot 21}$
$= 18\,220\,\text{cfm} \qquad = 8 \cdot 60\,m^3/s$

Room to supply air temperature $\quad \dfrac{430\,000}{18\,220 \times 1 \cdot 08} \qquad \dfrac{126 \cdot 5}{8 \cdot 6 \times 1 \cdot 21}$
$= 21 \cdot 9\,°F \qquad = 12 \cdot 16\,°C$

The reduced air volume would allow a reduction in air handling plant sizing, and a selection of a unit size smaller than conditions A

14 TERMINAL DEVICES AND AIR CONDITIONING SYSTEMS 271

and B reveals a face velocity of 511 fpm ($2\cdot60\,m/s$). This would result in a revised fan operating level of:

	in. w.g.	N/m^2
Roll filter	0·35	87
Preheater	0·16	39
Cooling coil	0·78	193
Ducting	0·94	233
Terminals	1·50	374
	3·73	926

The fan selection would result in an absorbed power of 15·8 bhp ($11\cdot8\,kW$).

REFRIGERATION PLANT REQUIREMENTS

Each of the foregoing conditions has a different operating level with respect to refrigeration plant. On the basis that for the application considered one water chiller is required to serve two identical air plants, then the following selections can be made.

	Condition		
	A	B	C
Ref. duty required	112·7 tons	111·6 tons	113·1 tons
	397·8 kW	393·8 kW	399·4 kW
Apparatus dewpoint	50 °F	50 °F	47·1 °F
	10 °C	10 °C	8·4 °C
Entering coil chilled	40·0 °F	42·2 °F	40·3 °F
water temp.	4·4 °C	6·8 °C	4·6 °C
Water flow	197 igpm (14·93 litres/s)		
Chiller selection size	140	120	140
Absorbed power	101·1 kW	90·2 kW	102·1 kW
Condensing temperature	98·6 °F	97·7 °F	100·0 °F
	37·0 °C	36·5 °C	37·7 °C
Entering condenser water	75·0 °F	74·8 °F	75·2 °F
	23·9 °C	23·8 °C	24·0 °C
Leaving condenser water	86·1 °F	85·6 °F	86·4 °F
	30·1 °C	29·8 °C	30·2 °C
Condenser water flow	254 igpm (19·25 litres/s)		
Cooler tower requirements	Identical		

Using the above relative information which has kept chilled water and condenser water pipework and pumps, and cooling tower selection identical, a direct comparison of capital cost and design absorbed power for air handling plants and refrigeration plants can be made as follows:

Equivalent hours run:
Air handling plants
—52 weeks × 5 days × 12 hours × 0·8 = 2496 hours
Refrigeration plants
—20 weeks × 5 days × 12 hours × 0·5 = 600 hours
The resulting comparative kWh p.a. would therefore be:

	Condition		
	A	B	C
Absorbed fan power × 2	32·0 kW	28·0 kW	23·6 kW
Absorbed chiller power	101·1 kW	90·2 kW	102·1 kW
Fans	79 872	69 999	58 906
Refrigeration	60 660	54 120	61 260
Total kWh p.a.	140 532	124 008	120 166

Similarly, capital costs can be compared as follows:

	Condition		
	A	B	C
Air handler selection	130 (4-rows)	130 (6-rows)	120
Chiller selection	140	120	140
2 air handlers	£18 750	£18 950	£14 800
Chiller	£20 600	£19 000	£20 600
	£39 350	£37 950	£35 400

Quite obviously for the selections considered, condition C would be the best selection, both in terms of energy and capital cost considerations. A simple guide to VAV economy is to maintain the highest possible temperature difference between supply and room air and to provide good air side heat exchangers. There should be one note of caution with this statement, i.e., that some VAV air terminals cannot provide proper room air distribution at high temperature differences. The designer should take this into consideration if he has already pre-selected a certain type of terminal for physical, aesthetic, control or economy reasons.

At most partial load conditions it is the room sensible heat that is

14 TERMINAL DEVICES AND AIR CONDITIONING SYSTEMS

reduced and not the latent load. The only room latent gain to be considered in most applications is that from the occupants and it is fair to say that it is when the occupants are present that proper room conditions should prevail.

Therefore, at partial load the room sensible heat factor will reduce, calling for a lower coil dew point to maintain adequate room relative humidity conditions. Condition C as well as being the most economical approach, would also satisfy partial load conditions best. One factor which must be considered at partial load is the need to maintain adequate ventilation rates. If a system has been designed without a fresh air economy system to provide free cooling, then a 50 per cent reduction in supply air volume will probably result in 50 per cent ventilation rate. In such circumstances it is essential that some form of fan assistance is added to the fresh air connection to guarantee ventilation standards at partial loads.

Central plants using economy dampers will increase the ventilation rate at partial load, since the fresh air would be used as a free cooling medium and it is not normally necessary to provide a guaranteed fresh air supply. What will be essential is the need for a preheater on the fresh air inlet to protect chilled water coils when low ambient air is used for free cooling.

The principal rule in duct layout is that the ducting should feed the terminals rather than the terminal positions be restricted by the duct location. The terminal type will also dictate the design velocities of the ducting system, but generally the categories can be divided into low velocity systems, 1500 fpm *(7·5 m/s)* initial velocity, and medium velocity systems, 2500 fpm *(12·5 m/s)* initial velocity. With either system it is essential that the sizing of branch ducting is adequate to meet the individual maximum air volumes to the zone area it serves, whereas risers or main ducts need only be sized to cater for the diversified or block load air volume. Providing that these guide lines are adhered to, then no consideration to dampers for system balancing need normally be made.

By definition, the VAV system has a changing air volume and any pre-balancing of the air system will be effective at only one condition during the cycle of changing conditions that will exist minute by minute throughout a building.

The actual leaving velocity from the fan will be to a great extent dictated by the total system static pressure. Fans required to cater for pressures as high as 6 in. w.g. *($1500 N/m^2$)* would have outlet velocities as high as 4000 fpm *(20 m/s)* to take advantage of optimum fan efficiency. It is common practice to include silencers with such systems and it would be practical to use the silencer to create the transformation to lower the velocities of the system ducting.

It is good practice to size main duct risers on static regain principles, since this would have a considerable effect in reducing the overall system resistance. This approach will also provide equal pressures to run-outs, limiting the throttling of the air volume at the terminal to the variations in room load only, rather than having to take up large differences in system static pressure which could cause noisy terminals. Since VAV terminals are essentially confined to a location above the ceiling, then advantage of the ceiling void can be taken in lieu of local return air ducting. There are considerable advantages in using the ceiling void, since even at maximum air volumes the air velocities are extremely low, keeping noise to a minimum, and ensuring an easy passage for varying air volumes from different zones to return to the central plant. Unless local codes insist, it is not recommended that there is any return air ducting above a ceiling and that just a riser duct, sized for sound reasons rather than pressure considerations, connects the various ceiling voids to the main plant or return air/exhaust fan.

The use of the ceiling void will also provide economy of plant operation if return air light fittings are used. Since as much as 60 per cent of the heat from lights can pass directly into the return air system, this can substantially reduce the supply air volume to any particular zone. In addition to this, it enhances the opportunities for free cooling, since much of the room heat is picked up in the return air system, creating a temperature rise. This means that there will be a greater occurrence when the ambient conditions are lower than the return air conditions.

Return air light fittings require a certain air motion to be effective. Since the lights themselves can often contribute between 60 and 75 per cent of the room sensible heat gain, air motion can be guaranteed whenever they are switched on, since the terminal will react to this heat gain by passing more air to the room.

VARIABLE AIR VOLUME TERMINALS

The location of terminals must be consistent with the desired flexibility of partition layout for any given application. The terminals

14 TERMINAL DEVICES AND AIR CONDITIONING SYSTEMS

must integrate with lighting fittings within the ceiling, and should be capable of providing adequate air distribution within a zone throughout the spectrum of air volumes when the zone is occupied, and should operate within the desired sound level of the space. Terminals should also have the flexibility of controls such that single or groups of terminals can be controlled as building requirements change.

Since terminals are available in a wide selection of sizes, it is rarely the air volume to be handled which dictates the location of units, but rather the physical considerations of the building itself.

A fundamental factor in determining a final layout of terminals could well be the method of heating used in conjunction with the VAV system. Unlike the VAV system the heating requirements are normally met with a constant volume variable temperature system, whether by air or hot water, which operates to satisfy the transmission losses of a building.

The most common form of heating is a wet perimeter system having scheduled hot water temperatures compatible with the temperature difference between ambient temperatures and desired room conditions. It is probable that with existing buildings such a system is already installed, therefore the designer need only concern himself with the cooling requirements of the building, which is the major reason for VAV systems being so widely accepted for refurbishment projects.

The perimeter heating system has many advantages in that the heat is below window areas, thus preventing down-draughts, stratification of warmer air to the ceiling and cold floors. The practice of installing such systems is well known and they can with confidence be installed economically. No matter which heating system is installed, good outside wall insulation and double glazing should be common practice for energy conservation alone. Good insulation will also mean that the heating would not have to be switched on until ambient temperatures are quite low—perhaps 18 °F (*10 °C*) below room design condition. The room gains of lights and people will more than offset transmission losses at these temperature differences, and it is not until the heat loss at the perimeter is such that down-draughts occur that the heating is needed. Therefore the time that the heating system and cooling systems are working against each other could be substantially reduced.

The advent of good insulation has brought an alternative method of heating into use which can be integrated with the variable air volume terminals themselves. This is a constant volume system whose temperature is controlled on a reheat schedule in the same manner as an induction system, the quantity of air provided at each terminal being dictated by the transmission loss per degree of the zone being served. The constant volume air would normally be introduced through one half of the VAV terminal, with two diffuser slots—the outside one catering for transmission losses and the inside slot maintaining its normal VAV function. An advantage of such a system is that during the summer months cold air can be introduced to offset transmission gain and some of the other room cooling requirements. However, for such a system to be successful the air must be introduced close to the perimeter and should have the capacity to diffuse across the whole of the perimeter, otherwise down-draughts could still occur on the untreated perimeter surfaces. Good practice dictates that a minimum of 0.5 cfm/ft^2 perimeter (2.5 *litres/s/m^2*) be supplied by a constant volume system to offset down-draughts and to prevent condensation forming on the glazing.

Providing that both heating and cooling can be simultaneously achieved, there is a justification for integrating other air conditioning systems serving perimeter areas, with a VAV system serving the internal core. A good example of this would be an induction system serving a zone 12–20 ft (*4–6 m*) in from the perimeter. Such an arrangement will cater for all transmission losses during winter and provide cooling of lights, occupants and solar gain throughout the year. This logic can, however, prove expensive in terms of building space in that a ceiling void is required for the VAV system whilst the induction system requires a clearance around the perimeter, reducing lettable floor area. The integration of a wet perimeter system or constant volume air system with the VAV system could save some 3–5 per cent of lettable floor area of an office block.

An alternative heating system is that of VAV terminal reheat as shown later in Fig. 14.18. As the cooling requirements of an area reduce, so the air volume is reduced by the dictates of a room thermostat controlling a damper motor. Once the damper has reached a pre-determined minimum air volume, and the thermostat is still unsatisfied, then a reheater is actuated. The reheat system is normally hot water, which can be heat-reclaimed water from the refrigeration condenser circuit, although electric reheaters can also be used.

14 TERMINAL DEVICES AND AIR CONDITIONING SYSTEMS

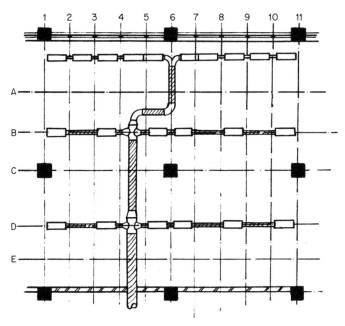

FIG. 14.12. VAV terminal layout arranged to suit possible partition lines.

Figure 14.12 shows a ceiling grid pattern for a typical office block showing how terminals can be arranged to suit various partition layouts. At the perimeter the units are sized to meet the associated high air volumes and it is good practice to allow one terminal per window mullion, such that irrespective of partition layout, maximum heat gain can be catered for.

With interior areas it is normal practice to alternate units. Since the air requirements per given floor area are reduced it should be possible to size units to meet the worst possible arrangement. For the example shown this would be two units per three window mullion widths, or an inbuilt system capacity of 150 per cent. In reality this condition rarely exists since often the load can be shared by more than two terminal units.

The example shown has placed terminals on centre lines B and D such that the two-way air distribution pattern can be utilised to serve areas either side of partitions located on such lines. This is good practice, providing the terminal has the facility to control either direction air flow, which in this case would serve separate areas.

278 APPLIED AIR CONDITIONING AND REFRIGERATION

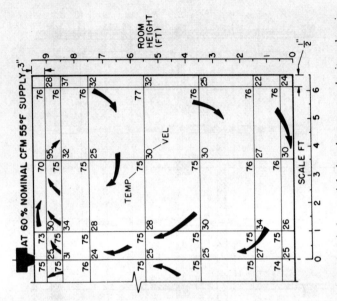

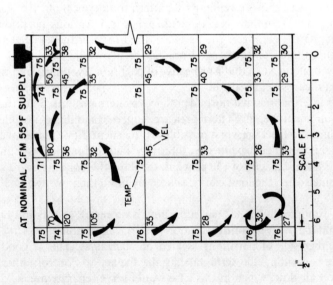

FIG. 14.13. Cross-sections of air patterns of a VAV system. Temperature values in lower right-hand corners; velocity values in upper left-hand corners.

14 TERMINAL DEVICES AND AIR CONDITIONING SYSTEMS 279

Fig. 14.14. VAV terminal unit suitable for two zone control. (Courtesy: Carlyle Air Conditioning Co. Ltd.)

The location of units will also be dictated to a high degree by the type of control used at the terminal. Figures 14.15 to 14.18 show various approaches to a common perimeter area. Figures 14.15 and 14.17 show systems of the self-powered type where only one thermostat has been shown located in the terminal unit itself. Figures 14.16 and 14.18 have either electric or pneumatic controls which are powered from a remote source and are controlled from a room thermostat which by necessity must be located on a partition or column. The use of remote controls imposes restrictions on the system flexibility since several terminals would normally be controlled from adjacent areas. Since the primary function of a VAV system is to provide the correct amount of air to satisfy the exact load in a given area, then a room without lights and in shade controlled from an adjacent area with lights and solar gain, could well be 9–12 °F

280   APPLIED AIR CONDITIONING AND REFRIGERATION

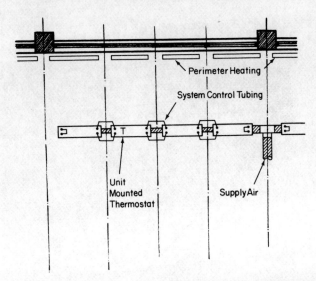

Fig. 14.15. Typical layout of a 'Moduline' system with wet perimeter heating.

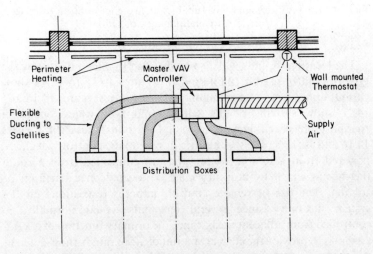

Fig. 14.16. Typical ceiling layout of master/satellite VAV system with wet perimeter heating.

14 TERMINAL DEVICES AND AIR CONDITIONING SYSTEMS   281

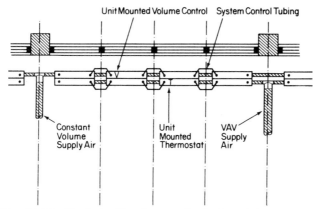

Fig. 14.17. Typical ceiling layout of a dual conduit system.

(5–7°C) lower than design requirements because it lacks a controller. In the opposite condition the room could well be 9–12°F (5–7°C) warmer. It is essential therefore that the designer considers whether the terminal system chosen has the capacity to change the thermostat controls and controls positions to meet the changing needs of his building—without the need to add master control boxes serving individual satellites. Conversely the system powered concept allows the designer to provide the minimum quantity of thermostats to control the building as though it were an open area. As the building

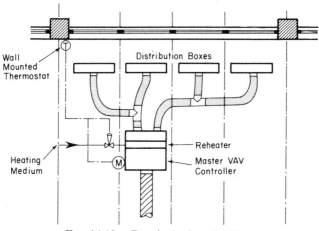

Fig. 14.18. Terminal reheat VAV system.

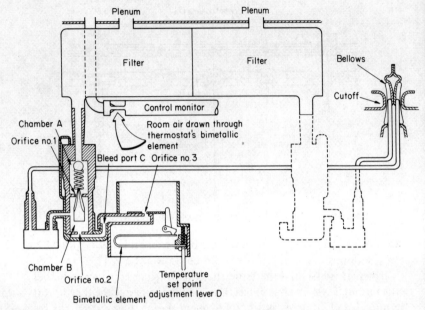

Fig. 14.19. VAV terminal direct acting control system.

requirements change, then by removal of the centre of the diffuser, thermostats can be added or removed and only control air tubing need be changed to suit the new partition configuration.

Figure 14.19 shows the schematic controls of a direct acting system. Air is taken from the air supply plenum, is filtered and introduced into a regulator control, which is used to dictate the maximum air volume of any given outlet. In operation, air passes into Chamber A where it bleeds through the adjustable orifice A to Chamber B. The adjustment is made to a vernier type scale on the outside of the regulator such that the factory calibrated control can be dialled for any desired air volume. Chamber B is connected directly to the bellows damper and provides the air to inflate the bellows. Orifice 2 is fixed and discharges through bleed port C. If a constant volume control was required then the air would be discharged to atmosphere. When the pressure in the plenum changes, a proportional change is produced in Chamber C, which inflates or deflates the bellows as required to maintain constant air flow through the diffuser slot. Such a system is capable of maintaining correct volume with fluctuations in plenum pressures as high as 5 in. w.g. ($1250\ N/m^2$).

14 TERMINAL DEVICES AND AIR CONDITIONING SYSTEMS 283

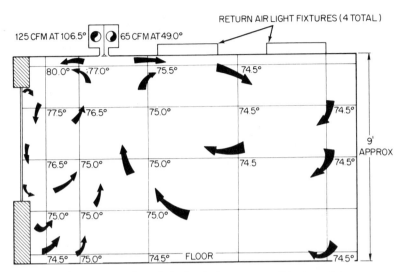

FIG. 14.20. Cross-section of air circulation pattern of dual conduit system. Single glass window, no drapes, room lights on. Outdoor air temperature 28·5°F DB. Black numbers indicate indoor air temperatures in °F DB. (Temperatures taken from laboratory test data.)

For variable air volume air slots, a thermostat package is added to the constant volume regulator. A control monitor connected to the terminal plenum ensures that the thermostat senses room temperature accurately by inducing room air through a perforated bimetallic element within the thermostat. The thermostat is set at the desired room temperature by adjustment of lever D which protrudes between the diffuser slot such that it can be adjusted from within the conditioned space. As the room temperature decreases, the bimetallic element moves to restrict orifice 3, thus increasing the pressure in Chamber B, which in turn inflates the bellows and reduces the air volume through the diffuser slot.

There is one further category of terminal, as shown in Fig. 14.22, which is the by-pass type. Essentially this consists of a constant volume central station system which dumps unrequired air into the ceiling void. Terminals are often fed from a master unit such as shown in Fig. 14.16. However, much of this provides variable air volume to the conditioned space, it does not have the desirable energy savings of a true variable air volume system in that there is no fan power reduction at partial load.

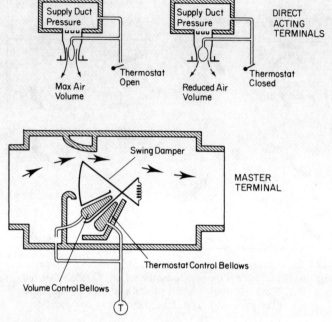

Fig. 14.21. Principles of direct acting control systems.

It is essential that the successful designer makes himself aware of the complete system concepts together with the hardware available to fulfil his client's defined flexibility.

Since, in its simplest form, the VAV system is a cooling only system, the basic central station control requirements are relatively simple. The control system may be modified to provide for requirements such as minimum ventilation air, heat reclaim and economy of operation, by use of free cooling at low ambient conditions.

The primary requirement of the central station plant is to provide cool air at a relatively constant dewpoint temperature to offset the individual room or controlled zone sensible heat gains sensed by the controlling thermostat. This implies that since room latent heat gain from occupants would be relatively constant as air volumes are reduced there would be an increase in room relative humidity at partial load. The control of the cooling coil therefore should be arranged to provide increased dehumidification per unit amount of air circulated, by providing a lower coil dewpoint as the air volume is

14 TERMINAL DEVICES AND AIR CONDITIONING SYSTEMS 285

reduced. It would be normal practice for the design engineer to establish the desired dewpoint to meet the job maximum load requirements and to establish what dewpoint is required for the necessary air volume when only lights and occupants contribute to the cooling load. This approach will dictate the necessary control band to meet the job partial load requirements.

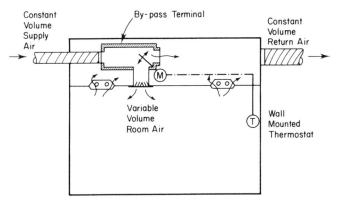

FIG. 14.22. By-pass type VAV terminal.

The control of the cooling coil on a central plant using chilled water could be achieved without control of the coil at all, by just using a 'wild' coil approach with constant flow, constant temperature chilled water. The flow and temperature of the chilled water will be dictated by the maximum cooling requirements, and any reduced air volume conditions will result in lower dewpoint or off-coil conditions (Fig. 14.23). Whilst this approach has been widely adopted and adequately provides good dehumidification at partial load, current thinking is that this could be an energy cost implication. There is a move whereby maintaining say 50 per cent RH at partial load is not considered so important, since it is considered that occupants cannot sense an increase to as high as 55 or even 60 per cent RH providing proper room dry bulb temperatures are maintained.

With energy in mind, designers are tending back towards a three-way valve control of dewpoint temperature with a narrow range to provide a relatively constant dewpoint. However, this could be considered a moot point, particularly if the designer has not previously considered good building insulation, adequately sized heat exchangers, or high performance fans, all of which will result in greater energy savings than can be achieved by using three-way valve control.

Direct expansion systems can only be controlled within the unloading characteristics of the refrigeration machinery. However the best approach is certainly to control the refrigerant evaporating temperature within the coil at the dictate of a suction pressure unloading system. Depending on the available number of capacity

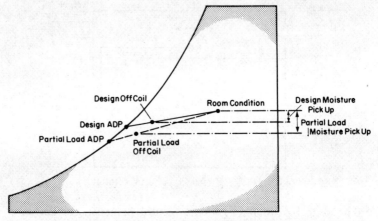

FIG. 14.23. Psychrometric principle of 'wild' chilled water cooling coil.

stages, a stable condition can easily exist and the correct dewpoints for the various loads can be simply governed by the pressures at which the compressor or compressors unload. With respect to energy, a direct expansion system has substantial savings over a chilled water system since it eliminates the necessary heat exchanged between refrigerant and water, which normally implies evaporating levels some 9–10°F (5–6°C) higher without the constant requirements of a chilled water pump. Over an operating year, energy costs of comparable systems can be reduced by as much as 25 per cent by use of a direct expansion system, which can be coupled with a reduced capital cost.

The control of the air volume itself is essentially confined to the terminal units. However, consideration should be made at the central plant as to whether a controlled discharge static pressure is desirable. Providing that terminals have the capacity to accept wide pressure variations without impediment to their control capabilities and

14 TERMINAL DEVICES AND AIR CONDITIONING SYSTEMS

without an adverse effect on the attenuation properties, then for small systems no static pressure control at the central station plant need be considered. Care should be taken in deciding whether a fan discharge damper controlled from duct static pressure is necessary by considering the terminals themselves. Most terminals of the remote controlled type cannot accept the wide static pressure variations which could exist. These can be as great as 2 to 3 in. w.g. (*500 or 750 N/m²*) with adverse effects on temperature control and noise regeneration.

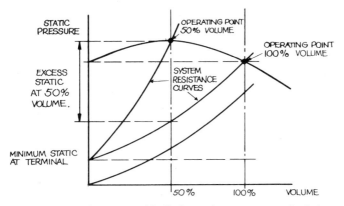

FIG. 14.24. Fan performance with discharge dampers or terminal pressure control.

The effect of reducing air volume by discharge damper or VAV terminal is the same in terms of energy (Fig. 14.24), and the energy costs, although much lower than a constant volume system, should be compared with those savings that occur when using variable inlet guide vanes (Fig. 14.25). A comparison of energy saving is shown in Fig. 14.26, and the designer should consider the variable inlet guide vane system for plants exceeding 10 000 cfm ($5 \, m^3/s$) capacity. An essential ingredient of VAV logic is that the more diversified areas that can be served from one plant, the greater the energy savings over conventional constant volume plants, therefore large systems are common.

Variable inlet guide vanes can only be associated with backward curved fans. However the larger the system the more static pressure required at the fan to overcome the longer ductwork runs, which makes the system tend towards backward curved fans in any event.

288   APPLIED AIR CONDITIONING AND REFRIGERATION

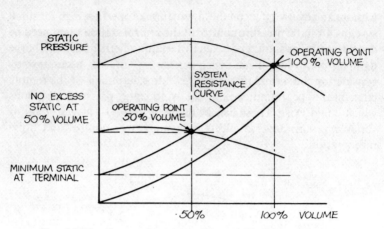

Fig. 14.25. Fan performance with variable inlet guide vanes.

There are even greater energy savings to be made if variable pitch axial fans are used. The only disadvantage in their use is that a built-up rather than factory-assembled central plant is necessary and designers often consider that only when the size of available factory assembled plants is exceeded, should built-up systems be considered.

Minimum ventilation rates may be laid down together with a conservation requirement of the use of return air for preheating the conditioned space during low ambient temperature conditions, with

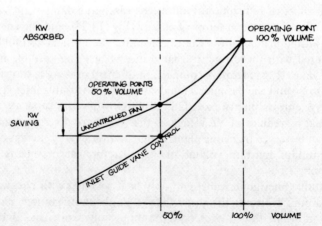

Fig. 14.26. Operating power requirements of VAV system fans.

14 TERMINAL DEVICES AND AIR CONDITIONING SYSTEMS

an all-air system the return air is predominantly extracted over the lighting fittings where it may attain temperatures in the region of 77 °F (*25 °C*). With winter design ambient at 32 °F (*0 °C*) and with a supply air temperature in the VAV system of 50 °F (*10 °C*), it is readily apparent that the fresh air quantity would be 60 per cent of the total.

Computation of the minimum operating load will determine whether this fresh air quantity will fulfil the ventilation requirement of the building. In cases where it does not, then a positive means of introducing this minimum will need to be adopted. This could take the form of a separate fresh air fan or introduction of a heater in the fresh air intake or control of the mixing dampers by means of volume measurement. All of these solutions imply the use of a separate heating medium rather than reclaiming heat from the return air.

A further possibility for dealing with the quantity of fresh air sometimes required, is the use of thermal wheels. However, the usefulness of these devices would only be reflected at extremes of design—there is some doubt as to the justification of additional costs in terms of fan kilowatts during intermediate periods when the wheel appears redundant.

During high ambient conditions, the control system may be further modified so as to select return air in preference to fresh air. This is as a conservative measure, since the return air is cooler than the fresh air. At this point a need to establish a positive fresh air volume may again exist. If a problem does exist, then the solution should be one that complements the winter situation also.

The complication of the control system in order to fulfil the ventilation requirements must expend energy. This situation begs the question, 'What price energy? Can fresh air be afforded?' Figure 14.27 shows the components of a central station variable air volume system and the location of its associated automatic controls. Superimposed are the added components which make up a dual conduit system.

The system shown is one with a fresh air, return air, exhaust air economiser system, which in conjunction with the chilled water control valve maintains a relatively constant dewpoint temperature under the dictates of thermostat T_1. Since it is the terminals which dictate the air volume to be supplied by their own local thermostatic control, the central plant itself should be capable of maintaining a constant duct static pressure in the interests of the terminals' performance, and when variable inlet guide vanes are used in the

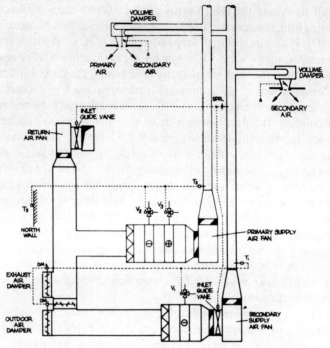

Fig. 14.27. Dual conduit VAV system.

interest of further energy conservation, the static pressure regulator SPR_1 would control the variable air volume fan, and, where economiser systems are used, the return air fan in harmony.

Because the physical size of central station VAV plants tends to be large, it is common practice to use pneumatic controls and a schematic control system is shown in Fig. 14.28.

When a dual conduit system is used, it is good practice to use 100 per cent return air on the primary supply air system. In the heating mode, maximum benefit will be achieved from the heat from light fittings, and only the minimum amount of reheating will be necessary. Therefore, during maximum recirculation no unnecessary waste in energy is expended in cooling outdoor air. During mid-season use when the constant volume system commences to act on its normal reheat schedule mode the return air would be close to the plant requirements.

There could be some justification on large systems to consider the

14 TERMINAL DEVICES AND AIR CONDITIONING SYSTEMS

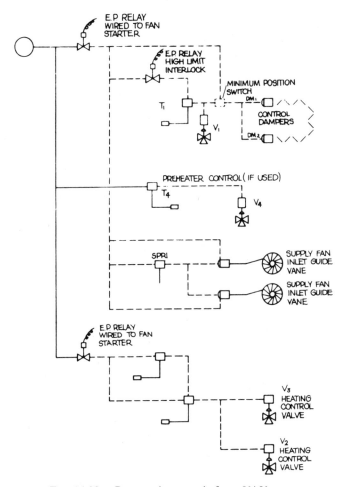

FIG. 14.28. Pneumatic controls for a VAV system.

use of separate fresh air, return air, exhaust air damper systems for this section of the plant. This has become feasible with the availability of reliable enthalpy sensors, whose use can maximise the use of the higher enthalpy of return or ambient air during a heating mode, or vice versa during a cooling mode. In addition, the dampers would adjust themselves to provide, whenever possible, an air mix to satisfy correct duct temperatures without the use of heating or cooling media in a mid season condition. This approach can be made to the VAV

Variable volume control box

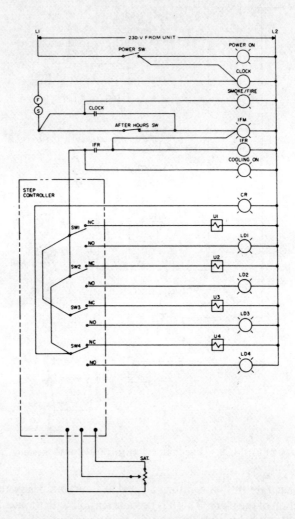

FIG. 14.29. Schematic wiring of a modular roof-top VAV plant. C, compressor contactor; CH, crankcase heater; CR, control relay; DM, damper motor; DR, damper relay; EMC, exhaust motor contactor; F, firestat (field supplied); Fu, fuse; HPCT, head pressure control thermostat; HPS, high-pressure switch; IFC, indoor fan contactor; IFR, indoor fan relay; IP, internal protector; LD, loaded (unloader indicator light); LPS,

14 TERMINAL DEVICES AND AIR CONDITIONING SYSTEMS

Main unit control box

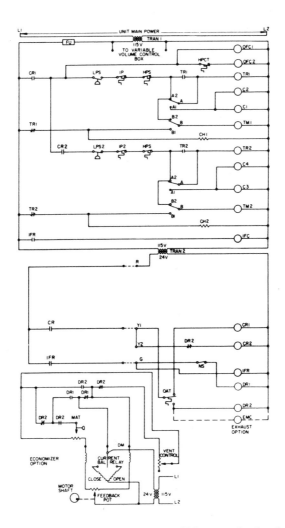

low-pressure switch; MAT, mixed air thermostat; NC, normally closed; NO, normally open; NS, night switch (field supplied); OAT, outdoor air thermostat; OFC, outdoor fan contactor; S, smokestat (field supplied); SAT, supply air thermostat; Sw, switch; TM, timer motor; TR, timer relay; Tran, transformer; U, unloader; ——, factory wiring; – – –, accessory or option wiring; ☐, optional.

section of the system, although because of ventilation requirements its use could be somewhat restricted. Figure 14.28 shows the normally accepted method of using fresh air for free cooling whenever the ambient dry bulb temperature is below a predetermined level, which should be at, or slightly above the return air design wet bulb temperature. The problem exists that with dry bulb temperature sensors of ambient air one cannot guarantee that the enthalpy of that air will be below that of the return air.

At first consideration there would seem to be an ideal case for heat reclaim from the condenser water of the refrigeration plant, to heat the primary supply of a dual conduit system. However, since a large economy would be achieved by the use of free cooling on the secondary supply air system, then at times when heating is required on the primary supply air system, there would be no refrigeration plant in operation. There would also be no benefit from using exhaust air as a heat sink, since by choice, return air is used for the constant volume section anyway. During warm-up prior to occupation it would be good practice not to use the VAV section, since it would require the heating of ventilation air and use unnecessary fan power.

On capital cost and physical space grounds, the use of modular roof-top plants is becoming more acceptable. In addition to the natural energy savings of the VAV system, such equipment can also provide for economiser operation and exhaust fans, and, where necessary, discharge duct dampers for system static pressure control. The use of discharge dampers in lieu of variable inlet guide vane dampers is considered against the use of forward curved fans, associated reduced capital and physical space requirements. By nature, the modular roof-top units are of the direct expansion type with inherent economy of running cost. As the name 'modular' implies, given that they are properly applied, only an external electrical feed is necessary as all controls are in-built. Figure 14.29 shows a composite wiring diagram of the electrically controlled system, having all possible accessories.

As the use of VAV becomes more widely acceptable, new applications are developing. What is becoming more apparent is that providing there is more than one occupied zone requiring individual and characteristically different operating air conditioning levels, then VAV may be applied. There is no limit to how small a VAV system can become. Providing that an air conditioning unit has sufficient external static pressure and has correct unloading characteristics,

14 TERMINAL DEVICES AND AIR CONDITIONING SYSTEMS

then modular systems as small as 2000 cfm ($1\ m^3/s$) can be considered. The ideal terminal for small systems would be either the by-pass type or the true VAV terminal having the ability to accept pressure variations without sound regeneration and without impairment to control function.

Although smaller systems cannot readily be adapted to economiser systems with return air/exhaust air fans, by nature they offer better control to individual zones than conventional all air systems, and also offer a substantial energy management possibility, particularly when compared with either individual units per zone, or terminal reheat systems.

DUAL-DUCT SYSTEMS

As mentioned in Chapter 7, a dual-duct system has similar features to a multizone system except that the mixing of hot and cold air is achieved at a terminal unit rather than at the apparatus. Physical differences mean that many more zones can be serviced by a dual-duct system using only two main supply air ducts, often at high velocity, whereas the multizone system requires one low velocity supply duct from the apparatus to each zone. The control features are very similar and the apparatus should have the facility to use primary air as a cooling medium at partial load or winter operation. Primary air can be introduced for heating purposes, but care should be taken, since this untreated air could add a substantial latent heat gain to the conditioned areas.

Figure 14.30 shows the return air system and fresh air mixing system common to all dual-duct arrangements. Fresh air for ventilation purposes is constantly introduced into the system through a preheater. This preheater can be omitted on systems using a single fan, but where stratification could occur when using an automatic filter or spray coil dehumidifier (for winter humidification) a preheater is recommended to avoid frosting. Exhaust, return air and outside air dampers are linked such that the return air can be introduced back into the apparatus at maximum load, or exhausted and replaced by all fresh air at partial or low load.

Figure 14.31 shows diagrammatically three dual-duct arrangements, each of which has desirable characteristics for system control. The first arrangement passes air across a reheater or dehumidifier at

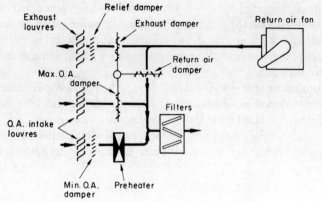

Fig. 14.30. Dual-duct return air system.

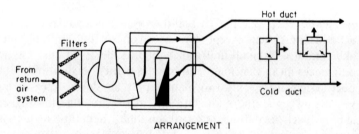

ARRANGEMENT 1

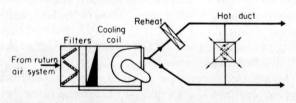

ARRANGEMENT 2

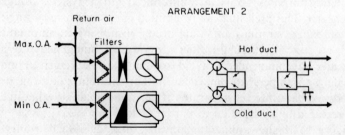

ARRANGEMENT 3

Fig. 14.31. Dual-duct supply systems.

the dictates of the individual demands of the terminal units. This arrangement is the simplest and most commonly used, but has the problem of allowing a percentage of the humid primary air into the hot duct and so causes a room latent heat gain to the room. A modification to this system would be to provide a dehumidifying coil in the minimum fresh air connection. Although overcoming the problem of high room relative humidity this adds balancing problems since the return air system must be dampered to create an equivalent resistance.

The second arrangement overcomes both these problems by dehumidifying all the air before it is sent directly to the cold duct or reheated in the hot duct. This arrangement, although providing excellent control, adds to the running costs of the system at partial load, the reheater having to put back the sensible heat removed at the dehumidifier.

A third arrangement utilises two fans, the cold duct handling the minimum fresh air at all times thus providing better humidity conditions. Return air, or all fresh air when required, is mixed into the cold duct system to maintain correct air volume.

No matter which arrangement is adopted all have the problem of maintaining good humidity conditions at partial load. This situation can be improved by the location of a humidistat located in the common return air duct sensing the average humidity condition within all zones. This will override the reheater control to provide warmer air, creating a requirement for greater quantities of cold dehumidified air resulting in lower humidity conditions at the expense of increased running costs.

ROOM FAN COIL UNITS

Applications of Room Fan Coil Units can be divided into several groups each providing varying degrees of adequate room conditions against installed capital costs. All units are supplied with filter, heat exchanger, fan and three-speed motor; accessories include cabinet for under the sill or ceiling suspension, fresh air connections, supplementary row or rows of heat exchangers for heating service only, two or three-way water control valves with optional automatic changeover features for single-coil operation, thermostat for fan speed or water valve control, and stamped or adjustable discharge grilles. Clearly these accessories when added to the basic unit to

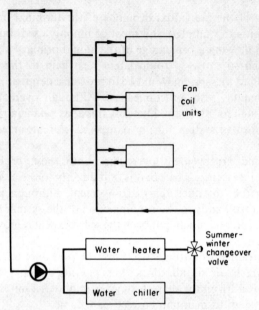

Fig. 14.32. Simplified two-pipe room fan coil system (see Fig. 14.34 for detailed water pipework and valves).

achieve improved room conditions also require extra services such as additional water pipework, additional electrical wiring, and primary air ducting.

All Fan Coil Systems provide individual room control, whether manual or automatic, and confined room air circulation. They can be located under the window to provide optimum room air distribution.

Figure 14.32 shows the most elementary system, the two-pipe system with changeover from heating to cooling at the plant room. This system is probably the most economically installed terminal system in that it eliminates the need for sheet-metal ducting and takes advantage of a diversified load at the refrigeration machine. It does not provide simultaneous heating and cooling such that adjacent modules, one requiring heating and the other cooling at mid-seasonal conditions, cannot be catered for, the result being that on the cooling cycle only a maximum temperature can be maintained and at the heating cycle only a minimum temperature can be maintained.

The simplest control system is the manual control of the three-speed fan switch which produces wide fluctuations in room

14 TERMINAL DEVICES AND AIR CONDITIONING SYSTEMS

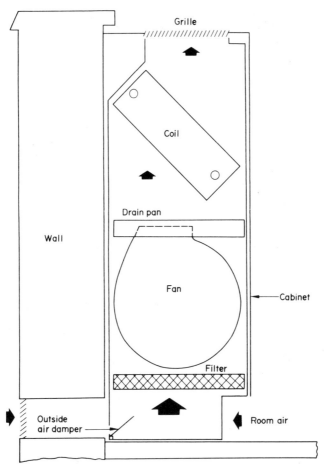

Fig. 14.33. Cross-section through room fan coil unit, showing fresh air damper.

temperature and humidity. The use of an automatically controlled water valve will maintain good temperature conditions because the coil surface temperature must be raised to cater for reduced sensible loads. The addition of electrical resistance coils controlled from a room thermostat, with constantly flowing chilled water during summer and mid-seasonal conditions provides excellent results of room temperature and humidity, since the system now becomes a terminal reheat system.

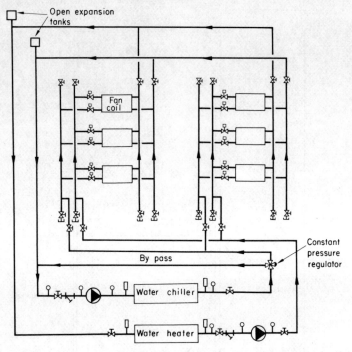

Fig. 14.34. Schematic four-pipe room fan coil system.

So far, however, no positive ventilation has been introduced into the system and stuffiness would result unless a separate ventilation system is introduced. Some systems rely on a ventilation system introduced through corridors to meet ventilation standards, this however, can cause air distribution problems since under the window unit distribution would act contrary to high sidewall distribution. Figure 14.33 shows a typical fan coil unit illustrated with an optional fresh air inlet from outside the building. This accessory economically overcomes the problem of ventilation, but could cause architectural problems since the inlet grilles can be seen on the building elevation.

The basic two-pipe system can be improved by using either a three-pipe system, one hot water and one cold water supply with a common return, or a four-pipe system using a separate hot and cold water supply and return. Both these systems provide simultaneous heating and cooling for mid-seasonal operation, the former having a more economical installed cost at the expense of increased running costs.

Figure 14.34 illustrates a four-pipe system, each fan coil unit having a thermostat able to select the operation of the heating or cooling valve without the need for an automatic changeover device. The advantages gained from multiple pipe systems are quick responses to temperature change since both hot and cold water are available immediately, the need for exposure zoning is eliminated and operational difficulties of the changeover system are avoided. As with the two-pipe system, primary air for ventilation purposes can be introduced through the outside wall or by using a separate ducted system.

The primary air fan coil system operates just as an induction system except that the motivation of secondary air movement is from the unit fan rather than the primary air itself. Figure 14.3 can be interpreted as this system, although the primary air is often distributed via low velocity ductwork. Simultaneous heating and cooling are provided with proper ventilation requirements. During summer design conditions, although some room load is catered for by the primary air cooling effect, the unit size selected for this system would probably be of equal size to the two-pipe system, because of the resultant lower temperature on to the cooling coil.

Providing the secondary water temperature to the fan coil unit is above that of the room dew point, drain connections can be eliminated. However, all other applications of room fan coil units require that drains be fitted, but since the drain pan is subjected to atmospheric conditions a drain trap can be omitted. Attention should be given to this point such that an adequate fall is maintained; this can be a difficult problem if there are many units on a long elevation.

The applications listed in Table 14.1 are again numerous and have been categorised as listed below to give an indication of the zoning and partial load requirements of each.

1. Residential
The architecture of residences is such that no one system can be advocated. However, in every case it is good practice to reduce any wide fluctuations in heat gain by the use of shading devices, etc. Common practice dictates that full effect of storage and temperature swing should be allowed for. When room air conditioners are used they should be so sized as to operate continually, so as to precool during the morning and allow the storage effect to overcome the heat gains of the afternoon and evening. A general rule is to undersize a room air conditioner to 60 per cent of the instantaneous heat gain at

TABLE 14.1
AIR CONDITIONING SYSTEMS AND THEIR APPLICATION

Category	System	Applications														
DX systems	Room air conditioners	1	2						9							
	Air-to-air heat pumps	1	2						9							
	Free-blow packaged equipment		2	3		5		7	9							
	Ducted packaged equipment	1	2	3		5		7	9							
All-air systems	Return air face and by-pass					5	6	7								
	Multizone					5	6		9	10	11					
	Double duct								9	10	11	12	13	14	15	
	Variable volume				4					10	11	12				
Air and water systems	Induction									10	11					
	Induction reheat								8							
	Reheat					5			9	10	11	12		14		
	Room fan coil (with primary air)				4				8	9	10	11	12	13	14	15
All-water systems	Room fan coil	1							8	9	10	11	12			

maximum load. Should the residence be of light structure the undersizing must be reduced.

2. Small shops
The operation of the air conditioning plant would normally be 8 to 10 hours, and the load within the space would be greatly affected by the frequency of occupation and ventilation requirements. It is good practice to carry out some precooling using merchandise and stock as storage, either floor mounted or ceiling mounted free blow units; similar units or roof mounted units with distribution ducting would be adequate, depending upon physical shape of area. Very small shops can use large room air conditioners located in the transom above the entrance.

Care should be taken with shops having special temperature or humidity requirements such as furriers, chocolate stores, etc.

3. Supermarkets
Generally, supermarkets tend to be single-floor buildings and are well suited to roof mounted air cooled packages. Where supermarkets form part of a multistorey building, split air cooled packages are common. As with small shops precooling should be used to advantage.

4. Department stores
The larger the store the more individual will be the air conditioning requirements per floor. However, this can be an advantage to the single water chilling plant which can take into account diversity of operation. It is also worth noting the frequency of occupation which could influence the design of the plant. (It is not normal practice to size the plant for an occupancy level which may only occur at a January sale for example.) Normally a central plant would provide dehumidified ventilation air to each floor, and a common chilled water system would serve multiple air systems on each floor. The location of air plants is critical because of the expected return per square foot of sales floor area.

5. Restaurants
Particular attention should be paid to the short periods of occupation which may occur at lunch time and/or in the evening or night. A major factor is the requirement of ventilation for the occupants and

exhausts from kitchens to prevent odours from food and smoking. Therefore, restaurant areas should be kept at a positive pressure. Because of the high latent gains return air face and by-pass or reheat systems are advantageous. Much of the heat gain from cooking appliances can be eliminated by good exhaust systems.

6. Theatres

The architectural shape of theatres or auditoriums, which have to cater for large occupation, plays an important role. Because of the frequency of operation high buildings can use the large air volume to act as a buffer to minimise ventilation requirements, and also allow for stratification. Low areas require careful treatment so that the air distribution does not cause discomfort to the patrons. Careful consideration should be made to outside conditions at the time of peak occupancy. As with restaurants the high latent load will demand a return air by-pass or reheat system.

7. Factories

Many factory areas require special treatment for temperature and humidity; however, where comfort takes the major role careful consideration should be paid to the architecture of the factory. Often the cost of an air conditioning system can be reduced, with subsequent running cost savings, if attention is paid to the insulation of a roof, which may well prove less expensive. Normally free-blow packaged units located at the perimeter of the factory will serve quite adequately, and high buildings allow large air volumes to be handled. Where low buildings or multistorey factories are being considered packaged units with sheet-metal ducting are necessary. Often partial cooling of a factory is effective if the cold air is directed towards the occupants.

8. Hotels

Because of the multi-room application terminal devices are economical. It is cheaper to pipe chilled and/or hot water and primary air around to the various rooms from a central plant, rather than tackle the problem with an all-air system. Occupancy is around the clock around the year. Diversity can be applied to a great extent since the majority of rooms are unoccupied during the day, although allowance must be made for occupancy at all times. Applying a diversity factor, for this event, lights and occupants gain at times of

14 TERMINAL DEVICES AND AIR CONDITIONING SYSTEMS

peak solar gains can be cut tremendously. Room Fan Coil Units having their own fan speed control or thermostat cater for this very well. Units are operated at low speeds during the night and are allowed to run at this level during the day unless the room is occupied. Where automatic controls are used they should be able to respond quickly and the system should be draught-free and quiet.

Where public rooms, restaurants, etc., are part of the hotel complex, they should be treated separately. However, it should be remembered that occupants cannot be in the hotel rooms and restaurant simultaneously, which can act in favour of diversity for the refrigeration plant.

9. Small offices

Free-blow or Ducted Packaged Equipment can be used providing attention is paid to partial load and zoning. Depending upon size, multizone systems or dual-duct systems could be used although these should be ruled out for multistorey applications where terminal systems prove more economical and flexible. For executive suites a reheat or induction reheat system would prove effective for close control of temperature and humidity.

10. Office blocks

Because of the need for zoning and partial load control high-rise office buildings lend themselves to terminal systems. Induction unit systems are commonplace for perimeter applications since they can meet the zoning requirements adequately and can meet the demand for heating and cooling; located below windows they prevent down-draughts and provide good distribution for areas within 15 to 20 ft (*4·57 to 6·10 m*) of the perimeter wall. Large floor areas having perimeter air conditioning are insulated from the external effects of solar radiation and transmission gain (except for the ground and top floors) in the centre sections and have a year-round cooling requirement. Variable volume terminal units or ceiling mounted, year-round cooling induction units can meet these requirements. Room Fan Coil Units can be used but consideration should be paid to the extensive wiring of individual units and the greater expense of such a unit compared to induction systems.

It is good practice to provide separate primary air plants to each face of the office building. The characteristics of the east face requiring cooling in the morning during summer months, the west face requiring cooling during the evening during summer months and the south face requiring cooling throughout the year, often greater during winter because of the low elevation of the sun, demand careful zoning consideration.

During mid-season operation perimeter area systems must provide both heating and cooling, and this can be achieved by the induction system with warm primary air and chilled water through the secondary coil. Unless all areas of a particular face have the same characteristics a room fan coil system may prove inadequate.

11. Hospitals

Patient rooms or wards have the same characteristics as office blocks, except their requirements are for 24 hour occupancy with no diversity. An important factor is that air circulation is confined to one room, which requires a positive exhaust creating a negative pressure. Because of the need for absolute cleanliness the rooms should be served with primary air adequately treated and filtered, which leans towards induction or induction reheat systems, although room fan coils with adequate primary air for cooling at mid seasons could be used.

Special areas such as operating theatres, therapeutic, surgical, mortuary, pathology, etc. must be specially treated with an all-air system. Because of their requirements it is common practice to use all-fresh-air plants with positive exhausts.

12. Schools

Even during 9- or 10-month occupancy of schools, high lighting intensity and occupancy demand year-round air conditioning. Ventilation and odour control are important in order to create a healthy environment for learning. The physical shape of school buildings plays an important role in air conditioning, current practice is single or two storey buildings of 'finger' design, spreading over large areas with large glass areas, and lightweight building construction. This reflects quick changes in load which could well be met with multizone units or return air by-pass systems. Latest trends are for schools to be windowless and of block construction so as to gain optimum enveloped area with minimum building fabric. This closed environment makes air conditioning mandatory—often year-round —and provides a comfortable atmosphere, distraction free, giving better pupil attentiveness resulting in better learning. Such systems can be met with variable volume systems much as the internal core of an office building.

13. Radio and TV studios
Electronic equipment and high lighting levels produce the need for air conditioning on a year-round basis. The central refrigeration plant serving many systems, each capable of wide variation in load to serve individual areas, can have large diversity factors applied, since all studios, etc. are not normally occupied simultaneously. Because of the varying size of the individual areas there may be a combination of equipment, such as multizone units, return air by-pass units for large areas, dual-duct systems, and room fan coil units in dressing rooms; a prime consideration must be sound and vibration-free equipment.

14. Laboratories
The nature of the laboratory will dictate the air conditioning system installed. Factors influencing selection are the accuracy in control of temperature and humidity, the amount of air extracted from fume cupboards, separation of individual areas, hygroscopic materials being tested or examined, diversity in the use of laboratories, whether 24-hour operation is required, and the degree of cleanliness. Laboratories can be grouped if their dew point levels are common. A single dehumidification plant for each area can maintain close levels of control. Care should be taken with areas having high exhaust rates through fume cupboards, as these exhaust rates may be in excess of the air volume required for air conditioning purposes. Where this case exists it may be feasible to introduce a separate untreated air system into the fume cupboard so that only a minimum amount of treated air is exhausted. Where areas require a high degree of filtration, the fine filter should be located at the discharge side of the air moving device. Care should be taken with corrosive elements within a laboratory as such areas would require all-fresh-air plants, with specially treated exhaust systems. Laboratories not requiring close humidity control could use multizone or dual-duct systems.

15. Computer rooms
The need for close control of temperature and humidity is questionable, the main criteria with most computers being that changes in temperature and humidity should be gradual. Because of the large amount of heat generated by computers and associated

lighting, etc., air conditioning is most often provided for the comfort of the occupants rather than the computer, although there are a few exceptions to this rule. Large sensible gains with little occupancy means high air volumes per ton of refrigeration (about 550 cfm/ton). This requirement rules out standard packaged plant designed for comfort applications. If packaged units are to be used, then they should be limited to those specifically designed for computer applications.

Standby plant tends to be synonymous with computer rooms, and care should be taken that the inclusion of standby does not result in unconditioned air constantly by-passing an unused evaporator section, which will result in unnecessarily high air volumes and the possibility of surplus latent removal from the operating coil(s). Often reheat systems are installed to provide the supposed accuracy of room conditions; in such instances care should be taken to ensure that the refrigeration plant working will not be always oversized. The high room sensible loads of a constant nature make variable volume systems a consideration for the larger installations.

Chapter 15

LIQUID CHILLING PACKAGES

The use of chilled water for air conditioning systems is confined to systems using multiple air handling plants, where it gives many advantages over a direct expansion system. The refrigeration plant is completely factory-assembled and requires only water pipework which requires less specialist treatment than refrigeration pipework and overcomes any oil return problems. Installation is simple and direct, and maintenance is confined to one location. Water pipework means less trouble with leaks and often simpler capacity control at the air handling unit resulting in better partial load operation. However, because of the extra heat exchanger required in the form of a chilled water cooler, the first cost per ton of refrigeration is higher, the evaporating temperatures are lower, the compressor is larger, and the operating costs are higher than a direct expansion system. However, the trend to central station cooling plants, and even district cooling installations puts the liquid chilling plant in the forefront of the larger air conditioning installation.

Basically, there are two types of cooler, flooded and dry expansion. The flooded cooler consists of a tube bundle and shell with the water flowing through the tubes. This arrangement permits cleaning of the water tubes where scaling can occur in such instances as spray type dehumidifiers or air washers. However, careful attention is required with the flooded cooler as it causes oil return problems, has a tendency to freeze up and requires liquid level control using a high or low side float valve. The trend in packaged chilling equipment is away from this type and only a few per cent are installed in comparison to the more favoured dry expansion cooler which has a similar arrangement but with refrigerant within the tube bundle.

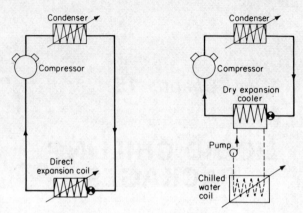

Fig. 15.1. Comparison of direct expansion and chilled water cooling and dehumidification.

The dry expansion cooler is available in two basic configurations, the straight tube with two header plates and the 'U' tube which requires only one header. Figure 15.2 shows the arrangements of these two types. The straight tube is often internally finned to give a better heat exchange and the tubes can be withdrawn individually from either end in the event of leaks. Chemical cleaning of the water side of a straight tube design is required if fouling becomes a problem. The 'U' tube design permits the withdrawal of the whole tube bundle

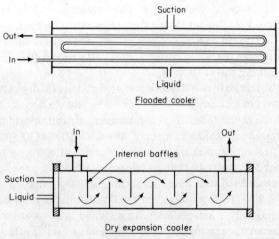

Fig. 15.2. Comparison of cooler types.

15 LIQUID CHILLING PACKAGES

so that inspection and, if necessary, cleaning can be made at regular intervals. The 'U' tube is preferred for low temperature brine applications or wide temperature ranges because of the floating action of the bundle allowing for expansion. In the event of tube damage as with the straight tube design a few tubes can be plugged without impairing the performance too much; however, if the damage is extensive it is normally necessary to replace the whole tube bundle.

Dry expansion coolers use expansion valves for regulating refrigeration flow and they are available for a wide range of refrigerants, although in package systems R.22 is commonplace. For improved part load performance multiple circuit coolers are often used with a thermal expansion valve and liquid line solenoid valve on each circuit. Because such units are factory-assembled and tested, the oil return problems are solved by the manufacturer.

The most common compressor used for liquid chilling application is serviceable hermetic; however, some manufacturers still use open machines normally direct driven, but this is losing favour. The hermetic compressor is ideally suited to the application which has relatively constant evaporating and condensing levels and has the advantage of smaller size, making a compact unit. The trend in larger chillers is to use a multiple compressor design which gives many advantages.

The performance of multiple compressor design gives advantages in lower starting currents since each is started in sequence, and although each is started direct-on-line, providing easier starting, the maximum inrush current overall is lower than what could be obtained from Star-Delta start, part-wound start or even stator–rotor start of a single compressor design. Because compressors are shut down at partial load the on-load compressors are working at full efficiency, whereas the single compressor design drops off in efficiency at light loads and could require power factor correction. In addition, where a nominal duty is required from the plant at times of peak electrical demand, compressors can be locked out to prevent high demand charges.

The use of multiple compressors gives standby protection in the event of a safety lock-out or equipment failure. By use of two standard compressor sizes a manufacturer can offer a wide range of equipment with standard components and low cost replacement. For example a 20-ton and 30-ton compressor can in multiples of 2, 3 or 4, give 10-ton increments from 40 to 120-ton capacity, resulting in a

Fig. 15.3. Hermetic reciprocating liquid chilling package. (Courtesy: Carlyle Air Conditioning Co. Ltd.)

minimum of 50 per cent standby. Because of the lower starting currents smaller power lines are required and the d.o.l. design makes starter costs lower. For further reduction in starting currents some manufacturers make available part wound accessories for each contactor, which in the case of a four-compressor unit would result in a maximum inrush current of about 150 per cent of the maximum running current.

The condensing medium is most often water cooled, the condensers being part of the whole chiller assembly. Where multiple compressors are used multiple condensers are supplied to match the compressor and split evaporator performance. Because dry expansion coolers are used with a flooded water design only one inlet and outlet connection is required for the chilled water. However, multiple condensers require a water manifold to provide single inlet and outlet connections. On larger chillers shell and tube condensers are more common, providing maintenance advantages, but many smaller units are supplied with shell and coil condensers which require chemical

15 LIQUID CHILLING PACKAGES

cleaning of the water tubes in the event of scaling, which is commonplace when using cooling towers. In the event of failure of a shell and coil condenser it is necessary to replace the whole assembly, whereas the shell and tube design provides facility for tube replacement and cleaning by the removal of the water box.

Chillers are available without condensers for remote air cooled or evaporative condensers and it is normal practice for the compressor to be provided with a larger motor to meet the higher power demands of the higher head pressures associated with air cooled condensing. The selection of air cooled or evaporative condensers must be made with care, and suitable head pressure control must be installed to prevent low suction pressures at start-up, which would cause nuisance trips on the low water thermostat to prevent freezing.

The water chilling plant requires many control devices which are generally supplied with the equipment within a control panel which should be complete with the following: starting equipment with overload protection; suction and discharge pressure gauges for each circuit; chilled water temperature control; on–off switch; high and low pressure switches for each circuit; chilled water safety thermostat; motor winding high temperature cutout; control circuit fuses; timer device to prevent rapid cycling; compressor crankcase heater relays; electrical interlocks for chilled water pump, condenser water pump and cooling tower fan.

In addition, oil safety switches for each compressor could be fitted and in the case of multiple compressors a sequence controller to alter the starting sequence of the compressors to provide average compressor usage for longer life.

The most common method of controlling the chilled water temperature is from a thermostat located on the entering side of the cooler. This takes advantage of the storage effect of the water system and realises a change in operating conditions before the water enters the chiller; in addition the thermostat can be controlled over a wide throttling range equivalent to the design split between entering and leaving water temperatures. However, this system cannot be used where the chilled water flow can change, such as the case when throttling two-way valves are used on chilled water coils. The operation of the plant of say 50 tons (*176 kW*) capacity, cooling 100 imp. gal/min (*7·6 litres/s*) between 55 and 45°F (*12·8 and 7·2°C*) means that at 55°F (*12·8°C*) entering water to the chiller, full compressor load will be called for. In the event that the circulated volume of water was dropped to 50 imp. gal/min (*3·8 litres/s*), the leaving temperature would drop to approximately 35 °F (*1·6 °C*) and cause nuisance trips on the low limit thermostat to prevent freeze up of the cooler.

Figure 15.4 shows the effect of return water temperature to leaving temperature against load conditions for a four-equal-stage water

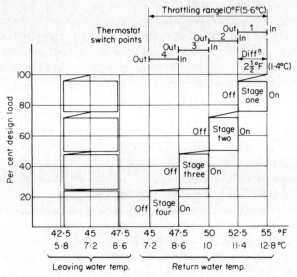

FIG. 15.4. Chilled water temperature control from entering (return) water.

chiller controlled from entering water temperature. At 100 per cent load the chiller will cool a specified water quantity through the design water temperature split, in this case considered at 55 to 45 °F (*12·8 to 7·2 °C*). Having four equal stages means $2\frac{1}{2}$ °F (*1·4 °C*) between each unloading step; it could be anticipated that the leaving temperature would also fluctuate by this amount. However, at the lower leaving temperatures the chiller is not capable of achieving design capacity because in this case it would approximate to 96 per cent capacity. Therefore, as the load drops from 100 to 96 per cent the leaving water temperature would drop by 2·1 °F (*1·2 °C*), to bring the chiller in balance with the load. However, a further drop in load will result in the return temperature falling more than 2·5 °F (*1·4 °C*) or to below 52·5 °F (*11·4 °C*), causing the unloading of one stage. At this

15   LIQUID CHILLING PACKAGES   315

condition the leaving temperature will rise to design, 45 °F *(7·2 °C)*, and circulate through the system. The system water volume and the increase in load over chilled capacity will dictate the duration of this higher temperature water returning to the chiller to be sensed at a higher temperature than the cut-in set point of the thermostat, calling for four stages of cooling again. Therefore between 96 and 75 per cent load the return and leaving temperatures will fluctuate between 55 and 52·5 °F and 47·1 °F and 42·6 °F *(12·8 and 11·4 °C and 8·4 and 5·9 °C)* respectively. The larger the system water volume, the less frequent the cycling which will also be affected by the increase of compressor capacity to cooling load.

As the load falls below 75 per cent so the above sequence of operation repeats itself, the capacity of the chiller being matched to load from 75 per cent, and the compressor cycling between 75 and 50 per cent duty as the load drops from 72 to 50 per cent and similarly for the lower stages.

A common misconception is that the return water temperature is a complete function of the load and that the stages of compression will only cycle as the load drops below the equivalent compressor capacity. However, as demonstrated above, this is far from being the case.

Figure 15.5 shows the return and flow temperatures against load for the machine discussed above, but controlled from leaving chilled water temperature. To minimise cycling it is necessary to provide a 4 °F differential on each step with a 0·75 °F *(0·4 °C)* differential between steps. The narrow differential in temperatures between cut-in and cut-out means more rapid cycling of the compressors, which makes this method less preferable than the return thermostat control. However, this control does give the advantage that at lighter loads the leaving water temperature drops, providing greater dehumidification at the coil which is a desirable effect.

Figures 15.3 and 15.4 indicate that for a four-stage chiller the leaving water temperature can be 2·4 to 6 °F *(1·3 to 3·3 °C)* below the design leaving temperature, depending on the control method used. Since water has an anomalous expansion resulting in an increase in specific volume below 39 °F *(4 °C)* to freezing at 32 °F *(0 °C)*, it is dangerous to design lower than 40 °F *(4·5 °C)* leaving, for the return water temperature control, and 44 °F *(6·7 °C)* for the leaving temperature control. Any design temperatures leaving the chiller below this would necessitate a narrowing of the thermostat differential, resulting in rapid compressor cycling, or worse, a dangerously low-set safety thermostat, normally set at 36 °F (2·2 °C) to avoid nuisance trips.

Often multiple machines are used, either for machine standby or because economics dictate two of a manufacturer's larger machines would prove more economical than a centrifugal chiller.

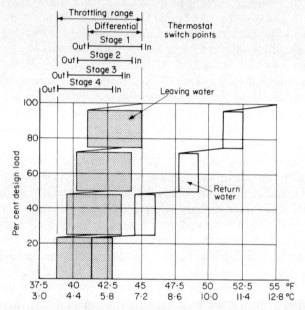

FIG. 15.5. Chilled water temperature control from leaving water.

Reciprocating machines are available up to approximately 150 tons R (528 kW) capacity in a single machine, and above this, say up to 200 tons R (704 kW), two reciprocating chillers could be cheaper in capital cost than a single centrifugal machine, although this break-even point differs from manufacturer to manufacturer.

Multiple machines may be installed in two arrangements, parallel or series flow. Parallel arrangement gives the advantage of low water pressure drops for a given cooling load through a fixed temperature range, but has the problem of mixing at light load when one machine is shut down for economy. It is commonplace to use return water temperature control with parallel machines, each machine having identical set points. The natural tolerance of control will ensure sequence of operation, resulting, in the case of two four-stage machines, in eight capacity stages. Examination of the part load performance of chillers will dictate when one machine should be closed down. In the case of single-compressor machines this would approximate to 35 per cent full load of 70 per cent single-unit capacity with a cut-in setting of approximately 80 to 90 per cent single-unit

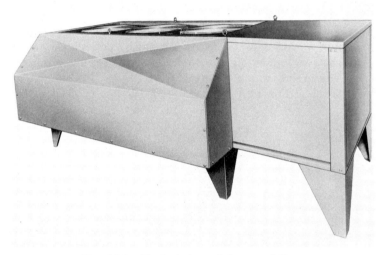

FIG. 15.6. Typical air cooled water chiller.

capacity. Multiple compressor design dictates that the economical point of stopping one machine is a point just below the single-compressor capacity. This stopping of one machine at light load is achieved simply from a thermostat in the common return line before splitting to each chiller. Lag-lead control can be brought into operation by means of a changeover switch which will send the signal to either machine.

A similar approach can be made using leaving water temperature control of the two chillers if other than constant water flow can be expected.

In cases where the water temperature split is high, chillers can be installed in series. With constant flow both machines can be controlled from their respective thermostats. Proper attention must be paid to the setting of the controls since the first chiller will provide more cooling duty because of the higher entering and leaving temperatures; however, by use of counterflow series condenser pipework, if wide temperature split can be tolerated, the load on each machine will be nearer balance. It is normal to allow the first machine to unload fully before switching off and allowing the second machine to operate at its design depressed temperature levels. As mentioned previously, the addition of a thermostat in the return line with use of a changeover switch will provide a means of running each machine equally at partial load.

Series arrangement can be used for variable water flow by changing the control to leaving water as with the single-machine operation.

In all cases before setting up a control sequence the manufacturer should be consulted as most controls are factory preset, and have to be commissioned to suit the job requirements.

Chapter 16

PACKAGED EQUIPMENT

The term 'packaged' is used very loosely. Room air conditioners are called 'packaged units' and even centrifugal water chillers of many thousand tons capacity are called 'packages'. So far as the author is concerned, packaged equipment refers to complete systems, comprising:

(i) factory-matched refrigeration cycle and air moving components;
(ii) performance which is factory set and engineered;
(iii) factory-made assemblies of one or two finished cabinets designed for installation in the field with little or no internal piping or wiring.

Current trends are for this equipment to be contained within 50 tons (*176 kW*) capacity, but this limit will doubtless be exceeded in the fullness of time. Before using packaged equipment the engineer should be aware of the wide range of apparatus available to him.

Packaged equipment is often referred to as a compromise; it is then misapplied, but for comfort applications there is rarely a better compromise when the equipment is properly selected. A tailor-made, build-up system may suit a certain application better than packaged equipment for a maximum anticipated load demand, but equipment rarely runs at design, and therefore, unless the tailor-made equipment has a variation in its load capacity with suitable controls, the net result for possibly 90 per cent of its operation would prove no more beneficial, although it would probably cost much more than a packaged unit.

Packaged equipment can, for simplicity, be broken down into four categories, although further themes can be worked on these. They are: single-piece systems (air or water cooled); single-piece air cooled packages; single-piece water cooled systems; and split systems.

SINGLE-PIECE SYSTEMS (AIR OR WATER COOLED)

These units have a fixed refrigeration performance matched to an air handling unit which, over 3 tons ($10.5\,kW$) capacity, normally has a variable pitch pulley to adjust the air volume from 300 to 500 cfm/ton (0.14 to $0.23\ m^3/s/kW$) to match design room sensible heat ratios. However, applications with high sensible loads requiring high air volume output do not have a high external pressure available, and the unit should be checked to consider whether a special drive and/or motor is required.

Since below 7 or 8 tons (25 or $30\,kW$) capacity these units do not have capacity reduction features, care should be taken in selection so that the unit is not oversized. Too often, a piece of apparatus is selected on the conservative side against a load estimate, considering an instantaneous gain at ambient condition probably reached for only 1 per cent of the year, with a 10 or 15 per cent safety margin thrown in as well. For comfort applications this is very bad practice, since the unit will be cycling on the thermostat too often and give erratic control. Applications having a large outside air volume would exaggerate this effect.

Consideration should be given to the partial load operation of the plant where the unit being controlled only from a thermostat will not cater for a fairly constant latent load. The effect of oversizing will cause the temperature at, say, 50 per cent partial load to swing between the range of the thermostat, probably $\pm 2\,°F$ ($1.1\,°C$), or say 70 and 74 °F (21.1 and $23.3\,°C$) every 5 or 10 minutes with the moisture level increasing steadily over the day, since only when the unit is cooling is dehumidification taking place.

In contrast, an undersized unit based on an ambient condition which is not exceeded more than 5 per cent of the time, against a load estimate allowing for building storage without any safety factors and a realistic diversity on internal gains, will operate on a more continual basis. The net result would be a temperature swing from say 70 to 74 °F (21.1 to $23.3\,°C$) during the whole day, with constant dehumidification.

16 PACKAGED EQUIPMENT

To obtain better internal conditions while giving a saving in capital and running costs, not to mention a longer equipment life, must be a serious consideration. Some years ago in the United States, a survey proved that for residential purposes units with 60 per cent capacity of the calculated instantaneous cooling load produced the best year-round comfort conditions. In considering cooling loads at lower ambient conditions, lower sensible heat factors will be found, which will practically always be around 400 cfm/ton ($0.19\ m^3/s/kW$), or the manufacturer's nominal capacity.

The advantages of reaching this level are that standard drives could be used, a lower noise level would be obtained from the fan and the air volume would be reduced, giving more economical sheet-metal ductwork and simpler air distribution.

Refrigeration pipework is not required on site since the manufacturer has provided this in his unit. Standard accessories to be found in single-piece units are thermal expansion valves with external equalisation, liquid line strainer/driers, sight glasses, hot gas mufflers and insulated suction lines, together with an operating charge of refrigerant.

SINGLE-PIECE AIR COOLED PACKAGES

These are designed for external use and are fully weatherproofed. Smaller units can be located through a wall with the evaporator half internal; however, weight limitations usually call for the unit to be located on a flat roof or at ground level adjacent to or above the conditioned space (see Fig. 16.1). These units are provided with inlet and outlet spigots for sheet-metal ductwork which has to be made to suit the job requirements. Often filters are not provided and must be installed in the return air duct, but the manufacturer's catalogue should be checked on this matter. Most units are cooling only, although heating coils are available as accessories to be located in the discharge ducting.

Site electrical work consists of a three-phase four-wire supply to the unit which must be isolated (some units up to 2 tons are available for single-phase duty) and control wiring to a remote thermostat from a prewired terminal block in the unit. Where heating is required, the controls should be interlocked with the evaporator fan, which can be

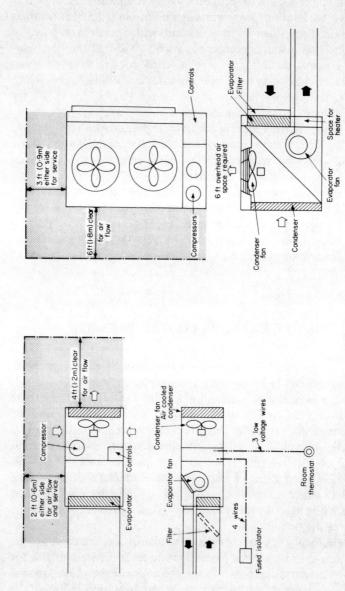

FIG. 16.1. Typical plans and cross-sections of single-piece packages, showing air flows and service requirements.

FIG. 16.2. Single-piece air cooled packaged unit. (Courtesy: Carlyle Air Conditioning Co. Ltd.)

achieved through the same control terminal block. Most manufacturers offer or recommend room thermostats with switch bases for selection of heating or cooling; fan, auto and off controls should also be included where possible. All too often air conditioning contractors have installed sophisticated control systems specifically designed for the job application on a one-off basis. In reality, such an approach could represent between 10 and 15 per cent of the equipment costs, whereas controls designed for the machine rather than the application would rarely exceed 2 per cent of the equipment costs. It is worth stressing that the use of 'customised' controls implies a special application for which the packaged equipment could well be misused.

In essence, up to 7 tons ($25\,kW$) single packages have a single-stage on/off control and any amount of sophisticated control equipment will not improve the performance.

The practical size limitation of single-piece air cooled packages is about 50 tons ($176\,kW$). This is not a limit imposed by the manufacturer but essentially by the physical size of the machine which has to be housed, usually at roof level. An indication of the physical size of such plant at about 25 ft × 8 ft × 5 ft ($7\cdot6\,m \times 2\cdot5\,m \times 1\cdot5\,m$) will give the designer some appreciation of potential site difficulties.

324 APPLIED AIR CONDITIONING AND REFRIGERATION

However, it should be appreciated that design trends are more and more favouring such roof-top equipment which not only is probably the cheapest form of air conditioning per ton, but also reduces contractual risk to a minimum, making the application favourable for both owner and contractor.

The trend towards single-piece equipment has allowed manufacturers to consider factory or field installed accessories to make a more flexible, energy conscious system. Accessories include roof curbs, discharge/return air electric or hot water reheaters, fresh air economisers, and exhaust fans, all of which can be combined to meet the exact job requirements. Compatible with such accessories are plug-in control components and room controls to achieve ultimate job flexibility. The state of the art is such that engineers have so much flexibility that they are designing jobs around the equipment, rather than making the equipment fit the job.

SINGLE-PIECE WATER COOLED SYSTEMS

These systems are designed for internal use with an external cooling tower (see Fig. 16.3). Units have been used which run water to waste, but this method should be frowned upon unless the water can be used for process purposes where the added heat rejection can be usefully served.

Because they are used internally, manufacturers attempt to make their design to pass through doorways; this means a vertical type unit giving the advantage of minimum floor area requirement. Larger units have a removable fan section to make handling easier.

Units up to 15 tons ($53\,kW$) capacity can be used for free-blow application within the conditioned space by use of a plenum distribution chamber mounted on top of the unit. Above this size, free-blow should not be used, as adequate distribution can rarely be met.

It is common to mount the units within the conditioned space without the need for return air ducting and remote controls.

The electrical wiring requirement is an isolated three-phase, four-wire supply to the unit with a terminal block for electrical interlock with a cooling tower and condenser water pump which require their own individual electrical supply. The controls are located on the unit and are prewired with the sensing element in the return air section of

16 PACKAGED EQUIPMENT

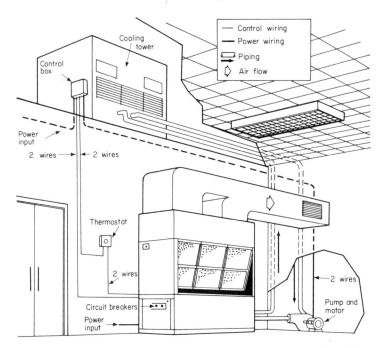

Fig. 16.3. Typical installation of single-piece water cooled package. (Courtesy: Carlyle Air Conditioning Co. Ltd.)

the unit. However, external controls can be fitted, as with the single-piece air cooled unit. Heating coils can be installed within the unit cabinet, and manufacturers offer these as optional accessories.

Units up to 7 or 8 tons (*25 or 30 kW*) are usually on–off control, up to 25 tons (*88 kW*), two-stage control, and above this, three-stage. This, however, is not a hard and fast rule, and larger units can often have more stages added by installing unloading compressors in addition to cycling multiple compressors which are used in the larger units. The use of multiple compressors is an advantage, since by use of a time delay relay reduced voltage starts can be obtained.

These units are also offered without the water cooled condenser for remote air cooled condenser application. The electrical wiring is basically the same, the condenser being interlocked in lieu of the cooling tower. Liquid and discharge refrigerant pipework is required in lieu of the condenser water pipework and pump. These applications often require a large winding on the compressor motor to cater for the higher head pressure attained with air cooled systems.

SPLIT SYSTEMS

These comprise a fan coil unit, often located within the conditioned space, and a remote condensing unit, normally air cooled. Water cooled condensing units can be used, although, like fan coil units, they are normally located internally. They have no advantage over the single-piece water cooled units unless a horizontal fan coil unit is required or more than standard or non-packaged duty is required.

The fan coil section of the package is complete with filter, evaporator coil, fan(s) and motor and drive in a finished cabinet. Both horizontal and vertical arrangements are available with many accessories for each such as heating coils, either hot water, steam or electric; sub-bases; return air grilles; distribution plenums and grilles; and suspension packages (see Fig. 16.4).

Up to 5 tons ($18\,kW$) capacity fan drives are normally direct with three-speed motors; above this size, variable pitch pulleys are supplied. Some manufacturers exclude fan motor and drives, which should be selected to match the air volume required against the calculated external pressure of the sheet-metal ductwork. Filter sections are always fitted or available as accessories, but some manufacturers do not supply the filters themselves.

The condensing unit is available in many sizes (see Fig. 16.5) to give a wide number of combinations with the fan coil units. For a given refrigeration duty the air handling capacity can be adjusted between 300 and 800 cfm/ton ($0\cdot14$ and $0\cdot38\ m^3/s/kW$) by mixing sizes. In addition, because the fan coil unit is capable of a variable volume, a larger or smaller unit could be used to make the most economical selection or the quietest unit. Because the fan coil unit is located internally the advantage over the single air cooled package is that the sheet-metal ductwork is reduced for the addition of suction and liquid refrigerant lines. The condensing unit is normally complete with hot-gas muffler, liquid line strainer/drier and sight glass and often thermal expansion valves. Up to 5 tons ($18\,kW$) capacity, precharged refrigerant tubing is available with quick coupled connections.

Electrical wiring required is an isolated three-phase, four-wire supply to the condensing units (single-phase available up to 2 tons

16 PACKAGED EQUIPMENT

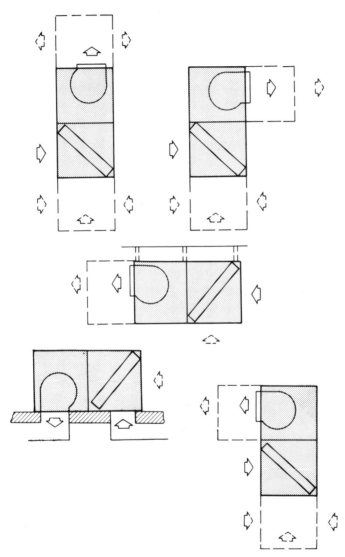

FIG. 16.4. A variety of arrangements for packaged fan coil units.

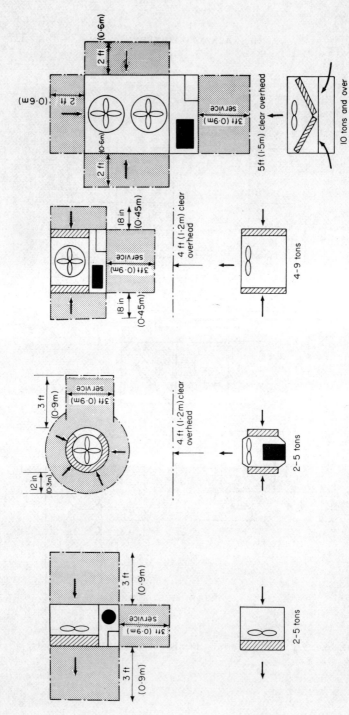

FIG. 16.5. Air cooled condensing units showing air flow and service requirements.

Fig. 16.6. Small air cooled condensing unit. (Courtesy: Carlyle Air Conditioning Co. Ltd.)

($7\,kW$) capacity) an isolated single-phase supply to the fan coil unit [over 1 hp ($0.75\,kW$) fan motor (approximately 8 tons or $28\,kW$ capacity) three-phase] and control wiring from a low voltage transformer at the condensing unit to a remote thermostat and fan relay.

It has already been said that oversizing of packaged air conditioning apparatus is undesirable, but really the situation should not arise if a proper site survey has been carried out. In fact, a proper site survey is essential no matter what the application, together with a good appraisal of the anticipated cooling load. The first consideration in sizing apparatus is the desired end result—the comfortable condition. This condition is often a subject of argument, so much so that current practice dictates no temperature graduation on single-piece packaged equipment. Every individual has his own preconception of an ideal temperature condition, which, needless to say, rarely coincides with his colleagues' in the same conditioned

FIG. 16.7. Air cooled condensing unit. (Courtesy: Carlyle Air Conditioning Co. Ltd.)

space. The situation of the Spartan athlete reaching the office first to set the thermostat to 65 °F (*18·3 °C*), to be followed by the dear spinster employed by the company for the past 40 years resetting to 75 °F (*23·8 °C*) is too often the case. Change the scale to read 1, 2, 3, 4, 5 and the situation ceases to exist. Notwithstanding this, the designer must consider the ideal comfort condition for his clime. It would be foolish to hold a room condition to 68 °F (*20 °C*), 50 per cent RH in an ambient condition of 110 °F (*43·3 °C*) DB, for, as many sea voyagers are only too aware, the quickest way to catch a cold is to leave the dining saloon, controlled at a condition compatible to the English Channel, for a stroll on deck in the tropics. It can safely be argued that in the United Kingdom a summer condition of 72 °F (*22·2 °C*), 50 per cent RH will satisfy most people. One unfortunate fact of life is that it is impossible to satisfy everybody at any one condition, so the statistical optimum is the best guideline.

Ambient conditions are often overestimated but the question one should ask is how often is the design condition reached or surpassed? The design condition for London is usually taken as 82 °F DB, 68 °F WB (*27·8 °C DB, 20 °C WB*), which can only be expected for 1 per cent of the year. A design condition of 76 °F DB, 65 °F WB (*24·5 °C DB, 18·3 °C WB*), is only surpassed for London for 5 per cent of the year, in Birmingham 2 per cent of the year, Cardiff 1 per cent, Belfast 1 per cent, and even more rarely in Glasgow and Edinburgh.

A cooling load estimate form (and a metric equivalent) suitable for comfort applications is printed here with an example. It takes into consideration an ambient condition of 76°F DB, 65°F WB (*24·5°C DB, 18·3°C WB*), with an internal room condition of 72°F (*22·2°C*),

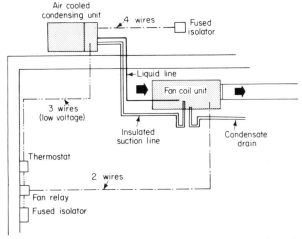

Fig. 16.8. Typical split package system showing external services.

50 per cent RH. This form will cover at least 95 per cent of all comfort cooling applications in the United Kingdom.

The constants used are for typical constructions, and provided a proper load survey is carried out it will produce answers suitable for comfort applications. This form, however, should be frowned upon for process calculations or zones requiring critical design conditions at all times of the year. The diagram (Fig. 16.10) shows a plan of a drawing office, which serves as an example for the use of the form printed.

From the notes given on the form, the correct factors could be obtained and, by adding the areas and fabric, together with lighting intensity and number of occupants, the form reveals the total cooling load, the supply air volume of the apparatus, the outside air volume required for ventilation and the entering wet bulb condition at the evaporator. From this information it is easy to select the correct piece of packaged equipment.

APPLIED AIR CONDITIONING AND REFRIGERATION

Fig. 16.9 'REFRIGERATION AND AIR CONDITIONING' COMFORT COOLING ESTIMATE FORM

ROOM CONDITION 72°F 50 per cent. RH AMBIENT CONDITION 76°F d.b 65°F w.b.

CLIENT: —	PROJECT: DRAWING OFFICE	ESTIMATE No. 1
	LOCATION: CROYDON	ESTIMATE BY: CTG
	AREA: 60' × 30' = 1800 sq. ft. × 10 = 18000 cu. ft.	DATE: 28 NOV 72

GLAZING

NOTES	ASPECT	FACTORS							QUANTITY AREA sq. ft.	COOLING LOAD B.t.u.'s/hr.
		SINGLE PANE			DOUBLE GLAZING					
		BARE GLASS	VEN. BLIND	AWNING	BARE GLASS	VEN. BLIND	AWNING			
	N OR SHADE	(21)	15	16	14	11	11	64		1 344
FACTOR × AREA = SOLAR AND TRANS. GAIN	NE OR NW	91	70	53	68	48	31			
SELECT ASPECT WITH HIGHEST GAIN ALLOW ALL OTHER ASPECTS EXCEPT SHADE, EXCEPT FOR HORIZONTAL	E OR W	125	(87)	59	99	66	38	128	11	136
	SE OR SW	107	74	53	84	56	31			
	S	78	56	40	60	42	25			
	HORIZONTAL	178	124		140	97				

EXT. WALLS

	ASPECT	11 in. CAVITY BRICK WITH PLASTER	9 in. BRICK WITH PLASTER	6 in. CONCRETE WITH PLASTER	BRICK OR CONCRETE WITH 1 in. INSULATION	COLOUR FACTOR		
						DARK	MEDIUM	LIGHT
	N, NE, SHADE	(1.4)	1.9	2.5	0.9	(1.0)	0.8	0.6
FACTOR × COLOUR FACTOR × AREA = GAIN ALLOW ALL ASPECTS SUNLIT UNLESS SHADED	E OR SE	6.0	8.1	11.0	4.0	1.0	0.8	0.6
	S	7.3	10.0	13.5	4.9	1.0	0.8	0.6
	SW OR W	(9.0)	11.8	16.0	5.1	(1.0)	0.8	0.6	(600−128)= 472	4 248
	NW	6.2	8.5	11.5	4.1	1.0	0.8	0.6

ROOF

		6 in. CONCRETE NO CEILING	6 in. CONCRETE WITH SUSP. CEILING	ASBESTOS OR WOOD WITH SUSP. CEILING	CONCRETE + CEILING WITH 1 in. INSULATION			
FACTOR × COLOUR FACTOR × AREA = GAIN	EXPOSED	15.0	6.3	8.0	3.6	(1.0)	0.8	0.6	900	3 240
	SHADED	2.0	0.8	1.5	0.5	1.0	0.8	0.6

INT. SURFACES

	CEILING	UNCONDITIONED SPACE ABOVE (0.8)						900	720
FACTOR × AREA = GAIN	FLOOR	UNCONDITIONED SPACE BELOW 1.0	GRD. FLOOR OR BASEMENT (NIL)					1800	—
	PARTITIONS	METAL OR METAL & GLASS 5.0	WOODEN 2.5	MASONRY (1.5)				900	1 350

(300−64)=236

16 PACKAGED EQUIPMENT 333

Fig. 16.9. (By permission of 'Refrigeration and Air Conditioning' Magazine.)

334 APPLIED AIR CONDITIONING AND REFRIGERATION

Fig.16.9a COMFORT COOLING LOAD ESTIMATE FORM (SI UNITS)

ROOM CONDITION 22.2 °C 50 per cent RH AMBIENT CONDITION 24.5°C d.b. 18.3°C w.b.

CLIENT: _____
PROJECT: DRAWING OFFICE
LOCATION: DAVIS HOUSE CROYDON
AREA: 18 × 9 = 162 m² × 3 = 486 m³
ESTIMATE No. 1
ESTIMATE BY: C.T. GOSLING
DATE: 6.9.73

NOTES	FACTORS									QUANTITY AREA m²	COOLING LOAD W
	ASPECT	SINGLE PANE			DOUBLE GLAZING						
		BARE GLASS	VEN.BLIND	AWNING	BARE GLASS	VEN.BLIND	AWNING				
GLAZING FACTOR × AREA –SOLAR AND TRANS. GAIN	N OR SHADE	(66)	47	50	44	35	35			6	396
SELECT ASPECT WITH HIGHEST GAIN ALL OTHER ASPECTS ALLOW SHADE, EXCEPT FOR HORIZONTAL	NE OR NW	286	220	167	214	151	98				
	E OR W	394	(274)	186	312	208	120			15	2 877
	SE OR SW	337	233	167	264	176	98				
	S	246	176	126	189	132	79				
	HORIZONTAL	560	390		440	305					
EXT.WALLS FACTOR × COLOUR FACTOR × AREA = GAIN	ASPECT	275mm CAVITY BRICK WITH PLASTER	225mm BRICK WITH PLASTER	150mm CONCRETE WITH PLASTER	BRICK OR CONCRETE WITH 25mm INSULATION	COLOUR FACTOR					
						DARK	MEDIUM	LIGHT			
ALLOW ALL ASPECTS SUNLIT. UNLESS SHADED	N. NE. SHADE	(4)	6	8	3	(1.0)	0.8	0.6		(27-6)=21	84
	E OR SE	19	25	35	12	1.0	0.8	0.6			
	S	23	32	42	15	1.0	0.8	0.6			
	SW OR W	(28)	37	50	16	(1.0)	0.8	0.6		(54-15)=39	1 092
	NW	20	27	36	13	1.0	0.8	0.6			
ROOF FACTOR × COLOUR FACTOR × AREA = GAIN		150mm CONCRETE NO CEILING	150mm CONCRETE WITH SUSP. CEILING	ASBESTOS OR WOOD WITH SUSP. CEILING	CONCRETE + CEILING WITH 25mm INSULATION						
	EXPOSED	47	20	25	(11)	(1.0)	0.8	0.6		81	891
	SHADED	6	3	5	2	1.0	0.8	0.6			
INT. SURFACES FACTOR × AREA = GAIN	CEILING	UNCONDITIONED SPACE ABOVE (3)			GRD. FLOOR OR BASEMENT (NIL)					81	243
	FLOOR	UNCONDITIONED SPACE BELOW 3								162	—
	PARTITIONS	METAL OR METAL & GLASS 16		WOODEN 8	MASONRY (5)					81	405

16 PACKAGED EQUIPMENT

INTERNAL	LIGHTS	INCANDESCENT 1.0	FLUORESCENT 1.25		32 × 146 watts (W)	6	480
	MACHINERY				746 ×	h.p.	
	PEOPLE	THEATRE 73	OFFICE 79	SEDENTARY 86	MANUAL 126	30 No.	2	370
						SUB TOTAL	14	838
				SUB TOTAL × 1.05 (5 PER CENT. FAN HEAT) = ROOM SENSIBLE HEAT		15	580
LATENT GAIN	PEOPLE	THEATRE 29	OFFICE 53	SEDENTARY 72	MANUAL 167	30 No.	1	590
					ROOM TOTAL HEAT	17	170
OUTSIDE AIR VOLUME - cfm USE LARGEST ANSWER	No. PEOPLE	30		× 7 FOR NON SMOKERS × 9.4 AVERAGE × 14 HEAVY SMOKERS =	282 litre/sec		
	FLOOR AREA	162		m² × 1.0 FOR 2.4m CEILING × 1.25 FOR 3.0 m CEILING × 1.5 FOR 3.6 m CEILING =	162 litre/sec × 10 W per litre/sec	2	820
					TOTAL COOLING LOAD	19	990

ROOM SENSIBLE HEAT / ROOM TOTAL HEAT = 15 580 / 17 170 = 0.91

RSH/RTH	1.00	0.95	0.90	0.85	0.80	0.75
litre/sec FACTOR	10.6	11.2	11.9	12.8	13.9	15.8
O.A. SUPPLY	0.1	0.2	0.3	0.4	0.5
E w.b. °C	15.9	16.1	16.4	16.7	17.0

O.A. VOLUME / SUPPLY VOLUME = 282 / 1310 = 0.215

SUPPLY AIR VOLUME × RSH FACTOR = 15 580 / 11.9 = 1310 litre/sec

AIR ENTERING EVAPORATOR = 16.1 °C w.b.

CHECK FIGURES	TOTAL COOLING =	12.3	W/m² FLOOR AREA	AIR CHANGE =	SUPPLY VOLUME × 3.6 / ROOM VOLUME =	9.7	EAC/hr.
	ROOM SENSIBLE HEAT =	9.6	W/m² FLOOR AREA	VENT CHANGE =	O.A. VOLUME × 3.6 / ROOM VOLUME =	2.1	EAC/hr.
	m² FLOOR AREA/PERSON =	5.4					
	watts (W) / m² =	32					

FIGURE 16.9a.

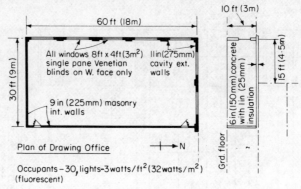

FIG. 16.10. Details of area considered in load estimating example Fig. 16.9 (see Fig. 16.9a for example in SI units using metric dimensions).

At the bottom of the form are check figures from which great advantage can be gained. After compiling several estimate forms the engineer can log his check figures so that quick budgeting can be given at the time of an initial inquiry to establish the approximate size of plant he would require, which in many cases can lead to a budget estimate at the time of a sales inquiry to establish its seriousness.

The selection of apparatus from manufacturers' data can be made after consideration of the physical requirements of the building.

The application considered lends itself to a split packaged installation with a condensing unit mounted on the roof and a room fan coil unit in the conditioned space below, or alternatively, a single-piece water cooled system. Because of the relatively low ceiling of 10 ft (*3 m*), free-blow would be inadvisable, and a simple run of sheet-metal ductwork down the length of the room would probably serve best. This approach also allows the designer to get more air to the west face where the main solar and roof gain occurs.

After an examination of the area and the respective loads, splitting the area into two zones along the 60 ft (*18 m*)-long centre line, the following break-down is found:

West half:

West glazing		11 136 BTU/h	2877 W
Roof		3 240 BTU/h	891 W
West wall		4 248 BTU/h	1092 W
	Total A	18 624 BTU/h	4860 W

16 PACKAGED EQUIPMENT

East half:

East partition $\left(\frac{60}{90} \times \text{total}\right)$ 900 BTU/h $\left(\frac{18}{27} \times \text{total}\right)$ 270 W

Ceiling 720 BTU/h 243 W

Total B 1 620 BTU/h 513 W

A + B = 20 244 BTU/h 5373 W

The room sensible heat is 56 100 BTU/h (*15 580 W*) and other than the 20 244 BTU/h (*5373 W*) above, the other loads are proportional across the centre line, i.e. 56 100 − 20 244 = 35 856 BTU/h (*15 580 − 5373 = 10 207 W*). The room sensible heat attributable to the west half is, therefore, 0·5(35 856) + 18 624 = 36 552 BTU/h (*0·5(10 207) + 4860 = 9963 W*). Since the air volume is proportional to the room sensible heat that required to the west half would be

$$2940 \times \frac{36\,552}{56\,100} = 1910 \text{ cfm} \quad or \quad 1310 \times \frac{9963}{15\,580} = 839 \text{ litres/s}$$

the balance of 1030 cfm (*471 litres/s*) being required for the east half.

This approach is good practice for single-zone areas, but should not be used for multi-zone areas, to which single packaged equipment cannot be applied, unless it can be guaranteed that the partial load characteristics of each zone are the same.

The load estimating form gives three pieces of information which enable the design engineer to select the correct packaged unit or units to meet his specific requirements. These are the air volume to be handled by the apparatus, the total refrigeration effect and the entering air conditions to the unit.

To establish the required air volume the load estimate has to make one assumption, that being a coil configuration resulting in a contact factor of 80 per cent, or as more commonly termed a by-pass factor (BF) of 0·20. If the by-pass factor of the apparatus to be considered is within the band 0·17 to 0·23, no correction need be made, since the resulting air volume would be within ±4 per cent of that selected. However, should the BF differ beyond this, a suitable correction should be made. A typical rating for a single-piece water cooled air conditioner is shown in Fig. 16.11. At 3000 cfm (*1415 litres/s*) the BF is 0·25. Since the required air volume is dependent on the amount of air by-passed around the coil, this volume is a proportion of (1 − BF) to 0·80, the assumed figure. For the example considered the volume would approximate to 3000 cfm (*1415 litres/s*), which for the apparatus under review is coincident with a BF of 0·25. The required volume would therefore be:

$$2940 \times \frac{0 \cdot 80}{1 - BF} = 3140 \, \text{cfm} \quad or \quad 1310 \times \frac{0 \cdot 80}{1 - BF} = 1410 \, litres/s$$

The table figure of 3000 cfm (*1415 litres/s*) could safely be used without interpolation.

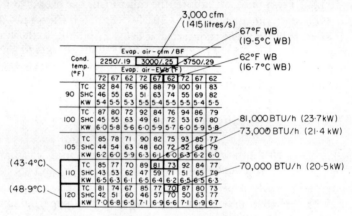

Fig. 16.11. Single-piece water cooled unit selection table.

On-site measurement of air volume can rarely be guaranteed to within 5 per cent, and often not to within 10 per cent. This is because instruments quickly become uncalibrated and, more important, readings can rarely be taken at ideal locations in the sheet-metal ductwork because of necessary transformation pieces and bends. In addition, normal duct sizing charts and tables make it difficult to distinguish between the narrow limits of, say, 5 per cent at any one air volume level. One other valid reason is in sheet-metal ductwork sizing. It is normal practice to manufacture ducting in 2-in (*50 mm*) size increments; therefore the size of the air distribution system for, say, 2940 cfm (*1390 litres/s*) the calculated air volume and 3000 cfm (*1415 litres/s*) would more than probably be the same.

When selecting apparatus the engineer should use his licence. It should be remembered that the range of equipment available to him is limited and any amount of correction within, say, 5 per cent limits will

16 PACKAGED EQUIPMENT

result in the same size apparatus, sheet-metal ductwork, water pipework, etc., once the basis of his design is laid down.

Correction is more important in the total heat of the air entering the apparatus and the condensing level of the compressor, which together control the evaporating level. At normal air conditioning levels 1 °F (0·55 °C) difference in the saturated suction temperature would result in about a 2 per cent load difference. As stated previously, the calculated entering wet bulb temperature was 61 °F (16·1 °C), which falls outside the published ratings. However, it would be permissible to extrapolate from such ratings within the limits of the load estimate form, the minimum being 60 °F (15·6 °C) WB, and the maximum 65 °F (18·3 °C) WB, which falls within the selection table.

Considering the probable operating level of the apparatus, say 110 °F (43·4 °C) condensing, the ratings show 81 000 BTU/h (23·7 kW) at 67 °F (19·5 °C) WB or 73 000 BTU/h (21·4 kW) at 62 °F (16·7 °C) WB or 8000 BTU/h (2·3 kW) per 5 °F (2·8 °C) WB difference. The difference for 61 °F (16·1 °C) WB would be 1 °F (0·55 °C) WB or 1600 BTU/h (0·46 kW). It is easier to consider a figure which occurs in the ratings than convert many rating figures to establish the requirement. Therefore, one should use a table figure of 71 000 BTU/h + 1600 BTU/h = 72 700 BTU/h at 62 °F WB (20·84 kW + 0·46 kW = 21·30 kW at 16·7 °C WB). Interpolating between 110 and 120 °F (43·4 and 48·9 °C) condensing temperature at 3000 cfm (1415 litres/s) and 62 °F (16·7 °C) WB the machine would result in 72 700 BTU/h (21·30 kW) at 111 °F (43·9 °C) condensing with an absorbed power of 6·2 kW.

The above selection follows the ratings given by some manufacturers. Most manufacturers, however, do not present selections in their sales literature and only publish the ARI rating. This method of rating is good, since it shows that the product is reliable to the figure given, but only at the specified level at nominal air volume, 80 °F DB, 67 °F WB (26·7 °C DB, 19·5 °C WB) entering air temperature and 105 °F (35 °C) condensing temperature—in the case of water cooled apparatus. The ARI rating is also a convenient way of comparing different products in establishing price for performance. However, the ARI rating should not be used for selection purposes, and this is particularly underlined in the fact that the ARI rating for the apparatus considered is 82 000 BTU/h (24 kW), when operating at 50 Hz, whereas the selection for the required operating level is only 71 100 BTU/h (20·84 kW).

The engineer would be well advised, when considering apparatus without selection tables, to seek confirmation of the unit's capacity at the operating level pertinent to his requirement. In the past, many units have been grossly overestimated in capacity on the assumption that the ARI rating was adequate for lower operating levels, and, even worse, these ratings could have been for 60 Hz operation, for use on 50 Hz electrical supplies.

Units installed on this basis have probably worked because the engineer in his turn has overestimated his cooling load requirement. However, if a proper load estimate is made allowing for building storage, diversification of load, etc., a serious equipment undersizing could occur.

At this point in the selection the sensible heat capacity of the unit should be checked. The refrigeration capacity of the packaged apparatus is dependent on the total heat of the air entering which can be approximated to the WB temperature, whereas the sensible heat capacity is dependent on both DB and WB temperatures. The ratings shown are the sensible heat contents assuming the entering conditions to be 80 °F (26·7 °C) DB coincident with the various published WB temperatures; therefore, some correction must be made.

A simple way of approximating the room sensible heat capacity of the machine makes the assumption that within the limits of entering and leaving conditions of the apparatus as shown on the psychrometric chart, Fig. 16.12, the increments of total heat are constant with change of WB temperature.

From the load estimate the room sensible heat is 56 100 BTU/h (16·45 kW), and the mixture of fresh and return air results in an on-coil condition of 72·8 °F DB, 61·0 °F WB (22·7 °C DB, 16·1 °C WB). The heat removed from the coil is 71 100 BTU/h (20·84 kW) at 3000 cfm (1415 litres/s), therefore the heat removed from the air (ΔH_1) could be found as follows:

$$\Delta H_1 = \frac{\text{total heat}}{\text{cfm} \times 4 \cdot 45} = \frac{71\,100}{3000 \times 4 \cdot 45} = 5 \cdot 33\,\text{BTU/lb}$$

or

$$\Delta H_1 = \frac{\text{total heat}}{\text{litres/s} \times 1 \cdot 19} = \frac{20\,840}{1415 \times 1 \cdot 19} = 12 \cdot 38\,kJ/kg$$

In this instance the by-pass factor was 0·25, therefore if the coil was

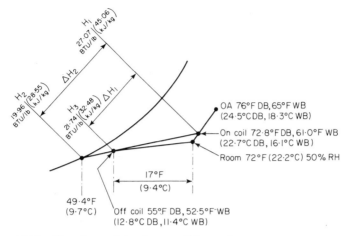

Fig. 16.12. Entering and leaving conditions of apparatus as plotted on psychrometric chart.

perfect the total heat of the apparatus dew point (H_2) would be the total heat of the entering air [H_1, 27·07 BTU/lb (*45·06 kJ/kg*)] minus the heat removed at zero by-pass factor, i.e.

$$H_2 = H_1 - \frac{H_1}{1 - BF} = 27 \cdot 07 - \frac{5 \cdot 33}{0 \cdot 75} = 19 \cdot 96 \, \text{BTU/lb}$$

$$= 45 \cdot 06 - \frac{12 \cdot 38}{0 \cdot 75} = 28 \cdot 55 \, kJ/kg$$

The total heat of the off-coil condition (H_3) would be

$$H_3 = H_1 - H_2 = 27 \cdot 07 - 5 \cdot 33 = 21 \cdot 7 \, \text{BTU/lb}$$
$$(45 \cdot 06 - 12 \cdot 38 = 32 \cdot 48 \, kJ/kg)$$

From the psychrometric chart the above points can be plotted where H_2 corresponds to a dew point of 49·4 °F (*9·7 °C*) and H_3 the leaving condition to 52·5 °F (*11·4 °C*) WB. By construction of flow lines on the chart, the off-coil DB temperature results in 55 °F (*12·8 °C*). The resulting room sensible heat therefore would be:

air volume (cfm) × 1·08 (room DB temperature − off-coil DB temperature) °F

$$= 3000 \times 1 \cdot 08 \, (72 - 55) \, °F$$
$$= 55\,100 \, \text{BTU/h}$$

or

air volume $(litres/s) \times 1 \cdot 21$ (room DB temperature
$\qquad - $ off-coil DB temperature)$°C$
$= 1415 \times 1 \cdot 21 \ (22 \cdot 2 - 12 \cdot 8) °C$
$= 16\ 200\ W$

This compares very closely with the desired figure of 56 100 BTU/h ($16 \cdot 45\ kW$). Being a little lower, it will result in a lower room RH and slightly higher DB temperature at design conditions. However, at partial load this is a good feature, since the latent heat requirements tend to be constant whereas the sensible heat reduces. Since the apparatus is controlled by temperature the only effect would be better humidity control. If the resulting sensible heat removal of the coil proves high, one could expect the room RH to be high at design and partial loads.

All that remains of the water cooled apparatus selection is the correct water quantity and temperatures for the condenser, which should allow for a scaling factor of $0 \cdot 000\ 5$ ($0 \cdot 000\ 088$) when using water to waste and $0 \cdot 001$ ($0 \cdot 000\ 176$) when using a cooling tower. The cooling tower selection should not be undersized and it is advisable that a selection is made at the highest normal expected WB temperature. A simple three-way valve control will ensure proper water temperatures to the condenser at partial load. This practice ensures that when conditions occur higher than $65\,°F$ ($18 \cdot 3\,°C$) WB and the apparatus is trying to do more duty because of possible higher entering conditions an overload condition does not occur.

The alternative suggested for the example was a split package air cooled system. Where combination ratings are available the operation is the same as the single-piece water cooled unit except that the selection is fixed by the design outside air temperature.

However, combination selections are not always available and the engineer must therefore mix–match selections to suit his requirements. As before, the first step is to consider the air side. Many fan coil arrangements are available and the easiest approach is to select a type of unit which most suits the physical requirements of the buildings, then find the correct size. Instead of finding the design condensing level, the coil refrigerant temperature is established from the ratings. Having reached this point a suitable condensing unit can be found at a suction temperature corresponding to the refrigerant evaporating temperature $-2\,°F$ ($-1 \cdot 1\,°C$), allowing for suction line pressure drop. The condensing unit can be selected for the saturated suction

16 PACKAGED EQUIPMENT

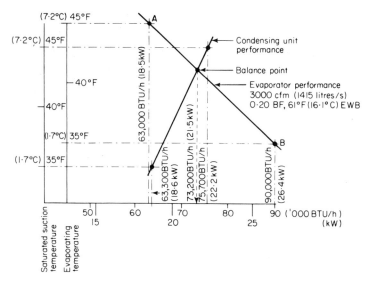

Fig. 16.13. Split package selection.

temperature and entering air temperature to the condenser to meet the required refrigeration effect.

Figure 16.13 shows a graphical selection method for the load estimate in question. Plotting evaporator performance for the fixed air conditions of entering WB temperature and air volume against refrigerant evaporating temperature is the line AB. Various condensing unit performances at fixed outside air temperatures plotted against saturated suction temperatures (2°F (1·1°C) lower than evaporating) give balance points of the available selections.

Having selected his apparatus the engineer should consider what the manufacturer includes in his package. Too often, equipment is purchased on price of the major components alone, without consideration of all the auxiliary components required to complete the installation. The running costs should also be compared.

Depending upon selection a manufacturer could oversize his compressor in relation to his heat exchange surfaces which means lower evaporating temperatures to achieve overall refrigeration effect out of a small evaporator; this would also result in over dehumidification, and higher condensing levels for a small condenser, the net result being high running costs and bad value. In comparing equipment the engineer should cost these features, and a guide for this is shown in Fig. 16.14 (see overleaf).

Selection.. Price..............................

Air volume.................(m^3/s) Total cooling capacity ....BTU/h

Component	Details	Extra cost if not included
COMPRESSOR		
Starter	No._____ Abs. power._____ Add 3 per cent to price if not serviceable D.O.L.[a] Star-Delta[a] Part-Wound[a]	
Thermal protection	Klixon[a] Single thermistor[a] Double thermistor[a]	
Unloading	Stages._____ Suction pressure[a] Solenoid valve[a]	
EVAPORATOR	_____ ft^2 (m^2) Face area_____ Rows_____ Fins/in._____ °F (°C) evaporating	
Metering device	Capillary[a] TX Valve[a]	
Fan	_____ hp (kW) _____ (rev/min)	
Fan drive	Fixed[a] Adjustable[a] _____ in. s.g. (N/m^2) external pressure	
Fan starter	D.O.L.[a] Split capacitor[a] Multi-speed[a]	
FILTERS	_____ ft^2 (m^2) Face area. Throwaway[a] Washable[a]	
AIR COOLED CONDENSER	_____ ft^2 (m^2) Face area_____ Rows_____ Fins/in._____ °F (°C) condensing	
Fan(s)	_____ No. _____ hp (kW) _____ (rev/min) Total cfm (m^3/s)	
Fan starter(s)	D.O.L.[a] Split capacitor[a]	
Head pressure control	Fan cycling[a] Ref. side[a] Solid state[a]	
WATER COOLED CONDENSER	Tube in tube[a] Shell and coil[a] Shell and tube[a]	
Cooling tower	_____ gal/min (litres/s) _____ ft hd (N/m^2). Press. Drop. Fouling factor _____ °F (°C) w. b. Water _____ °F (°C) entering _____ °F (°C) leaving	
Expansion device	Capillary tube[a] TX valve[a]	
H.P. and L.P.	Fitted[a] Supplied separately[a]	
Time delay		
Thermostat		
Control switch	Fan/auto control[a] Heating/cooling selection[a]	
Sight glass	Moisture indication[a]	
Strainer/drier		
Muffler		
Control transformer		
	Total	

[a] Delete where not appropriate.

Chapter 17

REFRIGERATION PIPEWORK

The design considerations of a refrigeration pipework system are the same as any other general fluid flow system. However, there are some additional factors that have a great influence on this design. Firstly, any excessive pressure drop not only requires more compressor power, but also has an adverse effect on system capacity. Secondly, the refrigerant exists in three conditions, as a hot liquid (liquid line), a cold, low pressure gas (suction line) and a hot, high pressure gas (discharge line). Thirdly, provision must be made to ensure that oil is returned to the compressor at the same rate at which it leaves and that the crankcase is free of liquid refrigerant at all times.

Consideration of the above points will achieve the design objectives of practical pipework sizes with economical pressure drops, proper feed to evaporators, and proper protection of compressors.

Pressure drops in refrigeration lines are usually sized to allow an equivalent of a 2 °F (*1·1 °C*) drop in saturated suction temperature, a 2 °F (*1·1 °C*) drop in saturated discharge temperature [i.e. condensing temperature $+2\,°F\ (1\cdot 1\,°C)$] and a 1 °F (*0·55 °C*) line loss in the liquid line. The pressure drop in the liquid line is not too critical since the system expansion device will drop the pressure ultimately, although the pressure loss should not be enough to cause flash gas before the expansion devices. A 2 °F (*1·1 °C*) suction line loss represents a 4 per cent drop in system capacity and a 2 °F (*1·1 °C*) discharge line loss a 1 per cent drop in system capacity. It is important, therefore, that particular attention is paid to the gas lines, with particular emphasis on the suction line.

Figure 17.1 shows a refrigerant pipework sizing chart which can be used by taking the following steps.

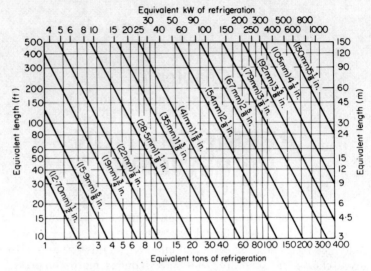

FIG. 17.1. Refrigerant pipework sizing chart (see Table 17.1 for corrections to above values).

1. Make the necessary correction factors to the system tonnage.
2. Measure the pipework length in feet and establish number of fittings.
3. Add 50 per cent to obtain a trial equivalent length.
4. Establish probable size of the pipework from chart.
5. Establish equivalent length of fittings from Fig. 17.2, add these to the actual length to establish the total equivalent length.
6. Check the pipe sizes, and repeat steps if necessary because of number of fittings.

Figure 17.3 shows a typical suction line from a direct expansion cooling coil, with the system using R.12 at 30°F ($-1°C$) saturated suction and 120°F ($48°C$) saturated discharge temperature, having a duty of 24 tons R ($84.5 kW$).

1. The correction factors are 0·5 for R.12 Suction, and 1·10 for 30°F ($-1°C$) Suction, 120°F ($48°C$) discharge, i.e. 0·66. The table duty must be reduced to 0·66 its value, or more conveniently a table duty of 24/0·66 or 36 tons R ($84.5/0.66$ or $128 kW$) should be used for pipe sizing.

17 REFRIGERATION PIPEWORK

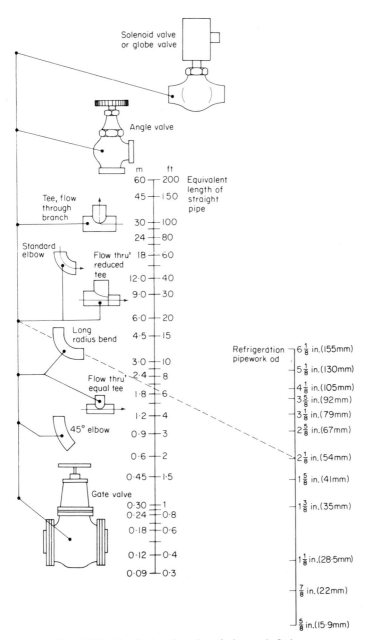

FIG. 17.2. Equivalent lengths of pipework fittings.

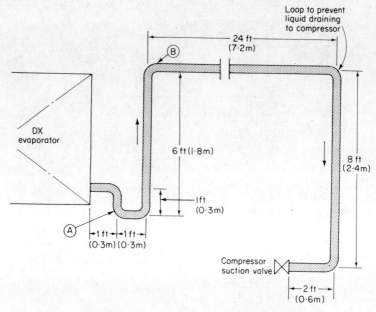

Fig. 17.3. Typical suction line pipework.

2. The total pipework run is 43 ft (*12·9 m*).
3. Trial equivalent length is 43 × 1·5 = 64·5 ft (*12·9 × 1·5 = 19·4 m*).
4. At 36 tons R and 64·5 ft (*128 kW and 19·4 m*) equivalent length pipe size is less than $2\frac{1}{8}$ in od (*54 mm*). Use this size for trial.
5. With $2\frac{1}{8}$ in (*54 mm*) pipework the actual run is

$$
\begin{array}{ll}
\phantom{+6\text{ elbows at }6\cdot5\text{ ft} = }43\text{ ft} & (12\cdot9\text{ m}) \\
+6\text{ elbows at }6\cdot5\text{ ft} = 39\text{ ft} & (6\text{ at }1\cdot95\text{ m} = 11\cdot7\text{ m}) \\
\hline
\text{Equivalent length} = 82\text{ ft} & (24\cdot6\text{ m})
\end{array}
$$

6. With 82 equivalent feet (*24·6 m*) $2\frac{1}{8}$ in (*54 mm*) suction line will carry 39 tons R × 0·66 or 26 tons R (*135 kW × 0·66 or 89 kW*).

Therefore the original size of $2\frac{1}{8}$ in (*54 mm*) od is satisfactory.

Figure 17.1 is based on R.22 having a suction line pressure loss equivalent to 2 °F (*1·1 °C*), at 40 °F (*4 °C*) evaporating with system

17 REFRIGERATION PIPEWORK

condensing at 100 °F (*37 °C*). For other refrigerants and other lines at varying temperatures divide required tonnage by corrections above and use that value in the table. Where other than the stated line loss is required adjust the tonnage to the ratio of line loss (i.e. 1 °F (*0·55 °C*) discharge line loss R.500, correction factor 0·75).

TABLE 17.1
CORRECTIONS TO FIG.17.1

Pressure drop		R.22 R.502	R.500	R.12
2 °F (*1·1 °C*)	Suction line	1·00	0·70	0·50
2 °F (*1·1 °C*)	Discharge line	2·00	1·50	1·00
1 °F (*0·55 °C*)	Liquid line	6·00	5·00	2·50

Evaporation temp. line	30 °F (*−1 °C*)		40 °F (*4 °C*)		50 °F (*10 °C*)	
	Suction	Discharge	Suction	Discharge	Suction	Discharge
80 °F (*27 °C*) Cond.	0·90	0·80	1·10	0·85	1·25	0·80
100 °F (*37 °C*) Cond.	0·85	0·95	1·00	1·00	1·10	1·00
120 °F (*48 °C*) Cond.	0·75	1·10	0·90	1·10	1·10	1·10

Having established a pipework size it is important that the velocity of the gas is high enough to carry oil around the system. Particular attention should be paid to the minimum capacity of a pipe size with unloading compressors which are very common in air conditioning systems or when multiple compressor systems are used. Table 17.2 lists these minimum tonnages for various pipework sizes. It can be clearly seen that a $2\frac{1}{8}$ in od (*54 mm*) suction line at 30 °F (*−1 °C*) has a minimum capacity of 8·5 tons R (*30 kW*) with R.22, or using R.12, 8·5 × 0·7 or 5·95 tons R (*30 × 0·7 or 21 kW*). For the example considered this is something less than 25 per cent of the maximum duty of 24 tons R (*84·5 kW*). It is possible therefore that the compressor could unload to 25 per cent without fear of the oil not returning to the compressor. In the event that the system could unload below this value, say to 16 per cent or 4 tons R (*14·1 kW*), the suction riser could not be more than $1\frac{5}{8}$ in od (*41 mm*). If the riser was sized at $1\frac{5}{8}$ in (*41 mm*) the pressure drop could well exceed the desired 2 °F (*1·1 °C*).

The example considered has a problem from points A to B (Fig.

TABLE 17.2
MINIMUM REFRIGERATION CAPACITY FOR OIL ENTRAINMENT R.22, FOR SUCTION AND DISCHARGE LINE RISERS

	Pipe outside diameter	in mm	$\frac{7}{8}$ 22	$1\frac{1}{8}$ 28.5	$1\frac{3}{8}$ 35	$1\frac{5}{8}$ 41	$2\frac{1}{8}$ 54	$2\frac{5}{8}$ 67	$3\frac{1}{8}$ 79	$3\frac{5}{8}$ 92	$4\frac{1}{8}$ 105
Suction lines	30°F sat. suction	Tons R	0.8	1.65	2.75	4.3	8.5	14.7	22.6	33.1	46.0
	−1°C sat. suction	kW	2.8	5.8	9.7	15.1	30.0	51.6	79.5	116	162
	40°F sat. suction	Tons R	0.9	1.8	3.0	4.6	9.1	15.8	24.3	35.4	49.4
	4°C sat. suction	kW	3.2	6.3	10.5	16.2	32.0	55.6	85.5	124	174
	50°F sat. suction	Tons R	1.0	1.95	3.25	4.9	9.8	17.0	26.1	37.8	52.7
	10°C sat. suction	kW	3.5	6.9	11.4	17.2	34.5	59.8	92	133	185
Discharge lines	80°F sat. discharge	Tons R	1.1	2.1	3.5	5.4	10.6	18.1	28.4	42.0	57.6
	27°C sat. discharge	kW	3.8	7.4	12.3	19.0	37.3	63.6	100	148	203
	100°F sat. discharge	Tons R	1.2	2.5	4.1	6.3	12.5	21.6	33.9	49.6	68.5
	37°C sat. discharge	kW	4.2	8.8	14.4	22.2	44.0	76.0	109	175	241
	120°F sat. discharge	Tons R	1.5	3.0	4.9	7.5	15.0	26.0	40.5	59.5	79.5
	48°C sat. discharge	kW	5.3	10.5	17.2	26.4	52.8	91.5	142	210	280

17 REFRIGERATION PIPEWORK

17.3), should the capacity fall as low as 4 tons R. The simplest answer would be to reduce the line sizes to $1\frac{5}{8}$ in if the pressure drop could be tolerated, and this could be calculated as follows.

Total run of $2\frac{1}{8}$ in *(54 mm)* od pipework 43 − 7 ft
$$= 36 \text{ ft} \quad \text{or} \quad 12 \cdot 9 - 2 \cdot 1 \, m = 10 \cdot 8 \, m$$
No. of $2\frac{1}{8}$ in *(54 mm)* elbows: 3 at 6·5 ft
$$= 19 \cdot 5 \text{ ft} \quad\quad 3 \times 1 \cdot 95 \, m = 5 \cdot 85 \, m$$
$$\overline{55 \cdot 5 \text{ ft}} \quad\quad\quad\quad \overline{16 \cdot 65 \, m}$$

Total run of $1\frac{5}{8}$ in *(41 mm)* od pipework
$$= 7 \text{ ft} \quad \text{or} \quad\quad = 2 \cdot 1 \, m$$
No. of $1\frac{5}{8}$ in *(41 mm)* elbows: 3 at 4·5 ft
$$= 13 \cdot 5 \text{ ft} \quad\quad 3 \text{ at } 1 \cdot 3 \, m = 3 \cdot 9 \, m$$
$$\overline{20 \cdot 5 \text{ ft}} \quad\quad\quad\quad \overline{6 \cdot 0 \, m}$$

At a table duty of 36 tons R *(128 kW)* (allowing for R.12, etc.):
$2\frac{1}{8}$ in *(54 mm)* pipework could be 90 ft *(27 m)* equivalent length for 2 °F *(1·1 °C)* loss.
$1\frac{5}{8}$ in *(41 mm)* pipework could be 24 ft *(7·2 m)* equivalent length for 2 °F *(1·1 °C)* loss.
Therefore the loss would be

$$\frac{55 \cdot 5}{90} \times 2 = 1 \cdot 23 \, °F \text{ for } 2\tfrac{1}{8} \text{ in} \quad \text{or} \quad \frac{16 \cdot 65}{27} \times 1 \cdot 1 = 0 \cdot 68 \, °C \text{ for } 54 \, mm$$

$$\frac{20 \cdot 5}{24} \times 2 = 1 \cdot 71 \, °F \text{ for } 1\tfrac{5}{8} \text{ in} \quad\quad \frac{6 \cdot 0}{7 \cdot 2} \times 1 \cdot 1 = 0 \cdot 91 \, °C \text{ for } 41 \, mm$$

Total $\overline{2 \cdot 94 \, °F}$ Total $\overline{1 \cdot 59 \, °C}$

The above line loss could probably be tolerated, although the loss could be reduced by increasing the $2\frac{1}{8}$ in *(54 mm)* od pipework to $2\frac{5}{8}$ in *(67 mm)*, which by the method above would reduce the loss to 1·91 °F *(1·06 °C)*.

Here a question of economics must be answered, and a further method, a double suction riser, could be considered. This is shown in Fig. 17.4, where it can be seen that as the compressor unloads and the refrigerant flow reduces, any oil not carried up the riser will fall back to the U-bend where it will block the passage, allowing the refrigerant to pass through the vertical by-pass leg only and so reach a sufficient velocity to ensure oil entrainment. The U-bend trap should be kept to a minimum size since a large volume could store sufficient oil to seriously lower the compressor crankcase level. In addition, when the trap clears out on an increased load operation large slugs of oil are not pulled back to the compressor.

Similar problems occur when using discharge lines, although since pressure drop is not so critical as with the suction line a reduced line

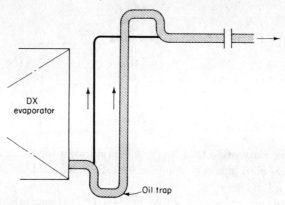

FIG. 17.4. Double suction riser for proper oil return at partial load.

size is often used. However, when multiple compressors are used the problem of light load can become serious and double discharge line users must be considered using the same treatise as with double suction users.

Good practice demands that refrigeration pipework runs are kept as simple and as direct as possible. Unnecessary complications add to cost, increase operating losses, and increase the possibilities of leaks. Horizontal lines should be kept level and hangers should be provided at frequent intervals to prevent sagging and pockets where oil and/or liquid can collect. It is essential that expansion and vibration are catered for either in the form of a loop or line vibration eliminators either side of the compressor.

Figure 17.5 shows recommended suction pipework to single evaporators. It can clearly be seen that all suction pipework includes loops and traps to ensure oil return to the compressor. Figure 17.6 shows how multiple evaporator coils should be connected. Double circuited evaporator coils can be treated as multiple evaporators.

17 REFRIGERATION PIPEWORK 353

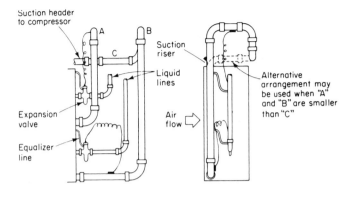

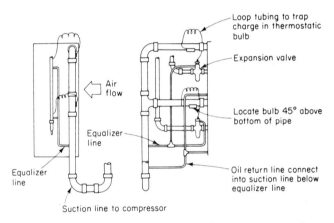

FIG. 17.5. Connections to direct expansion evaporator coils.

Figure 17.5 also shows multiple circuited coils which have oil return lines from the evaporators. Good oil return can be achieved by having oil return lines from the bottom of suction headers, which also ensures proper distribution from all sections of the coil without the danger of oil trapping.

Normally, only the suction line of a refrigeration system is insulated, essentially to prevent atmospheric condensation on the outside of the tube, which could result in a damaging effect caused by dripping. It is common practice, however, to use the same insulation —normally expanded closed cell foamed rubber with a vapour barrier

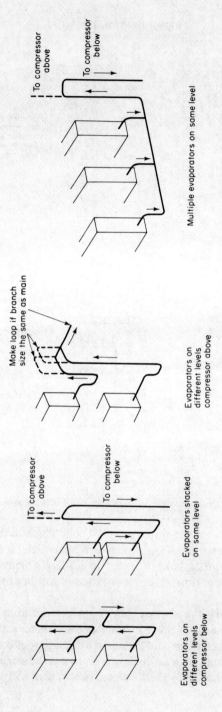

Fig. 17.6. Suction line pipework for multiple evaporator coils.

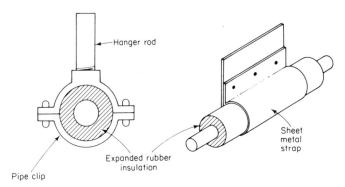

Fig. 17.7. Insulation and antivibration mounting of refrigeration pipework.

—on liquid and discharge lines at pipe clips. This practice avoids any vibration transmission and direct contact between pipe bracket and pipe, which could cause wear.

Discharge pipework, together with proper location of hot gas muffler, is shown in Fig. 17.8. Also shown is an oil separator, which may be an advantage in systems having sudden and frequent capacity changes, or systems having extensive pipe lines and/or numerous evaporators. There are some objections to oil separators, which must be considered. Firstly, they are not 100 per cent effective and oil is still allowed to circulate through the system and proper care in pipework design must still be maintained. Secondly, during the plant shut-down the separator acts as a condenser and allows liquid refrigerant to

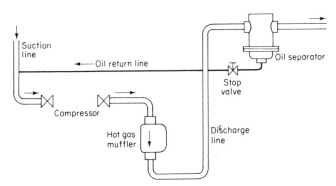

Fig. 17.8. Discharge line pipework and oil trapping.

return to the crankcase. Such applications should always use a crankcase heater in order to avoid poor lubrication on start up.

The liquid line does not have the same problems as discharge and suction lines, but care should be taken to keep the pressure drop within operable levels. Any undue blockage or liquid lift will cause flashing with subsequent malfunction of the TX valve, or pressure reducing device. Most condensers provide 5°F (2·8°C) subcooling,

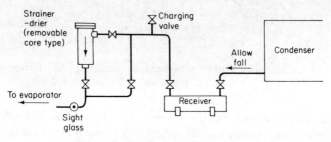

FIG. 17.9. Liquid line pipework.

although larger systems with subcooling circuits provide as much as 15°F (8·3°C). Additional subcooling to ensure liquid at the TX valve can be provided by a suction/liquid interchanger although careful attention must be paid to R.22 and R.502 systems which could result in excess discharge temperatures caused by high suction superheat temperatures (see Chapter 2, The Refrigeration Cycle).

Figure 17.9 shows a typical liquid line with various accessories required to ensure proper system operation.

Particular care must be given to multiple compressor systems to ensure that at partial load with compressor(s) stopped oil is not drained from one compressor to another, and when all compressors are operating they do so at constant evaporating and condensing levels. This is achieved by the use of equaliser lines as shown in Fig. 17.10.

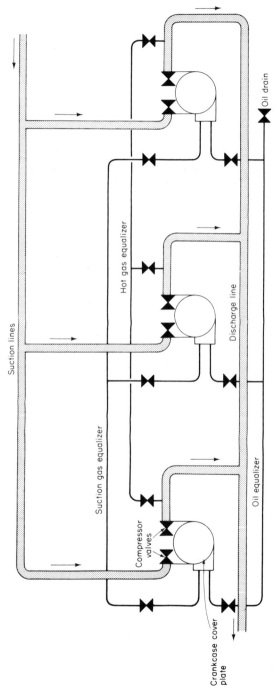

Fig. 17.10. Equalisation of multiple compressor installations.

Chapter 18

WATER PIPEWORK

The high thermal capacity, cheapness and safety of water make it an ideal fluid to convey energy in the form of heat in air conditioning systems. It serves a prime use in distributing cooling throughout a building to many terminal devices and/or air handling units and as a conveyance for heat rejection in water cooled condenser applications. The application of water piping in these two applications can be described as closed systems and open systems. Care has to be taken with the latter system since evaporation of water causes residual solids to remain in suspension, creating problems of scale build up.

Figures 18.1 and 18.2 show friction loss rates for both closed and open water pipework systems. The pressure loss in equivalent length of pipework can be assessed from Fig. 17.2. The calculation of any water system pressure drop must include the head loss through water chiller cooling coil; condenser, cooling tower nozzles and the static head loss through the cooling tower.

Recommended water velocities through various services are shown in Table 18.1. The recommendations are made taking into consideration the service for which the pipe is to be used; the maximum acceptable noise levels and erosion. Erosion can cause severe deterioration by the velocity of the water and the inclusion of solid matters will cause damage, particularly at the bottom of tubes and at elbows. Erosion is a function of these considerations and time and careful attention must be paid to proper pipe sizing.

Increased water velocities increase the friction rate of the pipework with resultant increase in pump and pumping costs. These costs must be considered against the installed pipework costs, which may be several times as high as the pump costs, and in air conditioning

18 WATER PIPEWORK

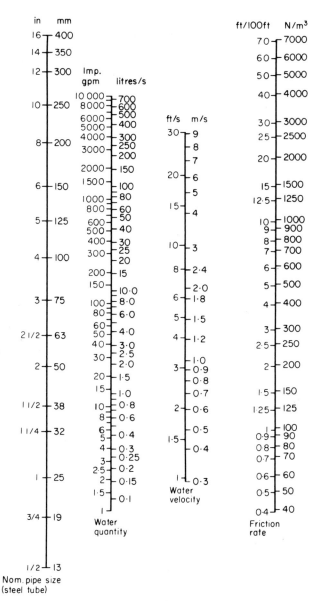

FIG. 18.1. Water-pipe sizing chart—closed systems.

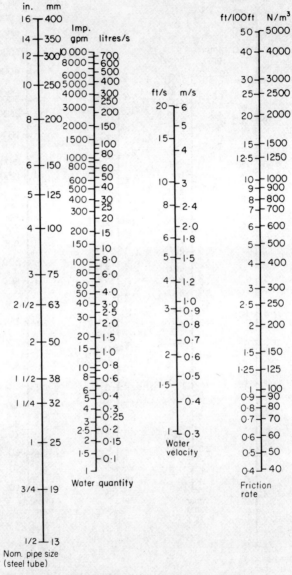

FIG. 18.2. Water-pipe sizing chart—open systems.

systems it is frequent practice to operate at the highest tolerable velocity and friction rate to minimise the installed cost.

Any system using water as a heat conveyance must cater for the change in water temperature, with resultant expansion and contraction of the total water volume. Every system requires the usage of an expansion tank to compensate for both expansion and contraction, and to act as a source for water make-up and often

TABLE 18.1
RECOMMENDED WATER VELOCITIES

Application	Water velocity (ft/s)		Water velocity (m/s)	
	Recommended	Range	Recommended	Range
Pump suction	5	4 to 7	1·5	1·2 to 2·1
Pump discharge	10	8 to 12	3·0	2·4 to 3·6
Main headers	8	4 to 12	2·4	1·2 to 3·6
Risers	6	3 to 10	1·8	0·9 to 3·0
General services	6	5 to 10	1·8	1·5 to 3·0
Drains and overflows	4	4 to 7	1·2	1·2 to 2·1
Make-up	4	3 to 7	1·2	0·9 to 2·1

system fill. In open systems where cooling towers are used, the cooling tower tank suffices as the expansion tank. Figure 18.3 shows an expansion tank which allows for overflow and quick fill. Attention must be paid to the expansion line such that the expansion tank can also accept any air which has been introduced into the system.

When piping to water pumps, several important factors must be considered. The velocities associated with pump inlets and discharges often necessitate a reduction from the main pipework system; if this reducer is placed above the pump, air pockets can occur, and it is recommended that an eccentric reducer is used with the reducer on the underside. Figure 18.4 shows two pumps in parallel and illustrates the requirements of valves, pipe unions or flanges, and proper use of elbows and tees. It is necessary to provide gauges for pumps, such that their performance can be commissioned and properly maintained. Pulsations from pumps can cause damage to the gauges, and valves are recommended to close the water supply when they are not in use, and for servicing. In addition a pipe loop or pulsation damper (not shown) should be included to prevent damage when gauges are in use.

Distribution water piping to many coils causes many balancing problems, and attention should be given to the return water piping to

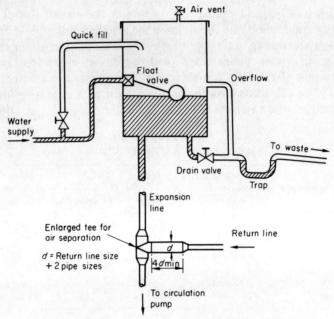

FIG. 18.3. Open expansion tank.

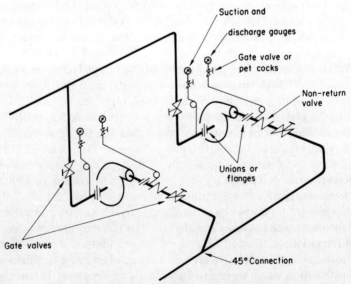

FIG. 18.4. Typical pipework—two pumps in parallel.

establish even or controlled flow to all units. Where many coils of identical performance are required, e.g. induction or room fan coil units, a reverse return system is advocated. This system provides for identical pipework runs from the main supply and return headers for each unit, and although more costly in pipework could avoid, providing the individual coil pressure drop is relatively high, the need

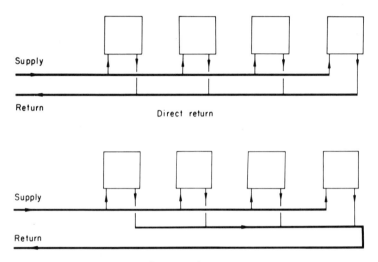

FIG. 18.5. Return water piping.

for balancing valves at each coil. When multiple coils of varying water quantities and pressure drops are required balancing valves are needed and little advantage is gained by using a reverse return system. Such applications could rely on a direct return system as shown in Fig. 18.5.

Individual cooling coils require that proper water flow is maintained, that the coils can be removed for cleaning, that the coils can be blown free of solid deposits and that the water pressure drop can be metered. Figure 18.6 shows a typical chilled water coil with automatic three-way control valve; manually controlled systems or systems using face and by-pass control would eliminate this valve and its service unions.

Figure 18.7 shows water piping used for a liquid chiller. The pump should always discharge into the chiller so that the heat gain to the water is eliminated before the water goes to the cooling coil. Provision

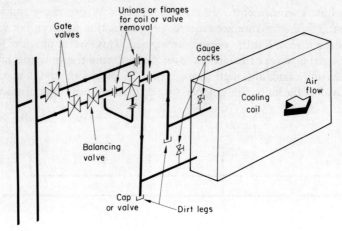

FIG. 18.6. Piping to chilled water coil with three-way control valve.

must be made for removal of the cooler water box for service to the cooler tubes, and this is achieved by the use of unions or flanges. The system shown provides for a gate valve in the leaving water; normally a further gate valve would be located on the inlet side for service. However, the example is close coupled and it is considered acceptable that the valve provided and that at the pump inlet would suffice for both pump and chiller service. Three gauges are shown providing measurement of pump head and cooler pressure drop. Since few manufacturers provide thermometers on their cooler vessels these should also be provided in the chiller return and flow lines.

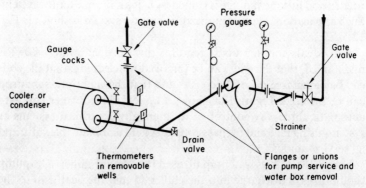

FIG. 18.7. Water pipework to chillers or condensers using cooling towers.

Condenser pipework must consider the same practice as that observed for a cooler. It should be stressed that the water tank in a cooling tower must be higher than the condenser to avoid, on plant shut-down, the water within the condenser from draining into the cooling tower tank. Figure 18.8 shows the water pipework for a once-through system. Attention must be paid to the return or waste water from the condenser. The example shows a loop to drain such that the

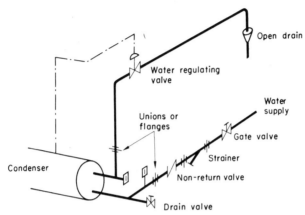

FIG. 18.8. Water pipework for once-through condenser.

condenser will remain full during shut-down or low load operation. Where a loop cannot be accommodated the water control valve should be located on the leaving water side. Attention must be paid to the free draining of the waste water and, although a recommendation, local authority codes must be observed.

Whenever multiple condensers are used it should be remembered that unless identical they are required to operate at different water quantities and pressure drops, and therefore balancing valves to each unit must be included.

Figure 18.9 shows how water temperature control is achieved at condensers to maintain proper condensing pressures. The three-way valve system is used when the cooling tower is at approximately the same level as the condenser. The control valve should be located in the position shown to avoid unnecessary head on the pump suction and to maintain constant flow through the condenser. Where cooling towers are located above the condenser, advantage can be made of the unbalanced head of the cooling tower and a simpler and less expensive

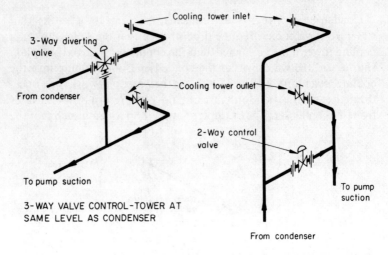

FIG. 18.9. Condenser head pressure control with cooling towers.

two-way valve can be used. The friction loss within the by-pass and control valve should be made less than the unbalanced head. The valve should be located close to the cooling tower to prevent cooling tower spill-over and motor overload when the valve is in a full by-pass position. As with the three-way valve system the by-pass is around the cooling tower and free flow is maintained through the condenser.

Probably the most frequently badly designed water pipework is that applied to drains. Whenever cooling coils dehumidify, the moisture must be collected and carried off as waste. Under operating conditions drain water is subjected to pressure conditions below that

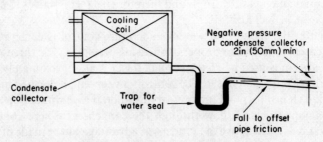

FIG. 18.10. Trapping of condensate drain.

of the atmosphere since in most applications the cooling coil occurs at the negative side of the fan. Therefore the line used to carry this water away must be trapped. The trap serves two functions; it enables the differential pressure to occur at the trap and so allow free drainage from the condensate collector, and in blow-through systems prevents unconditioned air from entering the air conditioning apparatus.

Figure 18.10 shows the proper trapping of a drain line from an air handling plant. The trap height must be determined from the expected negative pressure which can occur at the condensate collector which is the summation of the pressure drop of the cooling coil plus any other heat exchanger or air filtering device before the fan. The run-out from the trap must be pitched to allow for free-fall, and the pipework carefully designed to eliminate pockets and traps which will cause a build-up of the drain water. The drain system should provide for clean-outs and drain fittings to clear any blockages caused by sediment which may be collected at the drain pan.

Chapter 19

HEAT PUMPS AND HEAT RECLAIM DEVICES

The shortage of oil and limited resources of gas have caused a complete rethink for the system designer to the extent that he has had to turn upside down many of his previously conceived ideas about heating systems. On the bases that (i) oil can no longer be considered a suitable heating medium and logically resources will be conserved for transportation; (ii) that gas will probably soon be restricted—space heating is far down the list of priorities—and will probably be conserved for process work and possibly cooking; and (iii) that coal, nuclear energy and the yet to evolve alternative energy will all be used to generate electricity, the designer now has to consider electricity as his potential source of heating.

Electricity in a direct mode as resistance heating uses more basic energy resources than direct use of fossil fuels, but given the advantage of the refrigeration cycle the designer has the tools to conserve total energy resources, whilst in many air conditioning applications reducing capital cost as well.

Before consideration of any energy saving device, the designer should first assure himself that his building design has taken advantage of proper insulation on the premise that it is better to use as little energy as possible, no matter what the energy cost. In terms of both heating and cooling requirements building materials and insulation must be of prime consideration as well as double or even triple glazing with shading devices; the avoidance of air gaps and cracks causing infiltration; and a system flexibility that need not consider ventilation load during a warm-up period.

The heat pump cycle can be considered as a reversible refrigeration cycle, such that in the heating mode the condenser rejects its heat into the conditioned space whilst the evaporator uses a heat sink, usually the ambient air or a water source, to effect the refrigeration cycle. Figures 19.1 and 19.2 show an air to air system in both cooling and heating modes.

Considering this cycle in the heat pump with air available at both evaporator and condenser, the application has been air conditioning during the summer and in the winter the heat rejected is required for heating. The available heat source which is to be cooled is the ambient air. When the device was being operated as an air conditioner, air was entering from the room at, say, 72 °F *(22 °C)* whereas under winter heating conditions this could be as low as 32 °F *(0 °C)* which results in a very large capacity reduction of the machinery. However, instead of ambient air at, say 80 °F *(27 °C)* room air at about 68 °F *(20 °C)* is now entering the condenser, thus giving us an advantage in terms of condensing temperature.

This assumes that the locations of the condenser and evaporator can be changed from ambient to room air and vice versa. With other than the very smallest through-the-window units, changing the position of the coils is impractical because of associated sheet-metal ducting and fans. Therefore, to achieve the heat pump mode it is necessary to reverse the refrigeration cycle by an arrangement of valves. To operate the machine as a reversible cycle machine there is added to the system a four-way valve which is controlled by a thermostat. There is also an added expansion device, check valves around the two expansion devices, and also an accumulator at the suction entry to the compressor. This prevents any un-evaporated slugs of liquid from entering directly into the compressor.

Since the heat exchanger coils will be used for both condensing and evaporating they are referred to as the indoor and outdoor coils. Following the circuit in the cooling cycle, discharge gas from the compressor enters the four-way valve and is directed to the outdoor coil where it is condensed, then as a liquid the refrigerant goes through the check valve, by-passing the added expansion device. It is then expanded into the indoor coil where it evaporates, and is then drawn along the suction line to the four-way valve where it is diverted to the accumulator and into the compressor.

In the heating cycle the position of the four-way valve is changed by the action of a thermostat. The discharge gas from the compressor now goes to the indoor coil which is acting as a condenser and heats the room air. The condensed liquid by-passes the expansion device via

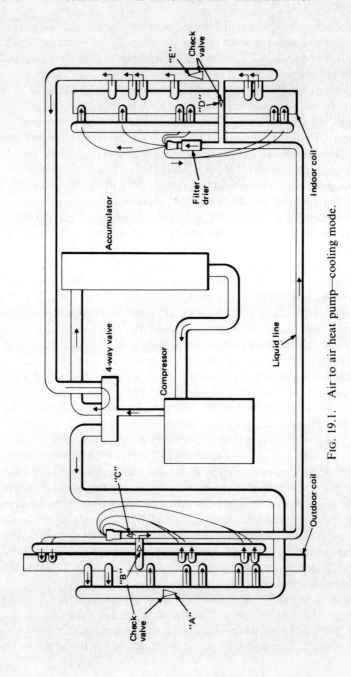

Fig. 19.1. Air to air heat pump—cooling mode.

19 HEAT PUMPS AND RECLAIM DEVICES

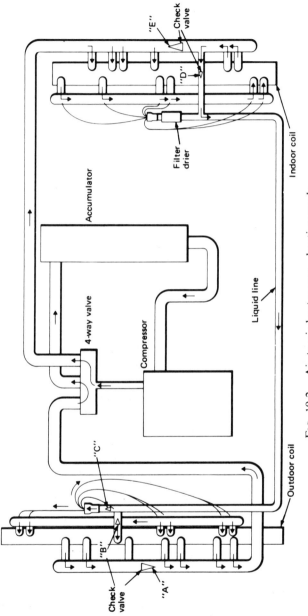

FIG. 19.2. Air to air heat pump—heating mode.

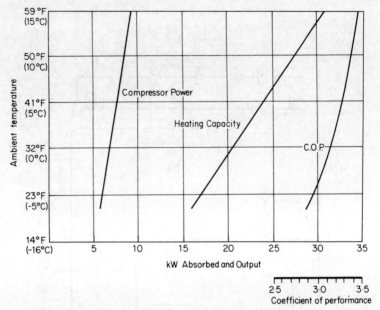

Fig. 19.3. Instantaneous heat pump performance.

the check valve and is then expanded in the outdoor coil where the ambient air is cooled by the evaporating refrigerant, and the gas is drawn through the four-way valve to the accumulator where it re-enters the compressor.

Full use of the benefits of COP (coefficient of performance) are now available to provide low-cost heating. There is, however, one major drawback to this very simple system. That is, when using ambient air at an already low temperature and further cooling it, the water vapour in the air condenses, and, if the coil temperature is already below freezing point, then this condensed vapour will form frost on the coil, in exactly the same way as a domestic refrigerator. The system will continue to operate until the coil fins become blocked with ice, and air can no longer pass in sufficient quantities to provide the necessary heat to evaporate the refrigerant. It should be said that all the time the coil is frosting so the evaporating temperature is falling, thus reducing the capacity of the plant.

The answer to this problem is quite simple: reverse the cycle and provide heat at the outdoor coil, to defrost the coil so that normal operation can be resumed once defrost has taken place. However, the

19   HEAT PUMPS AND RECLAIM DEVICES

effect inside the room can be dramatic because under the defrost cycle the room air is now being cooled.

Figure 19.3 shows the instantaneous performance of a typical air to air heat pump. Plotted are the required compressor power and heating capacity for given ambient temperature, and COP at that given temperature. It can be clearly seen that even at 32 °F (0 °C) there is still a great power saving. However, this only reflects the instantaneous performance and does not show the overall picture once the defrost requirements are taken into acount. To provide a true performance all factors influencing the heating capacity should be considered. The major factor affecting performance is the defrost cycle. Under normal UK operation a timer is incorporated to reverse the system's operation for 10 minutes every two hours, i.e., 10 minutes' cooling, followed by 1 hour 50 minutes' heating. Attached to the outdoor coil is a frost thermostat which would override the time cycle if no frost is present, i.e. at ambient temperatures normally above 45 °F (7 °C). Likewise, the power required at the indoor fan to overcome the resistance of the coil and sheet-metal ducting has the effect of adding heat. Also to be considered are the ambient temperature at which the heat pump is operating, the heat losses associated with the conditioned space itself, and consequently the actual running time of a heat pump to meet these losses.

Figure 19.4 compares the instantaneous performance of the heat pump with the integrated performance allowing for defrost, and a different picture emerges. Quite clearly the unit performance drops, and at 45 °F (7 °C) and below the kW input increases, which is reflected in the COP. The COP line no longer conforms to an even curve but now takes into account the added power required during the defrost cycles. It should be noted that the irregularity of the curve stems from the fact that at 45 °F (7 °C) there is a relatively large amount of moisture present in the ambient air which calls for a long defrost, whereas as the temperature lowers, the amount of moisture vapour in the air decreases, thus decreasing the period of defrost, although not the frequency.

As an example, the heating requirements for a building have been taken as 20 kW at 32 °F (0 °C) to maintain a constant space temperature of 68 °F (20 °C). The heat losses can be considered as a fraction of the design requirement versus the difference between ambient temperature and room temperature, e.g., at 50 °F (10 °C) ambient the heat loss would only be 10 kW.

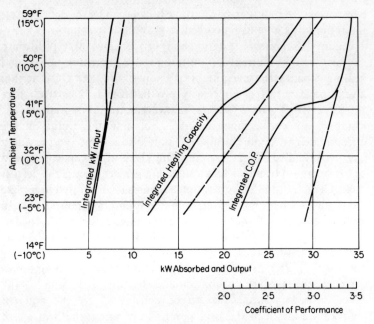

Fig. 19.4. Integrated heat pump performance.

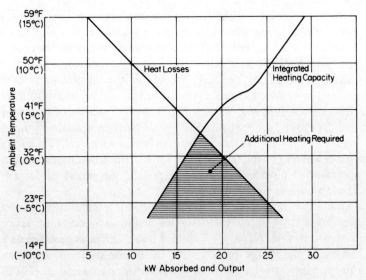

Fig. 19.5. Heat pump performance versus heat losses.

Using the integrated ratings the heat losses can be plotted to reveal the amount of additional heating required at various temperatures (Fig. 19.5).

The frequency of average dry bulb temperatures for London during the months November to April are shown in Table 19.1, and, for comparison to different applications have been broken down

TABLE 19.1
MONTHLY AVERAGE TEMPERATURES FOR LONDON (NOVEMBER TO APRIL)

Ambient temperature (°C)	No. of hours at ambient temperature			Integrated COP
	8 a.m. to 6 p.m.	6 p.m. to 8 a.m.	Total	
12	120	—	120	3·43
11	30	30	60	3·41
10	30	30	60	3·39
9	240	30	270	3·37
8	120	90	210	3·35
7	90	270	360	3·27
6	390	510	900	3·00
5	270	180	450	2·70
4	300	720	1 020	2·61
3	120	450	570	2·55
2	90	150	240	2·50

	Average COP
8 a.m. to 6 p.m.	2·96
6 p.m. to 8 a.m.	2·81
Total	2·87

against day and night-time operation. All the values are totalled and compared to the integrated COP shown against ambient temperature in Table 19.1. The resulting average day-time COP is 2·96, night-time COP 2·81 and the 24 hour average 2·87.

A direct comparison using the 24 hour average COP as a basis for calculation indicated a running cost for the heat pump of some 35 per cent of the cost for direct electrical heating.

Using this information, an owning and operating cost comparison can be made—but first it is necessary to consider the application. For residential use one can assume operation five days per week from 6 a.m. to 10 a.m. and 4 p.m. to midnight, and two days per week from 6 a.m. to midnight, that is a total of 96 hours of the 168 hour week. Conversely, an office application would only operate five days a week from 7 a.m. to 6 p.m. or 55 hours per week. The residential application can be considered to operate at the 24 hour average COP of 2·87 and the office application at the day-time average COP of 2·96.

To establish the viability of a heat pump one must consider the capital cost versus the savings that apply, these being affected by the power tariffs. In establishing this equation at any given time, the designer has the knowledge that trends are towards electricity use and can therefore be assured that any conclusion reached in favour of a heat pump system will give even greater savings in the future.

Using average ambient values for London, for a store application having a design heat loss of 30 kW at 32 °F ($0\,°C$) with electricity at £0·027 per kWh, and gas at £0·20 per therm (100·00 BTU/h), and assuming the constant cost of sheet-metal ducting and grilles which have been excluded from the capital cost values, the following emerges.

A.	Cost of heat pump installation	£3500
B.	Cost of heating and ventilating only	
	(i) with electric heating	£1500
	(ii) with new gas fired boiler	£3000
	(iii) with existing gas fired boiler	£1500
C.	Cost of air conditioning	
	(i) with electric heating	£3000
	(ii) with new gas fired boiler	£4500
	(iii) with existing gas fired boiler	£3000
D.	Energy cost of heat pump	£473
E.	Energy cost of electric heating	£1377
F.	Energy cost of gas heating	£533

On the basis of capital invested at 15 per cent and 10 years amortisation the owning and operating costs are as below and show comparison to A, the heat pump installation.

A.	£1092	
B.	(i) £1642—plus £550	
	(ii) £1084—minus £8	
	(iii) £819—minus £273	
C.	(i) £1908—plus £816	
	(ii) £1350—plus £258	
	(iii) £1084—minus £8	

Quite clearly the comparison shows a saving over even a ventilation system with electric heating; only the existing boiler installation with ventilation added would show any substantial saving. However, this argument could be turned around in that at £273 p.a. the heat pump system provides very economical air conditioning.

There is little doubt that the added capital cost of a heat pump more than justifies itself, given a system that requires air conditioning. It can also be considered that the inflation in fuel costs will outstrip the 15 per cent interest considered for the investment, if this was to be the case then the cost savings would be even more dramatic. For example, the £500 increase in heat pump cost over air conditioning with electric heating could be realised in about 7 months of operation.

Given the basic principles of a heat pump mechanism there are developments which are taking place on an almost hourly basis which are not only making the heat pump more reliable but are reducing capital and running costs. In the past the heat pump was a modified air conditioning system, but with the obvious trend towards energy saving devices, equipment designers are now putting the heat pump first and considering modifications to them for air conditioning systems.

This logic has already evolved some major design improvements such as a single-piece component which serves as both expansion device and by-pass valve, thus eliminating the service factor of the two previous components. The use of a sliding piston within a sleeve having a large orifice at one pipe connection and a small orifice at the other means that when the heat pump demands a by-pass flow the piston is positioned against the large orifice. The piston itself has a cross-section similar to a star such that the refrigerant can pass between the full area presented through the large orifice with little pressure drop. When the direction of the refrigerant reverses, the piston slides against the small orifice presenting a very restricted area to perform as an expansion device.

The defrost cycle cannot be avoided in an air to air system, but the use of two refrigeration cycles with phased defrost means that during defrost of one the second cycle can continue heating, thus avoiding the need for supplementary heating to prevent shock cooling loads to the system. Given that the integrated capacity of the heat pump system is in excess of the heating load, then supplementary heating can be avoided entirely. Similarly, applications in, for example, stores, which call for a year round daytime cooling cycle because of occupancy and lights can also arrange their system to provide warm-up only to bring the store up to minimum temperature under unoccupied conditions. Thus a defrost cycle will not cause discomfort and again, given that the heat pump is adequately sized, supplementary heaters can be avoided.

Where supplementary heaters are necessary it is good practice to include an outside air thermostat, such that above a dictated ambient temperature supplementary heaters are isolated for other than the defrost cycle. This means that at start-up when room conditions are lower than design, only the economical heat pump cycle is used to warm up the space. This logic can be extended to several stages of ambient sensing thermostats to match increments in supplementary heating.

There are opportunities to improve the heat pump COP, the simplest being to increase the heat exchanger sizes such that there is a low temperature difference between evaporating and condensing levels to the available heating and cooling media. However, because of energy requirements of fans, heat exchangers are limited to four rows, otherwise savings are offset by increased pressure drop. The use of exhaust air over the outdoor coil is of practical advantage for both summer and winter operation, but heat pump systems should have the facility to isolate exhaust air during warm-up since any efficiency improvement by using higher than ambient temperature would be more than offset by the cooling load imposed by the make-up air.

The designer should closely consider any application using large volumes of outdoor air. Theoretically, a low on-coil temperature to the indoor coil will improve the system COP. However, beyond a certain limit there would not be sufficient pressure differential to provide proper expansion valve operation and improved thermal efficiency would be more than offset by reduced system performance because of lowered evaporating conditions. The effect would be similar to that experienced in air cooled condenser systems at low ambient conditions (see Chapter 12).

There have been several attempts to convert existing air conditioning systems into heat pumps with little or no success. The conversion is far from simple and should be avoided, even if the added features of a second expansion valve, two by-pass valves, a dual flow strainer drier, a suction accumulator and four-way valve are added to the refrigeration cycle with additional control features. Standard air conditioning equipment does not have the compression equipment suitable for the frequent stop-start operation of the heat pump during defrost; neither does it have the surface area nor coil depth necessary for proper outdoor coil operation, but most significantly probably has a refrigerant charge and balanced heat exchanger sizing incompatible for reversed cycle operation.

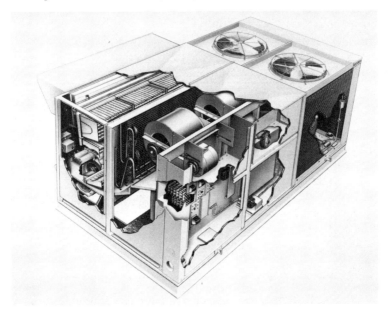

FIG. 19.6. Typical air to air single-piece heat pump.

In addition to the air to air heat pump available in a single-piece or split configurations as packaged air conditioning equipment (Chapter 16) other heat sources give the further combinations of air to water, water to water and water to air.

The air to water system is essentially a reversible air cooled water chiller and finds applications with fan coil systems. Ambient air is still used as a heat sink; however, the nature of the system is such that the defrost cycle rarely causes the potential shock experienced in air to air systems. Even considering the worst defrost of, say, 10 minutes, the reduction in water temperature will not be sufficient to fall below room conditions and therefore, given that the integrated heating performance is above or matches the heating load, no supplementary heating will be required (Fig. 19.7).

In terms of running and capital cost the air to water heat pump is more economical than the water chiller/boiler requirements of a two-pipe fan coil system. However, for a four-pipe system a close evaluation is necessary since the air side heating coil, probably having only one row available for heat exchange could well require elevated temperatures, giving reduced machine performance and reducing the

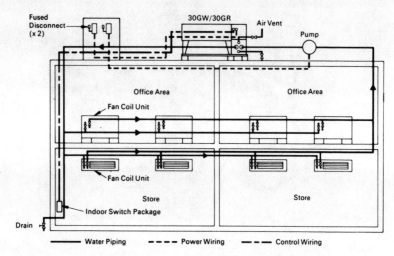

FIG. 19.7. Air to water heat pump with two-pipe fan coil system.

COP. Within practical terms the water temperature leaving the heat pump should not exceed 105 °F (40 °C) if maximum economy is to be realised. With a two-pipe fan coil system a simple arithmetic equation can be reached to compare heating and cooling performance.

COOLING

Room temperature	72 °F		22 °C	
		⟩22 °F		⟩12 °C
Leaving chilled water	50 °F⟨		10 °C⟨	
		⟩10 °F		⟩5·5 °C
Entering chilled water	40 °F		4·5 °C	

Given that the same coil is used for heating and cooling, and that the same duty is required from the coil for both heating and cooling, then the following results.

19 HEAT PUMPS AND RECLAIM DEVICES

HEATING

Room temperature	70°F ⟩ 22°F	21°C ⟩ 12°C
Leaving hot water	92°F ⟨ 10°F	33°C ⟨ 5·5°C
Entering hot water	102°F	38·5°C

Under such circumstances high COP values can be achieved, but given a single-row coil requiring a three-fold increase in temperature difference the entering hot water would have to approach 150°F (65°C) which is far from practical. The practical application of air to water heat pumps for four-pipe fan coils should therefore be confined to applications where heating loads are less than 50 per cent of cooling loads. There would be little merit in providing two deep coils for both heating and cooling since the fan power, required year round, would be excessive (see Chapter 20) and the potential of a changeover system in winter where the heating circuit uses the three-row cooling coil could cause system control problems; with the added cost of summer/winter changeover controls to both heating and cooling valves, that would achieve little better than the two-pipe system.

As yet the air to water heat pump has not found an application for heating only with the use of convectors or radiators, essentially because of the low grade heating available for system economy.

Water to water heat pumps have the same handicaps on the system side as the air to water units. However, given a heat source such as a well or river then the COP of such systems are very advantageous. The application of such units is rare and they are usually found in a heat reclaim mode without the potential for reversing the refrigeration cycle.

Water to air heat pumps however, are quite frequently used for multiple unit installation. The basic equipment tends to be small refrigeration cycle plants of up to about 3 tons (*10 kW*) capacity with a central source of heat rejection and heat sink on a ring main principle (Fig. 19.8). The manufacturers of such equipment claim low operating costs, but in the absence of a natural heat sink and the use of a boiler, the running costs would exceed fossil fuel burning during a heating season. However, during mid season or when there is a constant cooling requirement the system does have the advantage of utilising the effect of some units of the heat rejected on the cooling cycle as the heating source for reversed cycle units. The capital cost of such a system could be high if either a boiler installation is required or a high degree of partition flexibility, which calls for a large quantity of small units. The designer should consider water to air systems in consideration of the low efficiency of small machines together with their high maintenance cost and low recommended amortisation period. The best application for such systems would be a residential

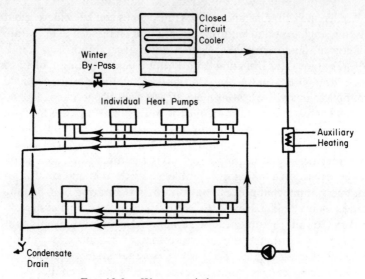

FIG. 19.8. Water to air heat pump system.

block where larger individual units serving multiple fixed rooms would be quite efficient, particularly in a sunny climate where the building could require cooling on one or two faces with heating on the others.

With water to air heat pumps it should be remembered that on the heating cycle the 'outdoor' section is acting as a water chiller and close attention should be paid to the heat source to ensure that the water leaving the 'outdoor' section does not fall below 40°F ($4.5°C$).

Heat reclaim systems differ from heat pumps in that their refrigeration cycle is not reversed. Essentially they rely on the building-wasted energy for conversion into useful energy for heating, and in some cases cooling. Before considering the merits of heat reclaim devices the designer should be aware of the design parameters and limitations associated with heat pumps, and primarily concern

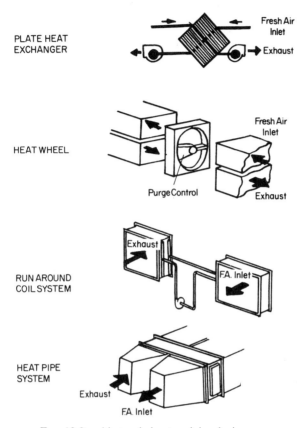

Fig. 19.9. Air to air heat reclaim devices.

himself with the seasonal load requirements of his application, always remembering that the prime objective is to conserve energy first, and wherever economically possible, reclaim any balance. Just as with a heat pump, each component of a heat reclaim system has its own COP, if defined as the ratio of the seasonal energy reclaimed to the energy used to conserve it.

Air to air systems are most effectively served by air to air heat exchangers, with the use of a refrigeration cycle only as an addendum. There are four basic heat exchange devices which are shown diagrammatically in Fig. 19.9. Such systems are restricted to applications requiring a positive exhaust, and find themselves confined to all-air systems associated with laboratories, swimming pools, and areas of high occupancy such as theatres and lecture rooms.

The air to air plate heat exchanger is by far the simplest but requires additional fan power on both supply and exhaust cycles. Should a very high efficiency plant with high pressure drop be considered then the additional fan power required year round could well detract from

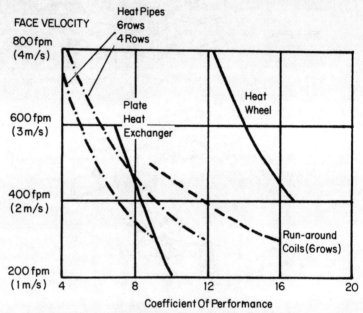

FIG. 19.10. Relative performances of air to air heat reclaim devices.

potentially limited heat savings in a mild climate. Figure 19.10 shows the relative performance of the various devices in terms of unit face velocity. It should be noted that for a given air volume the heat exchanger at 200 fpm (*1·02 m/s*) is three times as big as one at 600 fpm (*3·05 m/s*) and the designer should always consider the capital cost of such equipment versus potential savings before deciding the optimum selection.

In essence, the air to air heat exchanger is a series of flat plates, often glass or stainless steel, with supply and exhaust air in crossflow, in alternate air spaces between the plates. The physical arrangement of the plates prohibits the opportunities to use a counterflow arrangement.

Although generally higher in capital and maintenance cost, the somewhat more efficient Heat Wheel is very widely used and can be used as a sensible heat exchanger, or, given suitably absorbent material, can be used as a total heat or enthalpy exchanger. Such plant requires power to rotate the heat exchanger in addition to increased fan power. In addition, so that exhaust air should not be carried over from one section to another, it is equipped with a purge section such that some ambient air is always allowed to pass through the purge into the exhaust stream to avoid any carryover of exhaust air into the supply stream. With such a system, although the supply air fan can be positioned on either the system or weather side of the heat exchanger, to ensure a proper air flow through the purge it is essential that the exhaust fan is located on the weather side.

When using air to air heat exchangers there is always the possibility of frosting in the event of high humidity exhaust air being coincident with low ambient conditions. In this event it would be necessary to provide a preheater located on the weatherside of the system. This would slightly impair the effectiveness of the heat exchanger, but nevertheless would be adding heat necessary to offset the heat exchanger inefficiency, and building transmission loss. If ambient temperatures are expected to be very low then the preheater itself should be provided with frost protection (electric heating could also be used but is somewhat contradictory to an energy saving logic), or a by-pass arrangement allowing some of the already heated air to be mixed with the incoming air could be used.

Both the plate- and wheel-type heat exchanger require substantial ductwork connections and generally require quite large plant room areas. Under limited space conditions a run-around coil system can prove effective, although generally not as efficient. Essentially, brine filled heat exchanger coils are located in the exhaust, and supply air streams having a circulating pump in the coils should be selected such that freeze-up cannot occur even at the lowest recorded ambient temperatures at the application. Whilst the brine solution will impair heat exchange, it should be realised that the stronger the solution necessary the greater will be the potential savings because of the lower ambient temperature forecast.

An improvement over the run-around coil is the heat pipe. It has the advantage that it is unaffected by frost since refrigerant is within the tubes and does not require a pump to move the fluid, but has the disadvantage that both supply and exhaust systems must be close

386   APPLIED AIR CONDITIONING AND REFRIGERATION

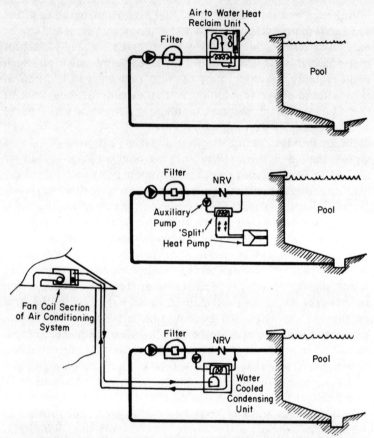

FIG. 19.11. Air to water heat reclaim for swimming pools.

since the heat pipe couples the two. The heat pipe consists essentially of individually sealed tubes arranged in a coil block formation with external plate fins. Within the tube is a refrigerant which in one half the length is evaporating, by the effect of waste heat, and in the other condensing and so imparting reclaimed heat into the air stream. The condensed fluid remains in the lower half of the tube flowing towards the evaporating section, whilst the evaporated refrigerant travels in the opposite direction, essentially as a convection current.

All the air to air heat exchanger devices can be augmented by an air to air heat pump provided the outdoor coil is located on the weather side and the indoor coil on the inside, thus overcoming expansion valve problems and in mild climates probably avoiding defrost problems. If, however, a close control of the supply air is required then since the heat pump tends to be an on–off performance, or at the best on the largest systems three-stage, then it must be the heat exchanger which provides the modulation. This would exclude the heat pipe and plate heat exchanger but could be achieved with the heat wheel by varying its speed of rotation, or on the run-around coil with a variable speed pump or three-way valve control.

Air to water heat reclaim devices find limited application because of the defrost problems associated with low ambient conditions. There are some devices used for domestic swimming pool heating but have the major disadvantage that they cannot be used below 45 °F (70 °C). A much improved solution would be the use of an air to water heat pump. However, given that an application has supplementary or alternative heating several opportunities exist, although possibly with the disadvantage of high capital cost. For example the swimming pool application could provide an alternative to an existing boiler whenever ambient conditions allow. Similarly the use of rejected heat from an air conditioning plant could serve to supplement pool heating.

Water to water heat reclaim has probably seen the widest application use, particularly with large office blocks and institution buildings. The use of a heat reclaim system is confined to the water or air/water air conditioning systems with a year round cooling demand, usually from the centre core of a building. Since all-air systems can utilise free cooling at low ambient conditions it is probably better not to use energy in the first place rather than attempt to reclaim energy. It is probable that the return air from lights will be used to mix with the required ventilation load at design minimum air temperatures to provide normal dew point temperatures, and there are few, if any, opportunities for heat reclaim unless the centre core area is of a large proportion to the perimeter (see Chapter 14). In the event that there is always a net cooling load because of lights and people, and there could be a potential heat reclaim situation, by definition it would not be required.

The heat reclaim principles are identical to the air to water heat pump on the system side. For maximum efficiency, low hot water temperatures should be used with deep air side coils which is generally associated with two-pipe fan coil systems, or the primary air side of a primary air fan coil system.

The heat rejection side of the equipment used for heating would normally be a cooling tower with the handicap of heat exchanger fouling. A common solution has been the use of a 'double bundle' condenser with sufficient heat exchanger tubes to do both heat rejection and heat reclaim within the same refrigeration circuit. Essentially, any refrigerant not condensed in the heat reclaim process would pass into the normal condenser used for air conditioning and

FIG. 19.12. Double bundle condenser system.

be rejected into the cooling tower. This procedure ensures the isolation of the cooling tower circuit from the heating circuit thus keeping a closed circuit free from water maintenance problems. It does however demand two sets of pumps and controls, and the designer should consider the capital cost of a second condenser and pumps added to the cost of a closed circuit cooling tower with a standard water chiller.

This totally closed circuit approach not only keeps the heating system water maintenance free, but also avoids fouling of the condenser, thus ensuring optimum heating and cooling performance, always assuming that the closed circuit cooler is adequately maintained.

In both water to water systems it has been common practice to require elevated condensing temperatures to provide hot water, often because of small air side heat exchangers. Any system should be designed such that whenever cooling is required the optimum condensing temperature is achieved for energy conservation within

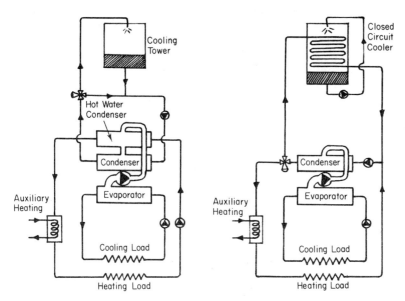

FIG. 19.13. Heat reclaim circuits for water to water systems.

the limits of expansion device operation, and these levels are always lower than the elevated temperatures required during the heating season. Subsequently a dual head pressure control is required, with the heating cycle preferably scheduled to ambient temperature. With the double bundle application, since the hot water condenser would take whatever heat it required, there would be no need to reschedule the normal condenser temperature control.

To establish the heat reclaim possibilities it is essential to evaluate the minimum cooling requirement of the building such that the heat rejection can be used for heating. It may be practical to install a heat storage vessel that will store enough energy for building warm-up before occupancy, providing the building load profile will provide sufficient residual heat. However, in the event that the residual heat is insufficient to meet the heating load, particularly during warm-up, then supplementary heating will be required. When considering a warm-up period, heat from lights should not be considered since if the lights are on, without their need for occupants, such energy equates to inefficient electrical resistance heating.

Although it has been established that reclaiming condenser heat at high temperature is inefficient, high grade heat can be achieved by the use of a hot gas desuperheater. Whether in an air conditioning or heat pump cycle the discharge gas from a compressor is usually at about 150°F (65°C) and such high grade heat can usefully be used for domestic hot water heating. The logic can be applied to any refrigeration plant providing either the hot gas is not condensed in the system or, as in the double bundle condenser any liquid refrigerant can free flow into the condenser properly. Since it is usually

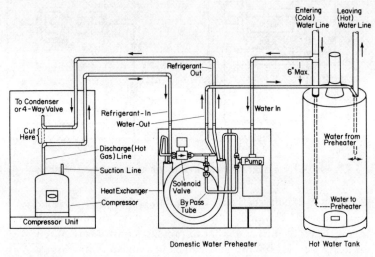

FIG. 19.14. Condenser desuperheater for domestic hot water heating.

impractical to mount the desuperheater above an air cooled condenser because of freeze-up problems, desuperheaters are equipped with a by-pass valve to ensure that condensation does not occur (Fig. 19.14).

The desuperheaters are widely used in residential and small commercial applications, and even with use in a heat pump system need not detract from system performance or application efficiency. The largest demand on any system is during warm-up, which by definition means the absence of people. Thus the hot water storage can be brought up to temperature economically during a prolonged warm-up period, and by the time the water is drawn off there will be some residual heat in the form of lights which could well offset the reduced heating performance of the heat pump. In the event that this was not the case then the same energy used for the water heating, often electric, could be used as supplementary heat in the heat pump, such that only at the very lowest ambient conditions could 'free' heating not be obtained. At all times during the air conditioning cycle advantage can be taken of 'free' heating.

To summarise, there are several elementary steps that the designer should follow before deciding on any heat pump or heat reclaim system.

1. Maximise insulation, provide multi-glazing and shading devices, and minimise infiltration.
2. Where possible, utilise free cooling rather than reclaim heat rejection.
3. Take whatever advantage possible of any exhaust air and/or heat from lights, either by heat exchanging with ventilation air or across heat pump outdoor coils.
4. Keep design condensing temperatures as low as practical by using large air side heat exchangers.
5. Minimise supplementary heating (including that required for defrost cycles).
6. Consider potential heat reclaim against added fan and pumping power.
7. Remember that any heat pump or heat pump system is not maintained by the designer and should always be kept as simple as possible.

Chapter 20

ENERGY COMPARISON OF AIR CONDITIONING SYSTEMS FOR MULTI-ROOM APPLICATIONS

Of the numerous air conditioning options available to a designer, current energy costs have already precluded systems such as dual-duct, all air terminal reheat, reheat induction, incremental room air conditions, etc., for applications not restrictive in building design, or where specific applications such as a hospital or laboratory demand them. Current trends are towards variable air volume, fan coil and induction systems, and in consideration of this an energy comparison has been prepared of the three basic systems commonly associated with multi-room office block air conditioning, compared to a dual-duct system, thus giving comparisons between two air/water systems and two all air systems.

The systems chosen are:

2-Pipe Non-Changeover Induction (see Fig. 14.3).
4-Pipe Fan Coil system (with primary air for ventilation) (see Fig. 14.34).
Variable Air Volume with wet perimeter heating (VAV) (see Fig. 14.10).
Dual-Duct system (D-D) (see Fig. 14.31).

To prepare a yearly energy operation budget, the office block as shown in Fig. 20.1 has been selected as typical of many buildings which potentially have an air conditioning requirement. The building consists essentially of ten floors, each having ninety-six 5×23 ft ($1 \cdot 5 \times 7m$) modules, which have been considered as the minimum controllable areas with respect to partition flexibility. The location has been assumed as London, with double glazing, light-coloured venetian blinds and an average level of insulation.

A block load estimate has been prepared on the basis of an ambient design condition of 82°F DB, 68°F WB (*27·8°C DB, 20°C WB*) and considering a room temperature of 72°F (*22·2°C*) DB, with 50 per

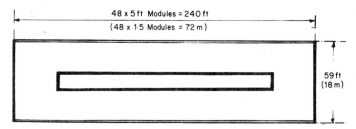

FIG. 20.1. Typical floor plan of building considered.

cent RH for both induction and fan coil systems and 45 per cent RH for the VAV and D-D systems. The results plotted in Tables 20.1 and 20.1a are for 1400 occupants requiring 35 600 cfm (*16·8 m³/s*) ventilation air, and consider by-pass factors of 0·1 for the air/water systems and 0·04 for the all air systems. At this point it should be stressed that the fan power to the various systems is substantially different and the values established later in the text have been inserted into the load estimates. It has been assumed that only 5 per cent of supply air is leaked via the warm air duct in the D-D system resulting in an effective BP factor of 0·09 (it should be stressed that the by-passed air volume can often exceed 10 per cent).

In summary the air requirements are:

2-pipe induction system	35 600 cfm (*16·8 m³/s*) primary air
4-pipe fan coil system	35 600 cfm (*16·8 m³/s*) primary air
VAV system	94 900 cfm (*44·8 m³/s*) supply air
VAV system	59 300 cfm (*28·0 m³/s*) return air
D-D system	109 500 cfm (*49·6 m³/s*) supply air
D-D system	73 900 cfm (*32·8 m³/s*) return air

For comparative reasons it has been assumed that the ventilation requirement will be exhausted via toilet extract systems and exfiltration.

TABLE 20.1
SYSTEM LOAD ESTIMATE SUMMARY (THOUSANDS BTU/h)

	System			
	2-Pipe induction	4-Pipe fan coil	VAV	Dual-duct
Fabric gain—solar	990	990	990	990
—transmission	333	333	333	333
Occupancy—sensible	325	325	293	293
Lighting	1 206	1 206	603	603
Net RSH	2 854	2 854	2 219	2 219
Fan gain	171	440	287	386
By-passed outdoor air	38	38	15	35
ERSH	3 063	3 332	2 521	2 640
Occupancy—latent	211	211	243	243
By-passed outdoor air	55	55	27	49
ERLH	266	266	270	292
ERTH	3 329	3 598	2 791	2 932
Outdoor air—sensible	347	347	369	350
Outdoor air—latent	490	490	639	617
Lighting (return air)	—	—	603	603
Return air fan	—	—	72	84
Chilled water pump	181	144	102	102
Grand total heat	4 347	4 579	4 576	4 688
ERSHF = $\dfrac{\text{ERSH}}{\text{ERTH}}$			0·90	0·90
Selected ADP			46·4°F	46·4°F
Effective coil Δt			24·6°F	23·3°F
Air volume			94 900 cfm	109 500 cfm

For equipment selection of the induction and fan coil systems a typical module will require 37 cfm (*17·5 litres/s*) of ventilation air, the induction unit selection would be with a nozzle pressure of 1·26 in. w.g. (*315 N/m²*) and a total secondary water system requirement of 1200 igpm (*90·9 litres/s*). Similarly, the fan coil units would require a total chilled water volume of 1152 igpm (*86·4 litres/s*) at 42°F (*6°C*) and each unit would require 107 W or a total secondary fan load of 102·72 kW. At this juncture it should be said that the fan power referred to is for a permanent split capacitor motor. Shaded pole motors are available as an alternative, but since their power requirements are between 60 and 100 per cent in excess of permanent split capacitor motors such a selection can be discarded since any capital cost reduction would be more than compensated for in increased running costs in less than one year. Table 20.2 shows the

20 ENERGY COMPARISON OF AIR CONDITIONING SYSTEMS

TABLE 20.1a
SYSTEM LOAD ESTIMATE SUMMARY (kW)

	System			
	2-Pipe induction	4-Pipe fan coil	VAV	Dual-duct
Fabric gain—solar	290·2	290·2	290·2	290·2
—transmission	97·6	97·6	97·6	97·6
Occupancy—sensible	95·3	95·3	85·8	85·8
Lighting	353·4	353·4	176·7	176·7
Net RSH	836·5	836·5	650·3	650·3
Fan gain	50·0	129·0	84·0	113·0
By-passed outdoor air	11·4	11·4	4·5	10·1
ERSH	897·9	976·9	738·8	773·4
Occupancy—latent	61·9	61·9	71·4	71·4
By-passed outdoor air	16·0	16·0	7·8	14·4
ERLH	77·9	77·9	79·2	85·8
ERTH	975·8	1054·8	818·0	859·2
Outdoor air—sensible	102·5	102·5	109·3	103·7
Outdoor air—latent	144·2	144·2	187·1	180·8
Lighting (return air)	—	—	176·7	176·7
Return air fan	—	—	21·0	24·5
Chilled water pump	53·0	42·0	30·0	30·0
Grand total heat	1275·5	1343·5	1342·4	1384·9
$ERSHF = \dfrac{ERSH}{ERTH}$			0·90	0·90
Selected ADP			8°C	8°C
Effective coil Δt			13·63°C	12·92°C
Air volume			44·8 m³/s	49·6 m³/s

component resistances of each system from which fan selections have been made.

From the load estimate for each system a water chiller can be selected. To minimise the quantity of selections it has been considered that condenser water flow and cooling tower sizing remain constant, with condenser water pumps at 18·5 kW and cooling tower fans at 30 kW. The resulting refrigeration plant absorbed power is:

2-Pipe Induction	362·2 tons R 330 kW
4-Pipe Fan Coil	381·5 tons R 350 kW
VAV System	381·3 tons R 350 kW
D-D System	390·6 tons R 358 kW

The heating requirements of each system are tackled quite differently. The VAV system has wet perimeter heating requiring a

TABLE 20.2
SYSTEM RESISTANCES (in. w.g.)

	System			
	2-Pipe induction	4-Pipe fan coil	VAV	Dual-duct
RESISTANCES				
Inlet louvres	0·25	0·25	0·25	0·25
Filters	0·35	0·35	0·35	0·35
Preheater	0·12	0·12	0·12	0·12
Spray humidifier	0·90	—	—	—
Dehumidifier	0·42	0·42	0·63	0·63
Eliminators	0·16	—	—	—
Reheater	0·36	0·16	—	—
Velocity conversion loss	0·57	0·41	0·42	0·42
Attenuator	1·00	0·50	0·50	0·50
Ducting	3·00	2·00	2·00	3·00
Terminals	1·26	—	0·96	0·96
Total	8·39	4·21	5·23	6·23
Supply air volume	35 600 cfm	35 600 cfm	94 900 cfm	109 500 cfm
Fan absorbed power (kW)	50·0	26·0	84·0	113·0
Return air volume	—	—	59 300 cfm	73 900 cfm
Fan absorbed power (kW)	—	—	21·0	24·5

TABLE 20.2a
SYSTEM RESISTANCES (N/m^2)

	System			
	2-Pipe induction	4-Pipe fan coil	VAV	Dual-duct
RESISTANCES				
Inlet louvres	*63*	*63*	*63*	*63*
Filters	*87*	*87*	*87*	*87*
Preheater	*31*	*31*	*31*	*31*
Spray humidifier	*226*	—	—	—
Dehumidifier	*106*	*106*	*157*	*157*
Eliminators	*40*	—	—	—
Reheater	*90*	*40*	—	—
Velocity conversion loss	*142*	*105*	*105*	*105*
Attenuator	*250*	*125*	*125*	*125*
Ducting	*750*	*500*	*500*	*750*
Terminals	*315*	—	*238*	*238*
Total	*2 100*	*1 057*	*1 306*	*1 556*
Supply air volume	*16·8 m^3/s*	*16·8 m^3/s*	*44·8 m^3/s*	*49·6 m^3/s*
Fan absorbed power (kW)	*50·0*	*26·0*	*84·0*	*113·0*
Return air volume	—	—	*28·0 m^3/s*	*32·8 m^3/s*
Fan absorbed power (kW)	—	—	*21·0*	*24·5*

20 ENERGY COMPARISON OF AIR CONDITIONING SYSTEMS

15 kW pump. The induction system requires heating only to the preheater and reheater coils, except for warm-up, when the secondary chilled water pump can be used with the secondary cooling coils acting as perimeter convectors, and the pump has been sized at 10 kW. The four-pipe fan coil system however, has to pump to 960 individual coils and has been sized at 23 kW, which is identical to the pump requirements for the secondary chilled water circuit to the

TABLE 20.3
INSTALLED POWER (kW)

	System			
	2-Pipe induction	4-Pipe fan coil	VAV	Dual-duct
Supply fan	50	26	84	113
Return air fan	—	—	21	24·5
Fan coil units	—	103	—	—
Refrigeration plant	330	350	350	358
Chilled water pumps				
Primary	30	42	30	30
Secondary	23	—	—	—
Cond. water pump	18·5	18·5	18·5	18·5
Cooling tower fan	30	30	30	30
Hot water pump	10	23	15	5
Total	491·5	592·5	548·5	579·0

induction units. The dual-duct system by contrast uses fan power to provide heating, including any warm-up periods. A summary of the installed power can be made as shown in Table 20.3.

Whilst at this stage the induction system would appear to have the lowest installed load, a detailed study of the operation of each component at each part load condition should be made.

To assist in the assessment and to avoid considerable arithmetical calculations for each component for each hour during each day of the year, the average mid-day wet bulb temperatures for each month of the year have been established, which will dictate the refrigeration load for the fresh air component of each system. Similarly, the monthly average dry bulb temperatures dictate the refrigeration and air side requirements for the transmission gain or loss; the average monthly sunshine hours give an average solar gain for air side and refrigeration loads, and it has been assumed that the occupancy and lighting again remain constant throughout the year's operation. It has been assumed that each month has 250 operating hours, with an additional 250 hours warm-up period for the five months between November and March.

The various values are plotted in Tables 20.4, 5 and 6. To establish these values an understanding of the part load characteristics of each system is necessary. Since the VAV system is an all air system, advantage can be taken of partial free cooling whenever the ambient air has an enthalpy below the return air temperature. Essentially the air volume is dictated by the Room Sensible Heat, which, taking the prior assumption of constant lighting and occupancy loads of 603 000 BTU/h and 293 000 BTU/h *(176·7 kW and 85·8 kW)* respectively against a design RSH load of 2 219 000 BTU/h *(650·3 kW)*, means that even during conditions of transmission loss and no solar gain, the minimum air volume would equate to some 41 per cent of design air volume. However, since the design return air temperature with heat from lights would be 75·2 °F DB, 60 °F WB *(24 °C DB, 15·6 °C WB)*, any ambient air condition below that would provide a considerable saving in refrigeration load.

Because of the mixing action of the dual-duct system (see Fig.

TABLE 20.4
SEASONAL AND PART LOAD CONDITIONS

	Month						
	April	May	June	July	Aug	Sept	Oct
Average DB temp. (°F)	52·0	60·0	66·0	70·0	69·0	63·0	56·0
Average WB temp. (°F)	45·3	52·5	58·1	60·8	60·3	56·1	48·6
Average enthalpy (BTU/lb)	17·75	21·66	25·11	26·91	26·57	23·81	19·47
INDUCTION FAN COIL							
Dew point enthalpy (BTU/lb)	19·32	19·32	19·32	19·32	19·32	19·32	19·32
Enthalpy difference (BTU/lb)	−1·57	2·34	5·79	7·59	7·25	4·49	0·15
VAV AND DUAL-DUCT							
Dew point enthalpy (BTU/lb)	18·37	18·37	18·37	18·37	18·37	18·37	18·37
Enthalpy difference (BTU/lb)	−0·62	3·29	6·74	8·54	8·30	5·54	1·10
Average solar (per cent)	46	58	69	76	75	64	53

TABLE 20.4a
SEASONAL AND PART LOAD CONDITIONS

	Month						
	April	May	June	July	Aug	Sept	Oct
Average DB temp. (°C)	11·0	15·5	19·0	21·0	20·5	17·0	13·5
Average WB temp. (°C)	7·4	11·4	14·5	16·0	15·7	13·4	9·2
Average enthalpy (kJ/kg)	23·51	32·59	40·57	44·77	43·90	37·66	27·46
INDUCTION FAN COIL							
Dew point enthalpy (kJ/kg)	27·05	27·05	27·05	27·05	27·05	27·05	27·05
Enthalpy difference (kJ/kg)	−3·54	5·54	13·52	17·72	16·85	10·61	0·41
VAV AND DUAL-DUCT							
Dew point enthalpy (kJ/kg)	24·84	24·84	24·84	24·84	24·84	24·84	24·84
Enthalpy difference (kJ/kg)	−1·33	7·75	15·73	19·93	19·06	12·82	2·63
Average solar (per cent)	46	58	69	76	75	64	53

TABLE 20.5
PART LOAD CONDITIONS—INDUCTION AND FAN COIL SYSTEMS
(thousands BTU/h)

	Month						
	April	May	June	July	Aug	Sept	Oct
Fresh air load	NIL	116	235	300	287	195	41
Solar	455	574	683	751	742	634	525
Occupants	325	325	325	325	325	325	325
Transmission	−333	−165	—	165	165	—	−191
Lighting	1 206	1 206	1 206	1 206	1 206	1 206	1 206
Sub total	1 653	1 940	2 449	2 747	2 725	2 360	1 731
INDUCTION SYSTEM							
Fan and pumps	352	352	352	352	352	352	352
Refrigeration duty	2 005	2 292	2 801	3 099	3 077	2 712	2 083
Absorbed power	148·5	151·8	204·6	224·4	222·7	196·4	141·9

Induction system total refrigeration plant power = 322 575 kWh p.a.

FAN COIL SYSTEM							
Fans and pumps	584	584	584	584	584	584	584
Refrigeration duty	2 237	2 524	3 033	3 331	3 309	3 044	2 315
Absorbed power	171·5	196·0	220·5	245·0	229·4	217·0	185·5

Fan coil system total refrigeration plant power = 366 225 kWh p.a.

TABLE 20.5a
PART LOAD CONDITIONS—INDUCTION AND FAN COIL SYSTEMS
(kW)

	Month						
	April	May	June	July	Aug	Sept	Oct
Fresh air load	NIL	34·0	69·0	88·0	84·0	57·0	12·0
Solar	133·5	168·3	200·2	220·5	217·6	185·7	153·8
Occupants	95·3	95·3	95·3	95·3	95·3	95·3	95·3
Transmission	−97·6	−48·8	—	48·8	48·8	—	−56·0
Lighting	353·4	353·4	353·4	353·4	353·4	353·4	353·4
Sub total	484·6	602·2	717·9	806·0	799·1	691·4	558·5
INDUCTION SYSTEM							
Fan and pumps	103·0	103·0	103·0	103·0	103·0	103·0	103·0
Refrigeration duty	587·6	705·2	820·9	909·0	902·1	794·4	661·5
Absorbed power	148·5	151·8	204·6	224·4	222·7	196·4	141·9
	\multicolumn{7}{l}{Induction system total refrigeration plant power = 322 575 kWh p.a.}						
FAN COIL SYSTEM							
Fans and Pumps	171·0	171·0	171·0	171·0	171·0	171·0	171·0
Refrigeration duty	655·6	773·2	888·9	977·0	970·1	862·4	729·5
Absorbed power	171·5	196·0	220·5	245·0	229·4	217·0	185·5

Induction system total refrigeration plant power = 322 575 kWh p.a.

Fan coil system total refrigeration plant power = 366 225 kWh p.a.

14.30) it is necessary to consider at each monthly average the mixed air temperature which results in a required air volume through the dehumidifier coil and a resulting refrigeration plant duty. A typical example of a mixture condition for April is shown in Fig. 20.2. To achieve the desired mixed air temperature of 63 °F (17·2 °C) with a fresh/return air mixture of 70·5 °F (21·4 °C) and an off-coil temperature of 47·5 °F (8·6 °C) would require 0·33 × 109 500 cfm or 36 500 cfm (0·33 × 49·6 m³/s or 16·5 m³/s) thus giving a refrigeration load of 74·9 tons R (263·5 kW) giving an absorbed power of 87 kW. Similar plots are necessary for other months to establish all other part load operating conditions.

Both induction and fan coil systems would have a proportionally reduced fresh air load as the ambient temperature falls, but the secondary circuit would be directly proportional to the room sensible heat gain. However, when the ambient temperature falls below the design coil dew point of the primary air plant, the system dehumidifier can be used to provide some free cooling for the secondary circuit, which means in effect the chilled water pumps will run continually.

TABLE 20.6
PART LOAD CONDITIONS—VAV AND DUAL-DUCT SYSTEMS
(thousands BTU/h)

	Month						
	April	May	June	July	Aug	Sept	Oct
Solar	455	574	683	751	742	634	525
Occupants	293	293	293	293	293	293	293
Transmission	−333	−165	—	165	165	—	−191
Lighting	603	603	603	603	603	603	603
Sub total	1 018	1 305	1 579	1 812	1 803	1 530	1 230
VAV SYSTEM							
Percentage air volume	45·8	58·8	71·1	81·7	81·3	68·9	55·4
Fan power	200	215	229	247	244	227	212
RSH	1 218	1 520	1 808	2 059	2 047	1 757	1 442
Net refrigeration load	Nil	966	2 202	3 140	3 004	1 759	382
Pumps	Nil	102	102	102	102	102	102
Refrigeration load	Nil	1 068	2 304	3 242	3 106	1 861	484
Absorbed power (kW)	Nil	99	173	240	230	142	72

VAV system total refrigeration plant power = 239 000 kWh p.a.

	April	May	June	July	Aug	Sept	Oct
DUAL-DUCT SYSTEM							
Fan power	287	287	287	287	287	287	287
RSH	1 305	1 592	1 866	2 099	2 090	1 817	1 517
Required off-coil temp. (°F)	59·5	56·0	54·7	52·7	52·7	55·0	57·7
Coil air volume (cfm)	50 700	70 800	78 000	86 700	86 500	75 700	61 300
Net refrigeration load	763	2 211	2 810	3 307	3 250	2 522	1 792
Pumps	102	102	102	102	102	102	102
Refrigeration load	865	2 313	2 912	3 409	3 352	2 624	1 894
Absorbed power	84	171	212	248	244	194	143

Dual-duct system total refrigeration plant power = 324 000 kWh p.a.

From Tables 20.5 and 20.6 the absorbed refrigeration plant power for each system can be summarised as

2-pipe induction system	322 575 kWh p.a.
4-pipe fan coil system	366 225 kWh p.a.
VAV system	239 000 kWh p.a.
Dual-duct system	324 000 kWh p.a.

To establish the absorbed power at partial load, Fig. 10.8 was used where it can be clearly seen for constant speed centrifugal water chiller

TABLE 20.6a
PART LOAD CONDITIONS—VAV AND DUAL-DUCT SYSTEMS
(kW)

	Month						
	April	May	June	July	Aug	Sept	Oct
Solar	133·5	168·3	200·2	220·5	217·6	185·7	153·8
Occupants	85·8	85·8	85·8	85·8	85·8	85·8	85·8
Transmission	−97·6	−48·8	—	48·8	48·8	—	−56·0
Lighting	176·7	176·7	176·7	176·7	176·7	176·7	176·7
Sub total	298·4	382·0	462·7	531·8	528·9	448·2	360·3
VAV SYSTEM							
Percentage air volume	45·8	58·8	71·1	81·7	81·3	68·9	55·4
Fan power	58·8	63·0	67·2	72·5	71·4	66·1	62·0
RSH	357·2	445·0	529·9	604·3	600·3	514·3	422·3
Net refrigeration load	Nil	283·0	645·0	950·0	910·0	516·0	112·0
Pumps	Nil	30·0	30·0	30·0	30·0	30·0	30·0
Refrigeration load	Nil	313·0	675·0	980·0	940·0	546·0	142·0
Absorbed power	Nil	99	173	240	230	142	72

VAV system total refrigeration plant power = 239 000 kWh p.a.

	April	May	June	July	Aug	Sept	Oct
DUAL-DUCT SYSTEM							
Fan power	113·0	113·0	113·0	113·0	113·0	113·0	113·0
RSH	411·4	495·0	575·7	644·8	641·9	561·2	473·3
Required off-coil temp. (°C)	15·3	13·3	12·6	11·5	11·5	12·8	14·3
Coil air volume (m^3/s)	23·0	32·1	35·4	39·3	39·2	34·3	27·8
Net refrigeration load	223·0	648·0	824·0	968·0	952·0	739·0	525·0
Pumps	30·0	30·0	30·0	30·0	30·0	30·0	30·0
Refrigeration load	253·0	678·0	854·0	998·0	982·0	769·0	555·0
Absorbed power	84	171	212	248	244	194	143

Dual-duct system total refrigeration plant power = 324 000 kWh p.a.

with variable guide vanes duties above 40 per cent, there is less power required per unit refrigeration capacity. Particular advantage can be made of this when choosing the optimum switching of multiple chillers.

In establishing the fan power requirements for the operating year, the primary plants for the induction, fan coil and dual-duct systems would remain constant, whereas the VAV systems' supply and return air fans would modulate on the variable inlet guide vanes and provide a saving in part load power consumption, as shown in Fig. 20.2. Using

20 ENERGY COMPARISON OF AIR CONDITIONING SYSTEMS

the values of minimum supply air volume of 41 per cent for the November to March period and the values established in Table 20.4 for the months April to October, for the same number of operating hours per month the following results:

Induction System
   Primary air fan   50 kW × 3000 h p.a. = 150 000 kWh p.a.

Fan Coil System
   Primary air fan   26 kW × 3000 h p.a. = 78 000 kWh p.a.
   Fan coil units   960 × 107 W × 3000 h p.a. = 308 160 kWh p.a.
   Total   386 160 kWh p.a.

VAV System
   Supply air fan   84 kW
   Return air   21 kW
   Total   105 kW

MONTH	PERCENTAGE VOLUME	PERCENTAGE POWER
April	45·8	56
May	58·8	60
June	71·1	64
July	81·7	68
August	81·3	69
September	68·9	63
October	55·4	59
November	41·0	54
December	41·0	54
January	41·0	54
February	41·0	54
March	41·0	54
		709

VAV supply and return air fans
   709 per cent × 105 kW × 250 h/month = 186 113 kWh p.a.

Dual-Duct System
   Supply air fan   113 kW
   Return air fan   24·5 kW
   Total   137·5 kW
   137·5 kW × 3250 h = 446 875 kWh p.a.

404   APPLIED AIR CONDITIONING AND REFRIGERATION

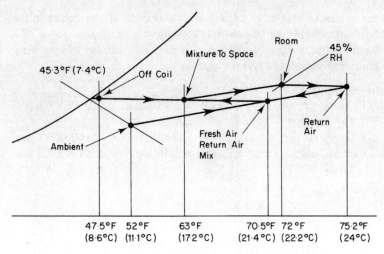

FIG. 20.2. Part load mixing condition for dual-duct system.

The auxiliary plants of each system comprising pumps and cooling tower fans can be summarised as follows:

Induction System
Primary chilled water pump   30 kW × 3000 h   90 000 kWh p.a.
Secondary chilled water pump   23 kW × 3000 h   69 000 kWh p.a.
Secondary chilled water pump,
   during warm-up period   23 kW × 250 h   5 750 kWh p.a.
Condenser water pumps   18·5 kW × 1750 h   32 375 kWh p.a.
Cooling tower fans   30 kW × 1750 h   52 500 kWh p.a.
Fan Coil System
Chilled water pump   42 kW × 300 h   126 000 kWh p.a.
Hot water pump   23 kW × 3250 h   74 750 kWh p.a.
Condenser water pump   18·5 kW × 3000 h   55 500 kWh p.a.
Cooling tower fans   30 kW × 2000 h†   60 000 kWh p.a.
VAV System
Chilled water pump   30 kW × 1500 h   45 000 kWh p.a.
Hot water pump   15 kW × 2750 h‡   41 250 kWh p.a.
Condenser water pump   18·5 kW × 1500 h   27 750 kWh p.a.
Cooling tower fan   30 kW × 1500 h   45 000 kWh p.a.

† Assuming that the cooling tower fans would cycle at low ambient temperatures.
‡ Includes 8 months heating and 250 h warm-up period.

TABLE 20.7
SUMMARY OF ANNUAL POWER REQUIREMENTS
(Excluding Heating Energy requirements)

	\multicolumn{4}{c}{System}			
	2-Pipe induction	4-Pipe fan coil	VAV	Dual-duct
Supply and return air fans	150 000	78 000	186 113	446 875
Fan coil motors	—	308 160	—	—
Refrigeration plant	322 575	366 225	239 000	324 000
Chilled water pumps				
Primary	90 000	126 000	45 000	52 500
Secondary	69 000	—	—	—
Condenser pumps	32 375	55 500	27 750	32 375
Cooling tower fans	52 500	60 000	45 000	52 500
Hot water pumps	5 750	74 750	41 250	13 750
Total kWh p.a.	722 200	1 068 635	584 113	922 000
Percentage comparison to VAV system	124	183	100	158

Dual-Duct System
   Chilled water pump   30 kW × 1750 h   52 500 kWh p.a.
   Hot water pump   5 kW × 2750 h   13 750 kWh p.a.
   Condenser water pump 18·5 kW × 1750 h   32 375 kWh p.a.
   Cooling tower fan   30 kW × 1750 h   52 500 kWh p.a.

Table 20.7 summarises all the data giving the resulting electrical power consumption for each system. One readily concludes the energy benefits of the VAV system against its contemporary alternatives, particularly when considering the dual-duct system, another all air system. However, the energy story does not stop here, since the heating requirements for each system have not been analysed in this exercise. The benefit of warm return air in the form of heat from lights minimises the required heating of primary air on a VAV system and only the net transmission loss can be considered as the heating requirement, whereas both induction and fan coil systems require the primary air to be raised to at least room dry bulb temperature. Whilst not a design feature of the induction system, the designer should also consider heat losses in the primary air ducting, which on bad installations have been known to incur a temperature drop of 20–30 °F (*10–15 °C*) between leaving the reheater and reaching the terminal.

In addition to heating, a true owning and operating cost analysis cannot be complete without considering maintenance. The maintenance cost of refrigeration plant, pumps, supply air fans and controls could be assumed similar for all three systems. Considering the balance of equipment for each system, as indicated below, certainly indicates another reason why VAV systems are enjoying the popularity they are.

Induction system: 960 lint screens—cleaned four times per annum
 960 sets of nozzles—cleaned annually
Fan coil system: 960 filters—changed twice per annum
 960 motors—to be checked annually
VAV system: 1 return air system to be serviced three times per annum
 960 terminals checked each five years
Dual-duct system: 1 return air system to be serviced three times per annum
 960 terminals to be checked annually

It should be stressed however that the example has considered only one building with its own peculiar characteristics; although every indication is that VAV systems are currently the most economical for multi-room applications, every application should be analysed on its own merits. The designer should appreciate that there are many applications for which VAV is not suited and should always keep an open mind when selecting any air conditioning system.

INDEX

Absorber, 161
Absorption
  cycle, 178
  refrigeration, 160
Accumulator, 369
Acoustic tile, 68–77, 82, 84
Adiabatic cooling, 13, 15
Aerofoil blades, 227, 246
Air
  by-pass, 18, 266, 394
  dehumidified, 20, 315
  distribution, 12, 32, 39, 94, 107, 125, 136, 225, 230, 257, 272, 275, 298, 300
  film, 46, 85
  flow, 39
  gap, 46
  handler, 236, 239, 272, 309, 320, 358, 367
  moist, 1
  outside, 19, 40, 130, 394
  primary, 96, 139, 242, 295, 392
  return, 132, 256, 259, 274, 289, 295, 321, 324, 394
    by-pass, 126, 128
  room, 8, 369
  secondary, 96
  supply
    condition, 11
    temperature, 12
    volume, 11, 131, 257, 274
  total, 96

Air—*contd.*
  velocity, 10, 99, 107
  volume, 15, 18, 20, 107, 129, 130, 135, 136, 229, 234, 236, 245, 267, 286, 289, 320, 331, 337, 341, 394
    constant, 259, 275, 282, 289
    variable, 124, 128, 256, 272, 392
  washer, 13
Air cooled condenser, 21, 29, 38, 39, 141, 186, 187, 225, 313, 325, 378
Air cooled condensing unit, 189, 194, 208, 326
Air cooled liquid chiller, 189, 207, 317
Air to air heat pump, 370
Air to air heat reclaim, 383
Air to water heat pump, 379
Air to water system, 242
All air system, 15, 242
All water system, 242, 262, 289
Altitude, 17, 18, 19
Aluminium
  coil, 140
  fin, 140, 237
Ambient condition, 40, 43, 46, 47, 50, 190, 201, 208, 257, 320, 330, 369, 385
Ammonia, 170, 181
Amortisation, 382

INDEX

Anti-vibration mounting, 153, 355
Apparatus dew point, 9, 18, 19, 129, 131, 132, 137, 237, 267, 271, 273, 394
Approach, 187, 216
ARI rating, 339
Asbestos, 66, 67, 72–5, 81, 83, 186
ASHRAE, 5, 40
Aspect ratio, 118, 257
Asphalt, 82, 84
Atmospheric pressure, 22
Attenuator, 245, 274, 396
Automatic control, 38, 289
Auxiliary water sprays, 15
Average temperatures, 375, 397
Axial flow fan, 225
Azimuth angle, 43, 63

Backward curved fan, 227, 234, 246, 287
Bank, 139
Barometric pressure, 17
Belfast, 49, 331
Belt drive, 145, 225, 236
Birmingham, 49, 331
Bleed-off, 210, 223
Block load, 264
Blow through system, 127, 239, 367
Boiling, 23, 24
Brickwork, 66, 67, 81, 83
British Thermal Unit, 2
Broadcasting studio, 106, 307
Builders' work, 35
Burn out, 153, 177
Butterfly valve, 164, 222
By-pass
  air, 18, 266, 394
    return, 126, 128
  control, 244, 377
  factor, 10, 130, 131, 132, 134, 139, 265, 337, 340
  VAV terminal, 285

Capacity control, 143, 163, 173, 180, 204, 320
see also Unloader

Capillary tube, 143
Capital cost, 107, 376
Cardiff, 49, 331
Carnot efficiency, 29
Carpet, 82, 84
Ceiling, 68–77, 259, 274
  effect, 96
  height, 136, 336
Central station system, 92, 124, 256, 273, 284, 309
Centrifugal compressor, 141, 154, 170
Centrifugal fan, 38, 225, 227
Changeover, 246, 264, 297, 301, 381
Charging valve, 356
Check
  figures, 38, 336
  valve, 369
Chemical
  cleaning, 212, 310, 312
  dehumidification, 16
Chilled water, 126, 132, 137, 163, 271, 299, 309; *see also* Water chiller
  coil, 137, 138, 237; *see also* Cooling coil
  flow switch, 166
CIBS, 7
Closed circuit cooler, 224, 382
Co-efficient of performance, 29, 30, 146, 161, 372, 374
Coil, 9, 12, 25
  cooling, 9, 12, 16, 39, 124, 126, 129, 132, 236, 252, 254, 267, 358, 367
  face velocity, 130, 132, 234, 384
  heating, 20, 39, 238
  indoor, 370
  outdoor, 370, 391
  secondary, 242
  temperature, 9, 125, 126
  wild, 285
Colour, 45, 63
Comfort, 1, 126, 319, 330, 378
Compression
  exponent, 30
  ratio, 30, 161, 208

Compressor, 20, 23, 24, 26, 27, 126, 161, 163, 186, 208, 345, 370
  centrifugal, 141, 154, 170
  head, 157, 161
  hermetic, 31, 145, 158, 208, 311
  motor, 26, 168, 189, 195
  multiple, 311, 325, 352, 357
  open, 146, 158
  power, 191, 195, 345, 372, 374
  reciprocating, 24, 141, 154, 161
  rotary, 141
Compressor/condenser balance, 192, 194, 208
Computer room, 139, 307
Concrete, 66–71, 75–9, 81, 83
Condensate, 140, 237, 247, 331
  drain, 366, 382
Condensation, 1, 218, 236, 239, 276, 390
Condenser, 23, 26, 34, 38, 149, 154, 163, 172, 186, 356, 358, 367, 368
  air cooled, 21, 29, 38, 39, 141, 186, 187, 225, 313, 325, 378
  evaporative, 20, 29, 141, 186, 208, 223, 313
  fan, 152
  limitation, 207
  load factor, 215
  pipework, 198, 222, 365
  pressure control. *See* Head pressure control
  ratings, 191, 215
  shell and coil, 211, 312
  shell and tube, 161, 213, 312
  water, 39, 167, 179, 183, 216, 221, 271, 294
  water cooled, 20, 29, 38, 39, 141, 170, 186, 208, 210, 312, 325, 358
Condensing pressure, 149, 158, 167, 197
Condensing temperature, 30, 149, 165, 167, 187, 196, 201, 207, 208, 215, 271, 345, 391

Constant pressure, 259
Constant volume, 259, 275, 282, 289
Contact factor, 10
Control,
  automatic, 38, 209
  by-pass, 244, 377
  capacity, 143, 163, 173, 180, 204, 320
  head pressure, 197, 224, 292, 313, 365, 389
  lag–lead, 168, 317
  on–off, 128, 323, 325
  refrigeration, 38
  safety, 195, 208
  sequence, 313
  solid state, 203
  static pressure, 261, 290, 294
  step, 124
  system powered, 244, 281
Convection, 46, 217
Cooling
  capacity, 42
  coil, 9, 12, 16, 39, 124, 126, 129, 132, 236, 252, 254, 267, 358, 367; *see also* Dehumidifier
  load, 14, 19, 32, 41, 42, 46, 129, 257, 321, 329; *see also* Refrigeration capacity
  total, 129, 331
  tower, 20, 34, 35, 39, 186, 210, 216, 271, 313, 324, 342, 358, 365, 388
  fan, 152, 166, 222, 313
Copper
  fins, 137, 237
  tubes, 137, 161, 237
Corrosion, 174
Cost(s)
  capital, 107, 376
  comparison, 344
  maintenance, 189, 221, 263, 312, 382, 406
  operating, 165, 257, 263, 309, 406
  running, 32, 107, 123, 125, 127, 214, 246, 256, 297, 300, 343
Counterflow, 140, 384

410 INDEX

Crankcase heater, 152, 292, 313
Crystallisation, 178, 179, 181
Cylinder, 141
  head, 152, 153
  cooling, 153

Daily range, 43, 49, 50
Damper, 107, 202, 222, 236, 255, 261, 287, 289
Defrost cycle, 373, 391
Dehumidified air, 20, 315
Dehumidifier, 9, 16, 19, 124, 320, 366, 396; see also Cooling coil
  chemical, 16
Dehydrator, 159
Density, 17, 81–4, 161
Department store, 33, 93, 106, 108, 139, 303
Desiccant, 16
Design
  condition, 52, 342
  day, 42, 126
  temperature, 42, 49–51, 246
Desuperheat, 23
Desuperheater, 390
Dew point, 1, 16, 50, 125, 126, 242, 246, 261, 341, 398
  apparatus, 9, 18, 19, 129, 131, 132, 137, 237, 267, 271, 273, 394
Diffuser, 95, 96, 257
Direct drive, 146, 225
Direct expansion, 126, 131, 137, 141, 237, 242, 286, 309, 346, 353
Disc valve, 142
Discharge
  line, 27
    pressure drop, 187, 190
    thermostat, 208
  pressure, 147, 163, 198
  temperature, 27, 29, 30, 149, 186, 187, 191, 207, 208, 345, 390
  valve, 141
Diversity, 33, 46, 86, 260, 320, 340

Domestic hot water, 390
Door, 80, 90
Double bundle condenser, 388, 390
Double suction riser, 352
Downdraught, 105, 259, 275
Drain, 34, 35, 39, 247, 264, 301, 331, 364
  condensate, 366, 382
  trap, 301, 366
Draught, 94
Draw through system, 239
Drive
  belt, 145, 225, 236
  direct, 146, 225
Drop, 95, 125
Dry bulb temperature, 1, 14, 49, 50, 136, 149, 186, 340, 398
Dry expansion cooler, 310
Dry steam, 23
Dual conduit system, 289, 294
Dual duct system, 128, 241, 295, 392
Dual pressure switch, 152
Duct
  design, 107
  heat gain, 11, 47, 91, 130
  leakage, 11, 47, 107, 246
  pressure loss, 110, 111, 252, 267, 396
  sizing, 115
  velocity, 97, 108, 273
Ductwork, 35, 94, 107, 244, 260, 267, 273, 321, 336, 369

Economiser, 259, 273, 289, 294, 324
see also Free cooling
Edinburgh, 49, 331
Effective room latent heat, 133, 266, 394
Effective room sensible heat, 133, 266, 394
  factor, 18, 132, 133, 266, 394
Effective room total heat, 133, 135, 394
Effective surface temperature, 16

Effective temperature difference, 15, 238, 267, 394
Efficiency
  carnot, 29
  fan, 226, 228, 231, 274
  saturation, 13
  static, 226, 228, 231
  total, 231
  volumetric, 161
Electric motor, 20, 46, 89, 145, 160, 163, 203
Elevation, 49
Eliminator, 218, 223, 237, 396
  vibration, 352
Energy
  comparison, 392
  conservation, 285, 368
Enthalpy, 2, 5, 27, 29, 216, 398
  deviation, 5
  sensor, 291
Entrainment, 96
Entropy, 27
Equaliser line, 356
Equilibrium diagram, 178
Equipment selection, 32
Equivalent duct sizing, 114
Equivalent length, 112
Equivalent temperature difference, 43, 45, 64, 65
Evaporating pressure, 147, 158, 167
Evaporating temperature, 131, 136, 167, 190, 198, 208, 237, 309, 342
Evaporation, 46, 216, 218
Evaporative condenser, 20, 29, 141, 186, 208, 223, 313
Evaporator, 23, 25, 26, 29, 33, 35, 141, 154, 157, 163, 171, 179, 186, 197, 326, 345, 348, 353, 369; *see also* Dehumidifier
  dry expansion, 310
  flooded, 161, 309
  multiple, 352
  shell and tube, 161
Exhaust, 256, 259, 294, 324, 378, 384, 393

Expansion
  tank, 361
  valve, 25, 31, 150, 201, 321, 326, 345, 369, 377, 389
External pressure, 320, 326
Extract, 225

Face and by-pass, 126, 127, 239, 363
Face velocity, 130, 132, 384
  coil, 234
Factory, 52, 93, 106, 139, 304
Fan, 20, 109, 225, 257, 297, 326, 369
  axial flow, 225
  backward curved, 227, 234, 246, 287
  centrifugal, 38, 225, 227
  coil unit, 326, 379, 392
  condenser, 152
  cooling tower, 152, 166, 222, 313
  cycling, 200
  efficiency, 226, 228, 231, 274
  forward curved, 227, 234
  heat gain, 92, 130, 252, 266, 394
  laws, 232
  motor, 20, 38, 226, 236, 297, 326, 329
  power, 20, 226, 231, 252, 260, 263, 267, 289, 294, 381, 384, 391, 403
  propeller, 203, 225
  scroll, 38, 228
  size, 232
  speed, 232
  supply, 11, 256, 259, 384
  wheel, 228, 235
Film resistance, 46, 85, 187
  air, 46, 85
  refrigerant, 214
  water, 214
Filter, 20, 34, 38, 233, 237, 245, 252, 264, 267, 295, 297, 321, 326, 396
  suction, 153
Fin spacing, 9
Flash gas, 26, 28, 30, 150, 156, 345

Float
  chamber, 161
  valve, 155
Flooded evaporator, 161, 309
Floor, 76, 77, 82, 84
Forward curved fan, 227, 234
Fouling, 213, 215
see also Scaling
Four-pipe system, 300, 380, 392
Four-way valve, 369
Free area, 95
Free blow, 35, 123, 324, 336
Free cooling, 261, 273, 284, 294, 387, 391
see also Economiser
Freezing, 17, 197, 313, 385
Fresh air, 34, 124, 126, 129, 132, 273, 289, 295, 300
see also Ventilation
Friction
  chart, 20
  loss, 107, 112, 358
Frost protection, 189

Gas fired absorption, 181
Gauge, 167, 313, 361, 364
Generator, 171
Glasgow, 49, 331
Glazing, 54–63, 80, 90, 248, 368
Grille, 95

Head pressure, 200, 222, 313, 326
  control, 197, 224, 292, 313, 365, 389
Heat
  compression, of, 26, 28, 29, 186
  exchanger, 29, 39, 129, 154, 170, 172, 179, 187, 199, 216, 233, 252, 297, 309, 367, 369, 383, 391
    plate, 383
  flow, 45
  gain, 40, 363
    duct, 11, 47, 91, 130
    fan, 92, 130, 252, 266, 394
    pump, 254

Heat—*contd.*
  loss, 14, 32, 374
  pipe, 383
  pump, 189, 368, 390
    air to air, 370
    air to water, 379
    water to air, 381
    water to water, 381
  reclaim, 284, 294, 368, 382, 389
    air to air, 383
    water to water, 387
  rejection, 26, 186, 192, 215
    factor, 221
  transfer. *See* Heat Exchanger
    coefficient, 33
  wheel, 383
Heating
  capacity, 372, 374
  coil, 20, 39, 238
Hermetic compressor, 31, 145, 158, 208, 311
High pressure switch, 152, 166, 214, 292, 313
Hospital, 93, 108, 139, 263, 306
Hot gas
  by-pass, 163
  line, 27, 164
Hotel, 86, 93, 106, 108, 304
Humidifier, 16, 34, 35, 239, 252
  spray, 16, 246, 396
Humidistat, 144, 297
Humidity ratio, 1
Hygrometer, 179

Ice box, 22
Immersion heater, 222
Indoor coil, 370
Induction, 96, 276
  ratio, 96, 246, 257, 263
  reheat, 255
  system, 242, 392
Industrial applications, 14, 86, 108, 225
Infiltration, 19, 47, 90, 368, 391
Installed power, 397
Instantaneous performance, 374

Insulation, 66–79, 82, 84, 91, 107, 275, 355, 368, 391
Integrated performance, 372
Internal gain, 40, 46

Laboratory, 93, 126, 307
Lag–lead control, 168, 317
Latent heat, 8, 23, 46, 124, 126, 256, 295, 342
  evaporation, of, 22, 210
  fusion, of, 22
  room, 18, 47, 129, 132, 133, 242, 273
  effective, 133, 266, 394
Level switch, 176
Library, 108
Lighting, 33, 40, 46, 88, 246, 248, 257, 266, 274, 331, 394
Linoleum, 76, 77
Liquid
  chiller, 154, 170, 309, 363; *see also* Water chiller
  air cooled, 313, 317
  line, 27, 198
  solenoid valve, 311
Lithium bromide, 170, 178, 181
Load
  estimate, 32, 39, 40, 120, 129, 132, 332
  profile, 258
London, 49, 330
Low pressure switch, 152, 198, 293, 313

Machinery load, 33, 40
Maintenance cost, 189, 221, 263, 312, 382, 406
Masonry, 81, 83
Metabolic rate, 46, 87
Metal roof, 70, 71
Metering device, 25
see also Expansion valve
Mixing process, 8, 9, 261, 295
Mixture condition, 129, 240, 257, 400
Module, 248, 263, 394

Moist air, 1
Moisture
  capacity, 1, 19
  content, 17
Motor
  cooling, 146
  winding, 166, 208
Muffler, 153, 321, 326, 355
Multiple compressor, 311, 325, 352, 357
Multiple evaporator, 352
Multi-zone system, 127, 128, 239, 337

Noise, 39, 94, 107, 123, 163, 225, 264, 274
Non-changeover system, 246, 392
Non-return valve, 365
Nozzle, 243, 255, 358, 394

Occupancy, 33, 40, 86, 87, 93, 124, 246, 248, 257, 273, 331, 378, 394
Off-coil condition, 129, 131, 132, 136, 237, 252
Office, 52, 86, 93, 106, 108, 120, 139, 277, 305, 387
Oil
  cooler, 152, 153
  entrainment, 350
  equaliser, 153
  filter, 153
  pump, 167
  return, 140, 311, 345, 352
  safety switch, 152, 313
  separator, 355
  sight glass, 152
On-coil condition, 131, 136, 237, 252, 340, 341
On–Off control, 128, 323, 325
Open compressor, 146, 158
Operating cost, 165, 257, 263, 309, 406
Orientation, 33, 120
Outdoor coil, 370, 391

Outlet, 94, 107
  velocity, 96, 106
Outside air, 19, 40, 130, 394
  thermostat, 378

Packaged equipment, 92, 124, 141, 236, 319, 379
Partial load, 120, 124, 127, 242, 272, 297, 301, 309, 317, 320, 337, 342, 356, 398
Partition, 78, 79
People, 86, 87, 93
see also Occupancy
Perimeter heating, 275, 392
Pipe loss, 254, 345
Pipework, 345, 348
  condenser, 198, 222, 365
  fittings, 347
  refrigeration, 39, 321, 325, 326, 345
  sizing, 345, 348
  water, 272, 309, 339, 358
Piston, 141
Pitched roof, 72–5
Plant room, 34, 260, 262
Plaster, 66–79, 81, 83
Plate heat exchanger, 383
Power consumption, 184, 187
Preheater, 238, 245, 261, 267, 295, 396
Pressure, 24, 27, 225
  atmospheric, 22
  barometric, 17
  condenser control. *See* Head pressure control
  condensing, 149, 158, 167, 197
  constant, 259
  discharge, 147, 163, 198
  evaporating, 147, 158, 167
  external, 320, 326
  head, 200, 222, 313, 326
    control, 197, 224, 292, 313, 365, 389
  loss, 20, 238, 246, 252, 345, 358, 384
    discharge line, 187, 190
    duct, 110, 111, 252, 267, 396

Pressure—*contd.*
  loss—*contd.*
    suction line, 190, 342
  static, 19, 108, 225, 273, 287, 294
  suction, 126, 143, 145, 147, 200, 313
  total, 109, 231
  vapour, 178, 179
  velocity, 97, 108, 112, 113, 225, 238
Pressure–enthalpy chart, 25–9, 155, 189, 207
Primary air, 96, 139, 242, 295, 392
Process, 33, 38, 126, 368
Propeller fan, 203, 225
Psychrometric chart, 1–8, 17, 18, 21, 50, 126, 136, 249, 341
Psychrometric process, 13, 216
Pump, 20
  heat gain, 254
  oil, 167
  power, 391
  refrigerant, 172
  solution, 172
  vacuum, 175
  water, 152, 166, 175, 245, 272, 286, 313, 324, 361, 388, 404
Purge, 159, 166, 173, 176, 385

Receiver, 159, 166, 193, 197, 222, 356
Reciprocating compressor, 24, 141, 154, 161
Recirculated air, 47
see also Return air
Reed valve, 142
Refrigerant, 24, 147, 150, 157, 161, 165, 170, 311, 321, 345
  charge, 194, 197, 213, 379
  cooled motor, 31
  film resistance, 214
  pump, 172
Refrigeration
  capacity, 253, 271
  controls, 38
  cycle, 22, 25, 155, 170, 319, 368

INDEX

Refrigeration—*contd.*
  effect, 24, 26, 29, 148, 150, 156, 161, 168, 186, 189, 192, 214, 254, 337
  pipework, 39, 321, 325, 326, 345
    sizing, 346
  plant, 34, 124, 129, 132, 189, 236, 257, 261, 271
Register, 95
Reheat, 16, 125, 127, 238, 245, 252, 276, 297, 324, 396
Relative density, 17, 21
Relative humidity, 1, 14, 124, 265, 285, 342
Residence, 86, 93, 106, 108, 301, 321, 390
Restaurant, 52, 93, 106, 108, 139, 303
Return air, 132, 256, 259, 274, 289, 295, 321, 324, 394
  by-pass, 126, 128
  grille, 105
  light fitting, 274
Reverse return, 363
Rise, 95
Roof, 64, 65, 68–75, 82, 84
Roof-top equipment, 260, 262, 294
Room
  air, 8, 369
    conditioner, 121, 319
    conditions, 40, 130, 132, 330, 380
  latent heat, 18, 47, 129, 132, 133, 242, 273
    effective, 133, 266, 394
  sensible heat. *See* Sensible heat
  temperature, 124
  velocity, 94
Rotary compressor, 141
Run around system, 383
Running cost, 32, 107, 123, 125, 127, 214, 246, 256, 297, 300, 343

Safety
  control, 195, 208
  relief valve, 152, 166

Saturated liquid, 29
Saturated vapour, 22, 23, 25, 154
Saturation, 16, 129
  efficiency, 13, 14
  line, 1
  temperature, 179
Scaling, 213, 221, 223, 313, 342, 358
  see also Fouling
School, 306
Sea level, 17, 19
Secondary air, 96
Secondary cooling coil, 242
Secondary water, 242, 394
Selection table, 338
Sensible cooling, 16
Sensible heat, 8, 11, 46, 124, 126, 252, 340, 342
  effective room, 133, 266, 394
  factor, 8, 19
    effective room, 18, 132, 133, 266, 394
  room, 8, 130, 132, 273, 320
  room, 8, 11, 18, 47, 124, 129, 132, 256, 259, 266, 272
Sequence controller, 313
Shade, 33, 43
Shading device, 33, 368, 391
Shaft seal, 146, 160
Shell and coil condenser, 211, 312
Shell and tube condenser, 161, 213, 312
Shell and tube evaporator, 161
Shops, 52, 86, 93, 106, 108, 139, 303, 377
Sight glass, 321, 326, 356
Silencer. *See* Attenuator
Single piece equipment, 189, 320, 321, 379
Single zone, 120, 126
Site survey, 32, 36, 40, 329
Slates, 72–5
Solar altitude, 43, 63
Solar gain, 19, 33, 40–3, 54–62, 245, 248, 257, 266, 394
Solenoid valve, 145, 311
Solid state control, 203
Solution pump, 172

Specific heat, 10
Specific humidity, 1, 8
Specific volume, 2, 10
Split system, 326, 379
Spray
  humidifier, 16, 246, 396
  pond, 210
  water, 14, 218
Spread, 96, 99
Standby, 311
Starter, 167, 311
Static efficiency, 226, 228, 231
Static pressure, 19, 108, 225, 273, 287, 294
  control, 261, 290, 294
Static regain, 274
Steam, 173
Step controller, 124
Storage
  effect, 41, 42, 313, 320, 340
  vessel, 389
Stores. *See* Shops
Strainer, 364
Subcooling, 27, 29, 150, 156, 189, 193, 213
Suction
  filter, 153
  gas, 147, 148
  line, 27, 164, 321, 345, 348, 352
    pressure drop, 190, 342
  liquid heat exchanger, 31, 152, 356
  pressure, 126, 143, 145, 147, 200, 313
  temperature, 27, 147, 149, 187, 196, 204, 342, 345
  valve, 141, 369
Summer, 49, 275
Superheat, 22, 27, 29, 31, 141, 148, 154, 208
Superheated steam, 22
Supermarket, 303
Supply
  air
    condition, 11
    temperature, 12
    volume, 11, 131, 257, 274
  fan, 11, 256, 259, 384

Surge, 162
Survey, 38, 40, 262
  site, 32, 36, 40, 329
Swimming pool, 383
System
  design, 32, 35
  powered control, 244, 281
  resistance, 396

Temperature, 27
  average, 375, 397
  condensing, 30, 149, 165, 167, 187, 196, 201, 207, 208, 215, 271, 345, 391
  design, 42, 49–51, 246
  differential, 96, 136, 189
    effective, 15, 238, 267, 394
    equivalent, 43, 45, 64, 65
  discharge, 27, 29, 30, 149, 186, 187, 191, 207, 208, 345, 390
  dry bulb, 1, 14, 49, 50, 136, 149, 186, 340, 398
  evaporating, 131, 136, 167, 190, 198, 208, 237, 309, 342
  room, 124
  saturation, 179
  suction, 27, 147, 149, 187, 196, 204, 342, 345
  supply air, 12
  swing, 42, 53
  wet bulb, 1, 49, 136, 149, 186, 208, 216, 221, 224, 331, 340, 398
Terminal, 242, 257, 267, 273, 358, 396
  reheat, 295, 299
  velocity, 95, 96
Textiles, 14
Theatre, 52, 93, 106, 108, 304, 384
Thermal expansion valve, 25, 196, 311
  see also Expansion valve
Thermal resistance, 46, 81–5
Thermal wheel, 289
Thermometer, 1, 2, 364
Thermostat, 121, 124, 144, 166, 208, 222, 240, 247, 257, 283, 289, 313, 321, 329, 369

Thermostat—*contd.*
  discharge line, 208
  outside air, 378
Three-pipe system, 300
Three-way valve, 222, 286, 297, 342, 363
Throw, 95, 99
Thrust bearing, 163
Tiles, 76, 77, 82, 84
Tool room, 33
Total
  air, 96
  cooling load, 129, 331
  efficiency, 231
  energy system, 160, 184, 368
  heat, 7, 8, 23, 29, 126, 253, 266 rejection. *See* Heat rejection
  pressure, 109, 231
Transformer, 293, 329
Transmission
  co-efficient, 45, 46, 66–80, 248
  gain, 40, 43, 45, 64, 65, 244, 257, 266, 276, 394
  loss, 242, 257, 275
Turbine, 160, 163, 185
Two-pipe system, 300, 380, 392

Unit heater, 225
Unloader, 126, 143, 204, 286, 292, 325, 349

Vacuum, 24, 157, 165, 170, 174
  pump, 175
Valve
  butterfly, 164, 222
  charging, 356
  check, 369
  disc, 142
  expansion, 25, 31, 150, 201, 321, 326, 345, 369, 377, 389
  float, 155
  four-way, 369
  liquid line solenoid, 311
  non-return, 365
  reed, 142

Valve—*contd.*
  safety relief, 152, 166
  solenoid, 145, 311
  suction, 141, 369
  three-way, 222, 286, 297, 342, 363
Vane, 98
Vapour, 25, 161
  line, 27
  pressure, 178, 179
  saturated, 22, 23, 25, 154
  water, 1, 16, 170
Variable air volume, 124, 128, 256, 272, 392
  inlet guide vanes, 159, 164, 166, 261, 287, 402
  terminal, 274
  by-pass, 285
Velocity pressure, 97, 108, 112, 225, 238
Venetian blind, 42, 54–63, 248
Ventilation, 8, 20, 34, 40, 47, 93, 129, 132, 242, 273, 288, 295, 300, 331, 368, 377, 391
Vibration eliminator, 352
Volume regulator, 263, 282
Volumetric efficiency, 161

Walls, 64–7
Warm up, 378, 379
Waste water, 210, 324, 342
Water
  chiller, 141, 154, 167, 198, 242, 247; *see also* Liquid chiller
  cooled
    condenser, 20, 29, 38, 141, 170, 186, 208, 210, 312, 325, 358
    head, 153
  film resistance, 214
  make up, 361
  pipework, 272, 309, 339, 358
    sizing, 359, 360
  pump, 152, 166, 175, 245, 272, 286, 313, 324, 361, 388, 404

Water—*contd.*
  regulating valve, 221, 365
  secondary, 242, 394
  spray(s), 14, 15, 218
    auxiliary, 15
  treatment, 189, 210
  vapour, 1, 16, 170
  velocity, 358
Water to air heat pump, 381
Water to water heat pump, 381
Water to water heat reclaim, 387
Weather data, 49
Well water, 210, 381

Wet bulb temperature, 1, 49, 136, 149, 186, 208, 216, 221, 224, 331, 340, 398
Wild coil, 285
Windage loss, 210, 223
Windows, 80, 90
Winter, 49
Wood, 68–77, 81, 83

Yearly range, 51

Zoning, 260, 301, 336